BIOTECHNOLOGY & NANOTECHNOLOGY

BIOTECHNOLOGY & NANOTECHNOLOGY

REGULATION UNDER ENVIRONMENTAL, HEALTH, AND SAFETY LAWS

B. DAVID NAIDU

UNIVERSITY PRESS

OXFORD
UNIVERSITY PRESS

Oxford University Press, Inc., publishes works that further Oxford University's objective of excellence in research, scholarship, and education.

Oxford New York
Auckland Cape Town Dar es Salaam Hong Kong Karachi Kuala Lumpur Madrid Melbourne
Mexico City Nairobi New Delhi Shanghai Taipei Toronto

With offices in
Argentina Austria Brazil Chile Czech Republic France Greece Guatemala Hungary Italy
Japan Poland Portugal Singapore South Korea Switzerland Thailand Turkey Ukraine
Vietnam

Library of Congress Cataloging-in-Publication Data

Naidu, B. David.
 Biotechnology & nanotechnology : regulation under environmental, health, and safety laws / B. David Naidu.
  p. cm.
 Includes bibliographical references and index.
 ISBN 978-0-19-534008-2 ((hardback) : alk. paper)
 1. Biotechnology industries—Law and legislation. 2. Biotechnology industries—State supervision.
3. Biotechnology—Law and legislation. 4. Biotechnology—Environmental aspects. 5. Biotechnology—Safety measures. 6. Transgenic plants. 7. Food law and legislation. 8. Cosmetics—Law and legislation.
9. Nanotechnology.
I. Title. II. Title: Biotechnology and nanotechnology.
 K3925.B56N35 2009
 343′.0786606—dc22   2009013012

1 2 3 4 5 6 7 8 9
Printed in the United States of America on acid-free paper

Note to Readers
This publication is designed to provide accurate and authoritative information in regard to the subject matter covered. It is based upon sources believed to be accurate and reliable and is intended to be current as of the time it was written. It is sold with the understanding that the publisher is not engaged in rendering legal, accounting, or other professional services. If legal advice or other expert assistance is required, the services of a competent professional person should be sought. Also, to confirm that the information has not been affected or changed by recent developments, traditional legal research techniques should be used, including checking primary sources where appropriate.

*(Based on the Declaration of Principles jointly adopted by a Committee of the
American Bar Association and a Committee of Publishers and Associations.)*

You may order this or any other Oxford University Press publication by
visiting the Oxford University Press website at www.oup.com

For Andrea, Ravi, and Kieran

CONTENTS

PREFACE

When I began this project in 2005, regulatory agencies in the United States offered only a limited amount of information as to how nanotechnology and nanomaterials would be regulated under environmental laws. Thus, at that time, it made sense to look at biotechnology as a model for how this other new and emerging technology would likely be regulated. But now, and especially in the last year, a number of different regulatory agencies have issued policy papers or guidance documents outlining their regulatory stance toward nanotechnology. Nonetheless, the policy positions of these agencies still need to be fully developed. Therefore, biotechnology—due to its longer regulatory history—continues to provide a good comparison for those seeking to understand the current and potentially future regulation of nanotechnology. Moreover, the controversies that have surrounded certain uses of biotechnology, such as genetically modified foods, offer insight into the pitfalls that can challenge a new technology in terms of rapidly shifting—and potentially severely negative—public opinion and the ramifications for regulatory action.

This book covers a diverse number of laws that affect a wide-ranging set of industries and that are administered by a variety of agencies. It can be correctly pointed out that the topics discussed are worthy of books themselves. And there are, in fact, books solely covering most of the topics discussed in this book, including, but not limited to, risk analysis, regulation of foods, regulation of drugs, regulation of medical devices, and laws dealing with chemicals. However, this volume is intended to offer the reader a uniquely comprehensive view of the overall scheme of environmental, health, and safety laws as they apply to both biotechnology and nanotechnology. From a practitioner's perspective, there are many sections of the book that detail the specific steps that must be taken in order to obtain approval for a product, while offering insight into the perspective that the applicable regulatory agency or agencies may adopt. In addition, and likely of interest to academics, the book discusses not only the regulations themselves, but also the historical context in which they were created, and offers an understanding of the political dynamics and economic realities that inevitably shape the laws that are developed.

While research and writing is a lonely exercise, many of my friends provided encouragement during the long process. If I were to list each person, I fear I would miss someone. Thus, I can only say, you know who you are. However, there are a number of people who directly assisted me with this project, and I am deeply grateful for all their efforts. I would like to thank Hetal Dhagat, Dawn Munson, Deborah Low, Mandy Lundstrom, Jonathan Barron, Phil Seliger, and

Nicole Behesnilian for their research and editing assistance. Barbara Tanzer was most helpful in finding journal articles and other secondary sources. I am especially grateful to Rebecca Halford Harris, Roger Pitt, Paul Stimers, Eric Stone, Suzan Onel, and Eric Kaminskas for their review of various sections of the book and their extremely helpful comments. I'm also thankful to the support that I received from my firm, K&LGates LLP, which allowed me the resources and time to complete this project, and especially to Woody Collins and Bill Hyatt for their unfailing support. I have special gratitude for Jean Reyes, who reviewed numerous rough drafts and provided insightful comments. Eugenia Frenzel receives my heartfelt thanks for all her tireless assistance with research and editing.

Naturally, this project would not have gotten off the ground were it not for the folks at Oxford, including Edward Burchell, Larry Selby, Ron Doering, Michelle Lipinski, Jessica Picone, and Sarah Bloxham.

Instrumental in almost every step of the process was my mentor and colleague Don Stever, who was generous with his time in providing numerous insightful suggestions on the non-copy-edited versions of the chapters.

Finally, I would like to thank my parents Karikutla R. Naidu and Elizabeth Naidu for their love of learning and teaching, which they instilled in me. And certainly the most important person in this entire process is my wife Andrea J. Stein, without whose unfailing love, support, and encouragement I could not have completed this book.

LIST OF ACRONYMS

ALJ—administrative law judge
ANDA—abbreviated new drug application
ANSI—American National Standards Institute
AOSCA—Association of Official Seed Certifying Agencies
APH(3')II—aminoglycoside 3'-phosphotransferase II
APHIS—Animal and Plant Health Inspection Service
ARS—Agricultural Research Service
ASTI—American Society for Testing and Materials International
ATP—adenosine-5' -triphosphate
BET—Biotechnology Evaluation Team
BLA—biologics license application
BRS—Biotechnology Regulatory Services
BSE—bovine spongiform encephalopathy ("mad cow" disease)
Bt—*Bacillus thuringiensis*
CAIR—Chemical Assessment Information Rule
CBER—Center for Biologics Evaluation and Research
CDER—Center of Drug Evaluation and Research
CEQ—Council of Environmental Quality
CESQG—conditionally exempt small quantity generator
CFSAN—Center for Food Safety and Applied Nutrition
CIB—Compliance and Inspection Branch
CIR—Cosmetic Ingredient Review
CJD—Creutzfedlt-Jacob disease
CNT—carbon nanotube
CPSA—Consumer Product Safety Act
CPSC—Consumer Products Safety Commission
CRT—cathode ray tubes
CSFAN—Center for Food Safety and Applied Nutrition
CTFA—Cosmetic, Toiletry and Fragrance Association, Inc.
CVM—Center for Veterinary Medicine
DEFRA—British Department for Environment, Food and Rural Affairs
DNA—recombinant deoxyribonucleic acid
EA—environmental assessment
ECB—European corn borer
EIS—environmental impact statement
ELA—establishment license application
EPA—Environmental Protection Agency

EU—European Union
EUP—environmental use permit
FIFRA—Federal Insecticide, Fungicide, and Rodenticide Act
FDA—Food and Drug Administration
FDAMA—Food and Drug Administration Modernization Act of 1997
FFDCA—Federal Food, Drug, and Cosmetic Act
FHSA—Federal Hazardous Substances Act
FIFRA—Federal Insecticide, Fungicide and Rodenticide Act
FONSI—finding of no significant impacts
FPLA—Fair Packaging and Labeling Act
FPPA—Federal Plant Pest Act
FQPA—Food Quality Protection Act
FSIS—Food Safety Inspection Service
GM—genetically modified
GMO—genetically modified organism
GRAS—generally recognized as safe
GRASE—generally recognized as safe and effective
HCS—Hazard Communication Standard
HT—herbicide-tolerant
IGF-1—insulin-like growth factor
INAD—investigational new animal drug
IND—investigational new drug
IRB—institutional review board
IRM—insect resistance management
ITC—Interagency Testing Committee.
IUR—Inventory Update Rule
IWGN—Interagency Working Group on Nanoscience, Engineering
          and Technology
LHAMA—Labeling of Hazardous Art Materials Act
LQG—large quantity generator
LVE—low volume exemption
MCAN—Microbial Commercial Activity Notice
MRI—magnetic resonance imaging
MSDS—material safety data sheet
MWNTs—multiwalled nanotubes
NADA—new animal drug application
NAS—National Academy of Science
NDA—new drug application
NEHI—Nanotechnology Environment and Health Implications
NEPA—National Environmental Policy Act
NF—National Formulary
NGO—nongovernmental organization
NIH—National Institutes of Health

NIOSH—Nanotechnology Research Center
NMFS—National Marine Fisheries Service
NNI—National Nanotechnology Initiative
NOC—Notice of Commencement
NRC—National Research Council
NSET—President's National Science and Technology Council
NSF—National Science Foundation
NSPS—new source performance standards
NTF—Nanotechnology Task Force
NTRC—NIOSH Nanotechnology Research Center
OCED—Organisation for Economic Co-operation and Development
OMB—Office of Management and Budget
OSH Act—Occupational Safety and Health Act
OSHA—Occupational Safety and Health Administration
OSTP—Office of Science and Technology Policy
OTC Monograph—over the counter monograph
PBN—premarket biotechnology notice
PEL—permissible exposure limit
PG—polygalacturonase gene
PHSA—Public Health Service Act
PIP—plant incorporated protectant
PLA—product license application
PMA—premarket application
PMN—Pre-Manufacture Notice
PPE—personal protective equipment
PPPA—Poison Prevention Packaging Act
PQA—Plant Quarantine Act
R&D—research and development
RAC—Recombinant DNA Advisory Committee
rBST—recombinant bovine somatotropin
RCRA—Resource Conservation and Recovery Act
rDNA—recombinant deoxyribonucleic acid
ROS—reactive oxygen species
SAP—Scientific Advisory Panel
SARA III—Superfund Amendments and Reauthorization Act of 1986
SNUN—Significant New Use Notice
SNUR—Significant New Use Rule
SPF—sun protection factor
SQG—small quantity generators
SWNTs—single-walled carbon nanotubes
TERA—TSCA Experimental Release Application
TME—test marketing exemption
TSCA—Toxic Substances Control Act

TSDF—treatment, storage, and/or disposal facility
UNEP—United Nations Environment Programme
USDA—U.S. Department of Agriculture
USP—United States Pharmacopeia
UST—Underground Storage Tank
UVA—ultraviolet A
UVB—ultraviolet B
VCRP—Voluntary Cosmetic Registration Program
WTO—World Trade Organization

1. INTRODUCTION TO REGULATION OF BIOTECHNOLOGY AND NANOTECHNOLOGY

The twenty-first century may be shaped in significant measure by the development of two different technologies: biotechnology and nanotechnology. A broad range of industries and products are impacted by both of these technologies, including medicine, medical devices, food and food additives, pesticides, cosmetics, other consumer products, chemicals, and many others. When it comes to the law, the fundamental question that each technology presents is the same: are the environmental, health, and safety laws that were formulated mainly during the latter half of the twentieth century both adequate to protect us from the risks associated with these technologies and permissive enough to allow us to obtain the benefits these technologies have to offer?

Because research into, and manufacture of, products using genetic modification began nearly two decades before nanomaterials reached the marketplace, regulators and stakeholders in biotechnology have had almost two additional decades to wrestle with the critical question of how to balance these competing interests while also developing a regulatory structure that is neither too permissive nor too restrictive. Nonetheless, as will be evident in subsequent chapters, regulators are still struggling with these issues in the biotechnology field. As technology advances, and new products that were previously only theoretical possibilities become reality, agencies must stake out policy positions in previously uncharted territory—and modify the regulatory structure accordingly.

As for those working in the field of nanotechnology, they too must now also confront the issue of whether the laws enacted decades ago are adequate and appropriate to regulate products that not only did not exist at the time of their enactment, but were unlikely to have even been contemplated. To their benefit, however, these people have the biotechnology experience from which to learn. That is, they can see how the regulations have shifted and changed in the face of new products made possible due to biotechnology advances as well as public concern about the risks of these new technologies.

To address this fundamental legal question, this book provides a review of the major laws and regulations that govern biotechnology and nanotechnology in certain key fields. These include a number of key environmental health and safety laws, as well as other laws that do not traditionally fall under the rubric of "environmental laws," such as the Food, Drug, and Cosmetic Act. Each of these laws has spawned a complex and detailed regulatory program administered by a number of different agencies. Any attempt to explain such a complex set of laws and regulations—especially one concerning a subject matter as technical as that addressed in this book—requires a degree of simplification. As a result, it is presumed that the reader, if he or she is interested, will in conjunction with reading this book also review as appropriate the relevant statute, Code of Federal Regulations, or agency policy position paper.

Another important point concerning the law governing these technologies relates to the use of guidance documents, agency policy statements, and even draft guidance documents in this book to outline an agency's position. These types of documents are not necessarily predictive of agency behavior, even though many agencies use them as "firm" policy, because it has been well settled that an agency is not bound by such statements. These documents merely serve as an agency's current interpretations of applicable statutes or regulations. An agency is free to choose different standards, procedures, or policies provided they meet the applicable statute and regulations. Yet despite the limitations of such documents, in those instances in which the agency has not released final rules or regulations but has, in fact, issued guidance documents or interpretative statements that specifically address the applicability of the underlying statute to biotechnology or nanotechnology, those guidance documents and interpretive statements are discussed. The reader is cautioned to remember the limitations of such documents.

The organization of the book reflects the view that, before reviewing the statutes and regulations applicable to these technologies, it is necessary to examine certain background issues. As detailed below, Chapters 2 through 4 lay out the groundwork for the chapters that follow—offering historical background and discussion of the theoretical underpinnings of the regulatory structure with respect to biotechnology and nanotechnology. Chapters 5 through 10 focus on specific industry sectors and the particular issues that arise in the regulation of each. This is not to suggest that statutes in one section are applicable only to that section. For example, the Occupational Safety and Health Act applies to a vast number of industries, and is not simply limited to the chemical manufacturing industry.

Chapter 2 offers a starting point for the discussion and provides those not familiar with the historical development of biotechnology or nanotechnology with some basic information, including some nontechnical definitions for specific nanomaterials referenced in other chapters. Additionally, it will offer examples of how these technologies are currently being used.

Chapter 3 offers a survey of the various factors (e.g., health and environmental risks, economic and ethical considerations, and public opinion) that have influenced discussions of biotechnology and nanotechnology. The risks discussed in this chapter serve as the basis for the way biotechnology and nanotechnology are currently regulated. The biotechnology discussion in this chapter focuses in particular on agricultural biotechnology because it is that area (as opposed, e.g., to medicines or medical devices) that has generated the most significant controversy.

The risks described in this chapter are divided into human health risks and environmental risks. In examining human health risks, the chapter focuses on two elements: toxicity (e.g., food allergens) and exposure (e.g., bioaccumulation and persistence in the environment and pathways into the human body). As to environmental issues, Chapter 3 examines the potential adverse consequences (e.g., impacts to microbial and aquatic communities) associated with products made from these technologies if they are released into the environment. The chapter will then focus on other factors that influence how a product is regulated— namely, the economic and societal impacts of the product, any uncertainty associated with what is known about the product, and public perceptions. Finally, the chapter will provide a brief discussion of an alternative mechanism for regulation: the application of a stringent version of the precautionary principle.

Chapter 4 provides a discussion on the evolution of the regulatory structure governing biotechnology and the current regulations governing nanotechnology. Because of the longer historical time line for biotechnology, it is possible to examine the early efforts at regulation in the late 1970s and compare them with the later regulatory structure imposed in the mid-1980s. As discussed in this chapter, the regulatory principles that were eventually arrived at for biotechnology are the same principles that are applicable to nanotechnology today. Additionally, the chapter will examine federal government policies designed to encourage the development of nanotechnology.

Chapter 5 extensively addresses the regulations applicable to genetically modified plants and animals. Specifically, the chapter focuses on the regulations and guidance documents issued by the Department of Agriculture (USDA), the Environmental Protection Agency (EPA), and the Food and Drug Administration (FDA) that govern the research into, and the manufacturing of, genetically modified plants (also referred to as "transgenic plants"). It also examines the specific EPA regulations applicable to genetically modified microorganisms as well as the guidance document issued by FDA for regulating genetically modified animals (also referred to as "transgenic animals") under its animal drug laws. Moreover, to emphasize how these regulations have been applied, three case studies are highlighted: the impact of genetically modified corn on the Monarch butterfly, the accidental mixing of genetically modified corn for animal feed purposes with corn for human purposes (the Star Link™ incident), and the request for approval of a transgenic fish. Finally, the chapter addresses the regulatory structure currently applicable to nanoscale products that have antibacterial or antimicrobial properties.

Chapter 6 examines how foods and food additives are regulated. This chapter addresses the regulation of transgenic crops (and it is only crops, as no transgenic animals have thus far been approved) once they are processed or otherwise made available for human consumption. This chapter covers how these foods are regulated both in the United States and in the European Union (EU). The regulation of genetically modified foods (or, in the opinion of some, the lack thereof) is a hotly contested topic, and this chapter aims to provide some understanding of the reasons for this controversy. The chapter also examines the way the current food regulatory structure may be applicable to nanoscale food additives and food packaging that contains nanoscale materials.

Chapter 7 discusses the regulatory structure governing bioengineered drugs, biologics, and medical devices. It also addresses the application of those regulations on drugs and medical devices that use or may use nanomaterials.

Chapter 8 focuses on the way cosmetics are currently regulated. The issue of cosmetic regulation is particularly relevant to nanotechnology as there are already a number of cosmetic products on the market that contain nanoscale materials. The chapter concludes with a discussion of a critical question: whether cosmetics containing nanoscale materials should be regulated as drugs.

Chapter 9 addresses the regulatory powers of the Consumer Products Safety Commission (CPSC) and its role in regulating nanomaterials in consumer products. Until 2008, the CPSC seemed to lack the funding and personnel to effectively carry out its mandate. However, with additional and enhanced enforcement powers, the CPSC may have a greater role in regulation of nanomaterials in the future.

Chapter 10 extensively discusses the statute that many originally believed was the appropriate vehicle for regulating nanomaterials—the Toxic Substances Control Act. It will also examine how the Occupational Health and Safety Administration's (OSHA) regulations are being interpreted and complied with by various companies involved in the manufacture of nanomaterials. Finally, the chapter examines a statute that may become more relevant as disposal of nanowastes becomes a more significant issue: the Resource Conservation and Recovery Act.

Chapter 11 concludes with a discussion of the lessons that can be learned by stakeholders in the nanotechnology field from the biotechnology experience. It also addresses the issue of how those lessons are shaping the actions and activities being undertaken by agencies, manufacturers, and nongovernmental organizations today.

Two final points should be noted about this book. First, given that it is intended for legal practitioners and scholars, a conscious attempt has been made to avoid being too technical in the discussions of either the technologies or the scientific data on toxicity, exposure, or other potential impacts. Second, the subject matter in this book continues to generate intense emotions. The author has sought to present an unbiased examination of the issues without interjecting his subjective opinions. Moreover, his aim has been to avoid support for or opposition to any particular policy, but rather, to fairly convey the comments of proponents and detractors of the particular policy.

2. WHAT IS *BIOTECHNOLOGY?* WHAT IS *NANOTECHNOLOGY?*

I. BIOTECHNOLOGY

A. The Evolution of Modern Biotechnology

For thousands of years, humans have been changing the inherent characteristics of plants and animals through cross-hybridization and selective breeding to create desirable traits that either have commercial value or perform a useful function. The late nineteenth and early twentieth centuries witnessed the use of microorganisms to produce vaccines and other medicines on a commercial scale. However, it was in 1973, almost two decades after Watson and Crick revealed the double helix confirming the belief others had postulated regarding the way genetic material was stored and copied, a team of California scientists announced they had conducted experiments using a technique that allowed for the combination of genetic material from different species.[1] Specifically, this process involved isolating a foreign gene, cloning that gene, inserting it into vectors (e.g., a bacterial plasmid), cloning that vector, and then introducing it into the host cells. This technology is referred to as recombinant DNA (rDNA) technology. At that moment it became possible to combine DNA from animal viruses with bacterial strains or to combine the DNA of different viral strains to create novel hybrids.[2] Subsequently, it was discovered that it was possible to clone DNA segments not only from bacterial or virus strains, but from virtually any organism.

Almost immediately, scientists understood the dual implications of this new technology. On one hand, it was potentially possible to create both precisely and commercially novel animals, plants, viruses, bacteria, and other microorganisms. On the other hand, potential dangers were associated with conducting experiments to create genetically modified organisms. The newly created organism could be pathogenic, and because it is genetically modified, no defenses might exist in the natural environment to restrict its growth; there also might be no adequate immunologic response. In addition, even if the newly created organism is not pathogenic in itself, if it were in the natural environment, it might be able to

---

1. M. Singer, *Where the Cloning Discussion Began.* Reflections from the Frontiers (Explorations for the Future: Gordon Research Conferences 1931–2006), GRC's 75th Anniversary. *Available at* http://www.frontiersofscience.org/reflections.aspx?category=2&essay=32.

2. M. Singer & D. Soll, *Guidelines for DNA Hybrid Molecules.* SCIENCE (Sept. 1973).

exchange information with bacteria or viruses that are pathogenic to humans, possibly leading to diseases for which there is no immunologic defense or that have allergenic properties for which again there is no immunologic response.

While most would recognize that modern biotechnology represents a fundamental technological advance over traditional selective breeding and cross-hybridization, a debate has nonetheless raged as to the implications of this technological advancement. This debate, as laid out in the following chapters, in the regulatory context has centered on the question of why a product created using modern biotechnology should be regulated more stringently or differently than other products. Are stricter regulatory controls necessary because of the *process* by which the product was created, or because the *product* itself has characteristics or qualities that render it potentially more risky to human health or the environment than products made using other techniques?

B. The Use of Modern Biotechnology Techniques

Modern biotechnology has been used in the manufacturing of commercial products for more than two decades. The late 1970s and early 1980s witnessed the creation of many different types of novel products. For instance, it was during this period of time that scientists at the University of California created a genetically modified bacterium (commonly referred to as *ice-minus*) that prevented ice crystal formation on the surface of plants and thus kept frost from destroying a crop.[3] The ice-minus bacteria could then be sprayed on crops.

This period also saw the first attempts at genetic modification of the plant itself, seeking to make the plant herbicide- and insect-resistant without applying a genetically modified spray. For instance, field tests began on tobacco, cotton, and tomatoes that incorporated genes from the soil bacterium *Bacillus thuringiensis* (Bt). Prior to genetic modification of the plant, Bt had been used as a spray-on pesticide. The insertion of the Bt genes in the plants themselves resulted in the development of proteins that were toxic to specific insects such as budworm, bollworm, and the European corn borer.

Also during this period, in the field of medicine the Food and Drug Administration (FDA) approved Humulin, the first genetically modified drug. Humulin is manufactured by genetically modifying *Escherichia coli* bacteria (a bacterium that resides in the human digestive tract) to create insulin that is identical to human insulin.[4] Prior to the development of this product, insulin was obtained from the pancreas of animals. Since these early years, biotechnology has expanded its reach in the agricultural and medical fields.

1. **Agriculture/Food Products** The use of genetic engineering in the field of agriculture has mainly focused on producing plants that are resistant to herbicides

---

3. *See* Foundation on Economic Trends v. Heckler, 756 F.2d 143, 149 (D.C. Cir. 1985).

4. *See* L. Altman, *A New Insulin Given Approval for Use in the U.S.*, N.Y. TIMES, Oct. 30, 1982 *available at* http://www.nytimes.com.

or insects, such as genetically modified corn, soy, and cotton. However, plants can also be genetically engineered to make them drought resistant, more nutritious, or capable of being grown in conditions of salinity. In the 1990s, there was an increase in both the acreage planted with genetically modified crops and the number of different types of crops that were genetically modified. However, two significant incidents—the Star Link™ incident and the Monarch Butterfly controversy—led to a severe curtailment in the introduction of new varieties of insect-resistant and herbicide-resistant seeds. As will be discussed in Chapters 3 and 5 respectively, the Star Link™ incident disrupted exports of corn from the United States while the Monarch Butterfly controversy generated intense adverse publicity and public outcry about the cultivation of such crops. The impact was widespread: Certain agricultural companies halted the developed of, or withdrew from commercial introduction of, certain genetically altered crops for fear of losing export sales. In fact, it was only during the 2008 growing season that for the first time since the 1990s, a new type of genetically modified crop was grown in the United States: sugar beets.[5]

Although the genetic modification of plants and their use in foods is a commercial reality, the genetic modification of animals for such purposes is not. Production of transgenic animals has primarily been limited to laboratory mice that serve as test subjects for drugs. With respect to food production, the exact number of companies that have sought federal government permission for genetically modified animals is unknown because this information is kept confidential. It is however known that a company attempted to obtain FDA permission for a transgenic salmon, but such permission has not been given.[6] Moreover, distinct from genetically modified animals are cloned animals, which could also be developed for food. However, the production of transgenic and cloned animals involves high costs, faces a skeptical public, and invokes ethical quandaries.

Finally, genetically modified hormones have been developed in order to be administered to domesticated animals. Specifically, the hormone recombinant bovine somatotropin (rBST) was approved in 1993 in order to improve milk production in dairy cows. As discussed in Chapter 6, the approval of this genetically modified animal drug was met with a firestorm of criticism, lawsuits, and state actions.

2. Drugs/Medical Devices Recombinant DNA technology has been and continues to be used to manufacture products that are intended for therapeutic, preventive, and diagnostic purposes. The technology has been used to produce enzymes, hormones (e.g., insulin), and interferons and their hybrids that have been tested and used to treat (among other things) diabetes, growth hormone deficiency, hepatitis B, Gaucher's disease, leukemia, multiple sclerosis, and anemia.

---

5. *See* A. Pollack, *Round 2 for Biotech Beets*, N.Y TIMES, Nov. 27, 2007.

6. *See* Office of Science and Technology Policy (OSTP), *Case Study No. 1: Growth Enhanced Salmon. Available at* www.ostp.gov/galleries/Issues/ceq_ostp_study2.pdf.

However, the development of such products is difficult. Unlike chemically synthesized drugs, products based on the genetic manipulation of a living organism or cell can result in complex mixtures that can be made only in small batches in order to ensure that the product is consistent; also, it may be targeted to address the condition in only a limited number of people. Nevertheless, there are now hundreds of products that are being manufactured or are in clinical testing based on the use of rDNA technology. Moreover, genetically modified bacteria have been used to detect pathogens, and there are hundreds of clinical diagnostic devices that use biotechnology.

II. NANOTECHNOLOGY

A. The Major Theoretical and Initial Breakthroughs in Nanotechnology

When it comes to nanomaterials, the discussion typically commences with the after-dinner speech given by the famous physicist, Richard Feynman, that is quoted below. However, the use of nanoscale materials predated Feynman's speech by hundreds of years. It's reported that in the medieval period, Arab ceramists applied thin layers of metallic silver and/or copper to glazed pottery to create the first luster potteries that would change color depending on the viewing angle.[7] Moreover, nanosized carbon black particles have been used in tires as a reinforcing agent for a hundred years.[8] However, because these uses do not represent intentional exploitation of nanomaterials, Feynman's speech is typically considered the starting point for a discussion on nanotechnology.

Feynman's after-dinner speech at the annual meeting of the American Physical Society at the California Institute of Technology on December 29, 1959 provided both a vision and a challenge to his audience. Feynman's analysis was influenced by the then-recently announced discoveries on DNA. Feynman said:

> This fact—that enormous amounts of information can be carried in an exceedingly small space—is, of course, well known to the biologists, and resolves the mystery which existed before we understood all this clearly, of how it could be that, in the tiniest cell, all of the information for the organization of a complex creature such as ourselves can be stored. All this information—whether we have brown eyes, or whether we think at all, or that in the embryo the jawbone should first develop with a little hole in the side so that later a nerve can grow through it—all this information is contained in a very

---

7. ALLIANZ, ET AL., SMALL SIZES THAT MATTER: OPPORTUNITIES AND RISKS OF NANOTECHNOLOGIES 6 (2005).

8. Consortium of Nanoscale Science and Engineering Controls, Environmental, Health and Safety Guidelines to NSF Nanoscale Science and Engineering Research Centers, available at www.cise.columbia.edu/nsecnetwork/health.php.

tiny fraction of the cell in the form of long-chain DNA molecules in which approximately 50 atoms are used for one bit of information about the cell.[9]

Feynman challenged his audience to develop a more powerful electron microscope in order for research to develop, but also presciently asked many of the same questions still being posed about the potential uses for—and consequences of—rearranging atoms at a molecular level:

> When we get into the very, very small world—say circuits of seven atoms—we have a lot of new things that would happen that represent completely new opportunities for design. Atoms on a small scale behave like nothing on a large scale, for they satisfy the laws of quantum mechanics. So, as we go down and fiddle around with the atoms down there, we are working with different laws, and we can expect to do different things. We can manufacture in different ways.[10]

It was not until 1974, however, that the term *nanotechnology* was coined by Norio Taniguchi of Tokyo University. He described nanotechnology as mainly consisting of the process of "separation, consolidation, and deformation of materials by one atom or one molecule."[11] What is covered under the rubric of nanotechnology remains a matter for scientific and regulatory discussion. A nanometer is one-billionth of a meter. While different regulatory agencies have provided different definitions of the term *nanotechnology*, typically, each of those definitions are a variation of the following understanding: *nanotechnology* refers to the creation, control, and use of materials at roughly 1 to 100 nm, in order to create and use structures, devices, and systems that have novel properties and functions because of their small and/or intermediate size. For example, the Environmental Protection Agency (EPA) defines *nanotechnology* as:

> [R]esearch and technology development at the atomic, molecular, or macromolecular levels using a length scale of approximately one to one hundred nanometers in any dimension; the creation and use of structures, devices and systems that have novel properties and functions because of their small size; the ability to control or manipulate matter on an atomic scale.[12]

At the level of 1 to 100 nm, as Feynman indicated, "different laws" are applicable. Because a nanoscale material has a relatively larger surface-area-to-mass

9. R. P. Feynman, *There's Plenty of Room at the Bottom* (1959), *available at* http://www.zyvex.com/nanotech/feynman.html. Originally published in the Feb. 1960 issue of Caltech's Engineering and Science.

10. *Id.*

11. *See* N. Taniguchi, *On the Basic Concept of Nano-Technology.* Proc. Int. Conf. Prod. Eng., JSPE, Tokyo, 1974.

12. U.S. Environmental Protection Agency, Nanotechnology White Paper, EPA 100/B-07/001, (February 2007) at 5.

ratio than conventionally sized or bulk matter, it can become more chemically reactive, and can manifest changes in its strength, toxicity, durability, and flexibility. "Moreover, below 50 nm, the laws of classical physics give way to quantum effects, providing different optical, electrical and magnetic behaviours."[13]

The next major event in the development of nanotechnology was the popularization of nanotechnology through the publications of K. Eric Drexler.[14] Drexler's approach referred to *molecular nanotechnology*, which he analogized to living organisms. Specifically, he argued that the structure and functions of proteins demonstrated how it would be possible to create mechanical devices on the molecular scale that performed similar functions. He envisioned versatile mechanical systems that had "structural members, moving parts, bearings, and motive power."[15] As Drexler noted, these "molecular assemblages of atoms can act as solid objects, occupying space and holding a definite shape."[16] Moreover, these "molecular assemblages" could then replicate themselves and create new molecular devices. The vision that Drexler articulated in his publications was of molecular-scale devices that could ultimately cure diseases and extend life. The appeal of Drexler's theory is aptly summarized by the following observation:

> Using nanotechnology, production would be carried out by large numbers of tiny devices, operating in parallel, in a fashion similar to the molecular machinery already found in living organisms. However, these "nanodevices" would not suffer from the constraints facing living organisms. . . . [T]hey could be constructed of whatever material, in whatever fashion, is most suited to their task. Known as "assemblers," these tiny devices would be capable of manipulating individual molecules very rapidly and precisely.[17]

Drexler's view, however, also came under criticism. Bill Joy, chief scientist at Sun Microsystems, wrote an article broadly covered in the popular press in which he hypothesized the potential for self-replicating nanomachines to destroy entire ecosystems. Accordingly, he articulated that humility and common sense in the face of powerful technologies requires the relinquishment of scientific pursuit of developing such technologies.[18] Others, such as Nobel Prize recipient Richard

13. ALLIANZ, ET AL., *supra* note 7, at 7.

14. E. K. Drexler, *Molecular Engineering: An Approach to the Development of General Capabilities for Molecular Manipulation.* 78 PROC. NATL. ACAD. SCI. USA. 5275–5278 (Sept. 1981), Chemistry section. *Available at* http://www.imm.org/publications/pnas/.

15. *Id.*

16. *Id.*

17. G. H. Reynolds, *Forward to the Future: Nanotechnology and Regulatory Policy*, PAC. RES. INST., Nov. 2002.

18. B. Joy, *Why the Future Doesn't Need Us*, WIRED (Apr. 2000), *available at* http://www.wired.com/wired/archive/8.04/joy.html. It should be noted that Drexler himself raised the prospect that, because of their superior capabilities, self-replicating nanobots could obliterate life, which he noted was referred to by the cognoscenti of nanotechnology as the

Smalley, argued that while creating nanomaterials for application is realistic, nanomachines are not, and thus Joy's fears will not be realized.[19] Meanwhile, as this theoretical debate raged, companies moved forward with testing and manufacturing products using nanotechnology.

There are two broad techniques for manufacturing nanoscale materials: the *bottom-up* and the *top-down* approaches. The bottom-up approach "starts with constituent materials (often gases or liquids) and uses chemical, electrical, or physical forces to build a nanomaterial atom-by-atom or molecule-by-molecule."[20] One mechanism to achieve the bottom-up method involves the atoms or molecules arranging themselves through a process of self-assembly, such as occurs in nature with salt crystals or snowflakes.[21] That is, natural forces would drive the atoms to self-assemble because of the need to create a thermodynamically stable state. However, in order to commercially exploit this process, it is necessary to understand and be able to control and manipulate the thermodynamic, gravitational, and electrodynamic forces on a nanoscale.[22] Another mechanism is through chemical synthesis. For example, one method involves beginning with a solution of nanoparticles, then using electrodes to apply a specific voltage. The charged particles would be attracted to the surface where they would bind themselves and create a film, which then could be applied as a coating to the particular production process.

The top-down approach is based on the traditional manufacturing view of taking a larger material and removing matter through etching, milling, or machining until only the nanoscale features remain.[23] The top-down approach can be achieved using techniques that have been used for decades in the semiconductor field, such as photolithography.

[This technique] involves the patterning of surface through exposure of light, ions, or electrons, and then subsequent etching and/or deposition of material on to the surface to produce the desired device . . . The main lithographic tools can be conveniently separated into methods that use a focused beam of electrons or ions to write patterns, and those that rely on the projection of light through a mask to define a pattern. . . .[24]

"gray goo" problem. Moreover, Joy advocated relinquishment not only for molecular nanotechnology, but also for certain genetic technologies. *See id.*

19. K. Chang, *Yes, They Can! No They Can't: Charges Fly on Nanobot Debate*, N. Y. TIMES, (Dec. 9, 2003).

20. *See* ALLIANZ ET AL., *supra* note 7, at 12.

21. The Royal Society & The Royal Academy of Engineering, *Nanoscience and Nanotechnologies: Opportunities and Uncertainties* (2004) at 27, *available at* http://www.nanotec.org.uk/finalReport.htm.

22. *See id.* at 26.

23. *See id.* at 28.

24. *See id.* at 29.

Regardless of the process, various types of nanomaterials are being tested and used in manufacturing. Among the nanomaterials that will be discussed in following chapters are:

- *Buckyballs* (also known as Fullerene or C60) are closed-hollow networks of small rings that form spherical particles. The name *buckyballs* refers to their resemblance to the geodesic domes built by architect Buckminster Fuller. Because the number of carbon atoms in each buckyball can be different, the shapes can also differ. Although the theoretical possibility of fullerenes was known for decades, scientists were first able to produce stable clusters of carbons in 1985. Though C60, referring to the number of carbon atoms that make up one sphere, is the most common fullerene, researchers have found stable, spherical carbon structures containing 70 atoms (C70), 120 (C120), 180 (C180), and others.[25]
- *Carbon nanotubes* are elongated versions of buckyballs. They are generally classified as single-walled carbon nanotubes (SWNTs), consisting of a single cylindrical wall that is hollow inside, or multiwalled nanotubes (MWNTs), which have cylinders within the cylinders.[26] Carbon nanotubes have various desirable quantities, including electric and heat conductivity, stiffness, and strength. The qualities are of significant importance in a number of fields, including energy production (in which solar cell efficiency can be increased through the application of a carbon nanotube coating), telecommunications (in which carbon nanotubes can be used to produce and detect pulses), and the medical industry (in which the potential exists to fill the carbon nanotube with a drug and thus serve as a drug delivery system).[27]
- *Quantum dots* are metal-based rather than carbon-based structures. They can be as small as 1 nm. Because of their small size, quantum effects are in play. As a result, there are changes in both the physical and optical properties in comparison to the bulk form. Specifically, the smaller the quantum dot, the higher the energy and intensity of the light that is emitted. Thus, quantum dots can find applications in the medical imaging field, in computer and television displays, and in lighting.
- *Dendrimers* are spherical, highly branched polymer molecules with numerous chain ends. They are assembled in discrete steps, which allows them to be constructed to perform specific functions.

---

25. U.S. National Institutes of Health. National Cancer Institute. NCI Alliance for Nanotechnology in Cancer. Nanotechnology Glossary. *Available at* http://nano.cancer.gov/resource_nanotech_glossary.asp.

26. *See* ALLIANZ ET AL., *supra* note 7, at 8–9.

27. *See* National Nanotechnology Initiative Grand Challenge Workshop, *Nanoscience Research for Energy Needs* (March 2004); T. Morton, *Nanotube Composites Help Fiberoptic Communication* (Sept. 26, 2008), http://www.arstechnica.com.

Dendrimers contain interior cavities and as such, the molecules could be placed in them such as a tumor-targeting or imaging-contrast agent.[28]

- *Composites* combine different nanomaterials or combine a nanomaterial with a bulk or conventionally sized particle.[29]

B. Uses of Nanotechnology

Virtually every single day, research is published that indicates a potential new use for nanoscale material. Every year more fields seem impacted by nanotechnology. Nanoscale materials have been used in a number of consumer products, including stain-resistant textiles, antimicrobial appliances, television displays, computer consoles, and sporting goods. With respect to many of these, nanosilver has been applied for its antimicrobial or antibacterial properties.[30] As discussed in Chapter 5, the application of an antimicrobial agent has caused consternation as to its impact on the environment. In the field of cosmetics, nanoscale materials have been used as delivery systems in anti-wrinkle creams and other beauty aids. These products have been touted as delivering the same anti-wrinkle results as Botox® at lower cost and with less pain. Moreover, nanoscale versions of titanium oxide and zinc oxide are being used in sunscreen products. Titanium oxide and zinc oxide have long been active ingredients in sunscreen.[31] On a conventional scale, titanium oxide and zinc oxide tend to appear as a white film when applied to the skin.[32] However, nanoscale titanium oxide and zinc oxide appear clear when applied to the skin and are claimed to be just as effective in blocking the sun's harmful ultraviolet rays.[33] Given that most consumers prefer a clear application, it's not surprising a myriad of companies are advertising that their sunscreen contains nanoscale titanium oxide and zinc oxide.[34]

In the field of medicine and drug devices, nanoparticles are seen as providing a mechanism by which time-released, targeted drug delivery can be achieved, thereby reducing the adverse impacts to healthy neighboring cells that may

---

28. *See* U.S. National Institute of Health, *supra* note 25.

29. *See* U.S. Environmental Protection Agency, *supra* note 12, at 10.

30. *See generally* The International Center for Technology Assessment et al., Petitioners, Filed With: Andrew C. Von Eschenbach in his official capacity as Acting Commissioner, Food and Drug Administration, May 16, 2006. *Petition Requesting FDA Amend Its Regulations for Products Composed of Engineered Nanoparticles Generally and Sunscreen Drug Products Composed of Engineered Nanoparticles Specifically* ("Petition"). May 16, 2006.

31. *Id.*

32. *Id.*

33. *Id.*

34. Woodrow Wilson International Center for Scholars, Project for Emerging Technologies, *New Nanotech Products Hitting the Market at the Rate of 3–4 Per Week* (Apr. 24, 2008), *available at* http://pewnanotechproject.us/news/archive/6697/.

result from conventional treatment.[35] Many types of nanomaterials are being considered for drug delivery, including SWNTs, nanoshells, and dendrimers. For example, carbon nanotubes have already been used to deliver drugs in a variety of cell culture systems to address cancerous tumors.[36] The nanotube is attached to a peptide or antibody on its outer surface; when administered, that peptide or antibody binds to its target.[37] Once the binding has occurred, the drug (which can be either inside or outside the nanotube) is released due to changes in pH or enzymes produced by the tumor, thereby delivering the drug directly to diseased areas of the body.[38] Nanoshells, on the other hand, are beads coated with gold that are linked to an antibody.[39] The antibody then binds to the target. Once the binding occurs, near-infrared light is applied, causing the light-absorbing nanoshells to generate heat and thereby destroy the targeted cells.[40] In either case, nanodevices are used to minimize the exposure of healthy tissues while more precisely targeting those areas that need to be treated.

Nanoparticles are also seen as offering tremendous benefits in the early and accurate detection of diseases. For instance, the current method for detection of cancer is through physical examination, symptoms, or imaging.[41] However, nanoparticles could be used to enhance imaging or to detect precancerous changes in cells.[42] For example, magnetic nanoparticles, such as manganese-doped iron oxide, can act as a powerful contrast agent in high performance magnetic resonance imaging (MRI).[43] Finally, researchers envision the possibility of linking both detection and drug delivery through a single nanoscale device.

Nanomaterials are also being tested or being used for both sewage treatment and as a tool for environmental remediation. Due to their enhanced ability to absorb and react with organic and inorganic materials in the environment, nanotechnology may improve environmental remediation and treatment by enabling quicker, more cost-effective cleanup of environmental contaminants in water, soil, and air as compared to current methods. To illustrate this point, EPA cites

---

35. *See id. See also*, National Cancer Institute, *Carbon Nanotubes Target Tumor Cells, Deliver Anticancer Drugs*, *available at* http://www.nano.cancer.gov/news_center/2008/aug/nanotech_news_2008—08-14b.asp; National Cancer Institute, *Nanoparticles Deliver DNA-Drug Combos to Tumors*, *available at* http://www.nano.cancer.gov/news_center/nanotech_news_2006-10-16a.asp.

36. Y. Liu & H. Wang, *Nanotechnology Tackles Tumours*, 2 NATURE NANOTECHNOLOGY 20.

37. *See id.*

38. *See id.*

39. National Cancer Institute, *Understanding Cancer and Related Topics: Understanding Nanodevices*, *available at* http://www.cancer.gov/cancertopics/understandingcancer.

40. *See id.*

41. *Id.*

42. *See id.*

43. Y. M. Huh et al., *Hybrid Nanoparticles for Magnetic Resonance Imaging of Target-Specific Viral Gene Delivery*, 19 ADVANCED MATERIALS 3109 (2007).

nanoscale zero-valent iron as being able to clean up chloro-organics in groundwater, metalloporphyrinogens and titanium-based nanomaterials as enabling degradation of chlorinated compounds, and other nanomaterials as being useful to increase retention or solubilization of contaminants for remediation.[44]

Nanomaterials may also have a significant role in increasing the efficiency of energy sources provided that technological challenges are mastered.[45] For example, the typical solar cell has a low efficiency of converting light energy to electrical energy. Carbon nanotubes and quantum dots may be able to increase those efficiencies. Because they are superior to copper at conducting electricity, carbon nanotubes could be used to send electricity through power transmission lines. Nanomaterials may also have the potential to enhance power storage in batteries and cells and to create improvements over incandescent or fluorescent lighting.[46]

Finally, there are many other fields in which the nanoscale materials are being used that at first blush would be considered surprising. For example, nanoscale materials are being used in food packaging or in food as additives, and even as a means of ensuring that money is not counterfeit.

The remaining chapters of this book will examine the various regulatory issues associated with the development of these technologies. The next chapter contains a discussion of the various health and environmental issues associated with these technologies. These risks will be a key part of the discussion in all of the remaining chapters because they determine the appropriate degree of regulatory control.

---

44. *See* U.S. Environmental Protection Agency, *supra* note 12, at 5.

45. National Science Technology Council et al., *Nanoscience Research for Energy Needs: Report of the March 2004 National Nanotechnology Initiative Grand Challenge Workshop,* (June 2005) at 6–12.

46. *See* National Nanotechnology Initiative Grand Challenge Workshop, *supra* note 29.

3. UNDERSTANDING THE RISKS FROM NANOTECHNOLOGY AND BIOTECHNOLOGY

I. INTRODUCTION

Products made using nanotechnology and biotechnology hold significant promise, yet questions and concerns have been raised about the risks they pose to human health and the environment, their economic impacts, and the ethical dimensions of using these technologies. As a result, this chapter covers a broad range of subjects that have shaped, and continue to shape, the debate about the way these technologies are to be regulated.

As this chapter will discuss, the impacts on human health are the driving force for the way products and technologies are regulated. The methodology used for evaluating such impacts requires an assessment of the specific risks associated with that product (i.e., risk assessment) and what mechanisms can practically be used to maintain those risks at an acceptable level (i.e., risk management). This methodology is referred to as the quantitative risk assessment paradigm. This chapter will examine the historical development of the quantitative risk assessment paradigm, the components of the paradigm, and the way the paradigm has been applied in the context of agricultural biotechnology (i.e., genetically modified plants and foods derived from such plants) and nanotechnology.[1]

The chapter will also examine other factors that will be used to determine how a product will be regulated. These factors include the ecological risks associated with, and the economic and societal impacts related to, agricultural biotechnology and nanotechnology. Then, the chapter will focus on two issues that have significant influence on how agricultural biotechnology and nanotechnology are regulated: uncertainty and public perceptions. Finally, the chapter will close with an examination of the precautionary principle. Some individuals and groups have argued that instead of the quantitative risk assessment model, a stringent application of this regulatory philosophy should be followed in order to prevent the manufacture and release into the marketplace of products whose risks are not fully understood.

---

1. There are a number of reasons for focusing only on the risks associated with agricultural biotechnology rather than all of biotechnology. For example, there has been an intense debate with respect to the risks posed by agricultural biotechnology while, generally speaking, biotech drugs and devices have not been subject to such intense controversy.

II. HISTORY OF THE RISK ASSESSMENT PARADIGM

The federal government's use of the modern quantitative risk assessment paradigm began almost 40 years ago and has continued to evolve. In the 1970s, regulators began to use risk assessment methodologies pertinent to carcinogens to assist in the formulation of appropriate regulatory policy.[2] In the prior decades, products that were potentially carcinogenic were dealt with in one of three ways: they were banned from the marketplace (e.g., Congressional authorization given to FDA to ban food additives believed to be carcinogenic); "permissible" levels were determined based on what was technically feasible; or the issue was simply ignored.[3] But, by the mid-1970s, these three options were being challenged. On the one hand, the public's growing awareness of the adverse health impacts from chemicals and other products in the marketplace, along with the development of technology that allowed detection of carcinogens at lower concentrations (e.g., carcinogens present in parts per billion were not able to be detected until the late 1970s), meant that ignoring them was no longer an option.[4] On the other hand, prohibiting the use or manufacturing of products was considered impracticable due to economic and political considerations.[5]

The third option—that of using technologically feasible limits—was addressed by the U.S. Supreme Court in *Indus. Union Dep't v. Am. Petroleum Inst.*, 448 U.S. 607, 100 S.Ct. 2847 (1980) (commonly referred to as the "*Benzene* decision"), which placed the Court's imprimatur on using quantitative risk assessment. In the 1970s, a link between exposure to high concentrations of benzene and leukemia was finally confirmed through epidemiological studies after such an association had been suspected as early as the 1920s based on individual reports.[6] As a result, OSHA set a standard by reducing permissible airborne concentration of benzene in the workplace from 10 ppm to 1 ppm.[7] The rationale for this position was that when no safe level can be determined, the exposure limit should be set at the lowest technologically feasible level that will not impair the viability of the industry.[8] While there was no majority opinion, the plurality of the Court held that before setting any standard, OSHA needed to establish that there were in fact, "significant risks" in the workplace due to exposure to benzene, and that

---

2. *See* NATIONAL RESEARCH COUNCIL, SCIENCE AND JUDGMENT IN RISK ASSESSMENT 31 (1994) [hereinafter *NRC, Science and Judgment*].

3. *See id. See also*, NATIONAL RESEARCH COUNCIL ET AL., RISK ASSESSMENT IN THE FEDERAL GOVERNMENT: MANAGING THE PROCESS 54–55 (1983) [hereinafter *Risk Assessment in the Federal Government*].

4. *NRC, Science and Judgment*, *supra* note 2, at 32.

5. *See id.*

6. *See id.*

7. Indus. Union Dep't v. Am. Petroleum Inst., 448 U.S. 607, 608, 100 S.Ct. 2844, 2847 (1980).

8. 448 U.S. at 613, 100 S.Ct. at 2849.

those risks could be eliminated or lessened by making changes to the permissible level.[9] While "significant risk" need not be determined by a mathematical straight-jacket or scientific certainty, the agency was required to demonstrate through the use of scientific evidence that it is at least more likely than not that long-term exposure to 10 ppm of benzene presented a significant risk of material health impairment.[10] In other words, the agency had to evaluate data in order to confirm that a risk existed and that a certain level of exposure would cause harm to the human body, not simply find the lowest point that contaminants could be reduced to and still be technologically feasible.

In the wake of the *Benzene* decision, and as criticism mounted from industry, consumer groups, and scientists regarding procedures used for estimating risk, Congress issued a directive to FDA to arrange for the National Research Council (NRC) to undertake a study of how federal agencies had been conducting risk assessment.[11] The NRC examined four federal agencies that had primary author-ity over environmental, health, and safety issues: OSHA, EPA, FDA, and CPSC.[12] The committee's report synthesized the scientific principles and policies that had already been adopted by these agencies while recommending certain grad-ual alterations to make the policies more effective. This report (commonly referred to as the "Red Book"), which remains the touchstone for the way risk is assessed, presented four analytic risk assessment steps: hazard identification, dose-response assessment, exposure assessment, and risk characterization. As noted in the Red Book, identification of hazards and determination of the dose at which there is a biological response are the components necessary for deter-mining toxicity. Exposure assessment involves evaluation of the pathways a par-ticular contaminant may travel and the routes it uses to enter the human body. The information garnered through the evaluation of toxicity and exposure (which combined are referred to as "risk characterization") is then placed in the context of political, social, and economic realities to create a practical policy prescription for controlling that risk (e.g., prohibition of an activity, labeling, etc.). This last step is referred to as *risk management.*

Following the issuance of the 1983 report, agencies (particularly EPA) devel-oped guidance documents that addressed a number of facets of risk assess-ment. For example, EPA issued guidelines for assessment of carcinogen risk, chemical mixture risk, ecological risk, neurotoxicity risk, reproductive toxicity risk, developmental toxicity, and mutagenicity risk.[13] In fact, EPA's Nanotechnology

---

9. 448 U.S. at 642, 100 S.Ct. at 2864.

10. 448 U.S. at 653, 100 S.Ct. at 2869.

11. *NRC, Science and Judgment, supra* note 2, at 33.

12. *Risk Assessment in the Federal Government, supra* note 3, at 5.

13. For a complete list of guidance documents, EPA's National Center for Environmental Assessment compiles the information at http://cfpub.epa.gov/ncea/cfm.

White Paper (February 2007) noted that the 1983 report served as the basis for its analysis of nanotechnology risk.

III. HEALTH RISK ASSESSMENT

A. The Basics of Toxicity: Hazard Identification and Dose Response

The evaluation of toxicity involves two steps: hazard identification and dose-response evaluation. Hazard identification involves the use of data to identify contaminants that are likely to pose health hazards.[14] Dose–response evaluation involves determining the conditions or levels at which the presence of a contaminant may trigger a biological response from the body.[15] Both hazard identification and dose–response assessment are based on using data from epidemiologic and animal bioassay studies.[16] There are significant advantages and disadvantages associated with each test methodology. Given that the data generated from such studies is used to justify certain arguments, it is necessary to appreciate the limitations associated with each methodology.

Epidemiological studies analyze whether a particular substance increases the risk of disease in groups of individuals. There are two types of epidemiological studies: prospective and retrospective. Prospective epidemiological studies follow a target group of individuals and a control group of individuals from a specific point in time forward while seeking to limit the potential for confounding variables (e.g., age, being a smoker). In contrast, retrospective epidemiological studies start at an endpoint and look backwards at the population. The latter are more common and significantly more prone to bias from confounding variables.[17] These confounding variables include: obtaining an appropriate population with a sufficient sampling size (both as the putative exposed group and nonexposed group) to avoid sampling errors; obtaining accurate information about the health of the subject individuals; obtaining accurate information on the duration, dose, and exposure of the substance; and identifying all the factors that may be influencing the health of the individuals being exposed.[18]

---

14. NRC, *Science and Judgment*, *supra* note 2, at 26.

15. *Id.* at 56.

16. Ideally, the best mechanism for determining whether a particular substance is related to the risk of developing a certain disease is to conduct a randomized experimental study of humans (*i.e.*, clinical study). *See* M. Green et al., *Reference Guide on Epidemiology*, *in* REFERENCE MANUAL ON SCIENTIFIC EVIDENCE (2nd ed., 2000) 338–339, *available at* www.fjc. gov/public/pdf.nsf/lookup/sciman06.pdf/$file/sciman06.pdf. However, precisely because clinical test subjects are humans, there are many experimental design problems in assuring that humans are not exposed to substances known or suspected to be toxic. *See* Ethyl Corp v. U.S. Envtl. Prot. Agency, 541 F.2d 1, 26 (D.C. Cir.), *cert denied* 426 U.S. 941 (1976).

17. NRC, *Science and Judgment*, *supra* note 2, at 57.

18. *See id.*

While epidemiologists use various techniques to deal with these issues, without further analysis the data from an epidemiological study only allows the researcher to draw inferences and associations.[19] Typically that conclusion is based on epidemiologists weighing the results from several studies (ideally involving different populations and investigative methods), and determining if there is a consistent pattern of findings, whether there is a statistical relationship between a particular disease and particular substance, whether the risk of the disease is higher with increased exposure to the particular product, and the degree to which other factors can be ruled out.[20]

The other test methodology involves experimentation on live animals and an examination of their tissues after death. Unlike epidemiological studies, the conditions under which the animals live are controlled and manipulated; thus, establishing causation is generally not as difficult.[21] Moreover, animal tissues can be dissected postmortem in order to make an accurate assessment of the precise parts of the animal's body that were adversely impacted. Because it is ethically and legally prohibited to test humans under conditions that one could test an animal (notwithstanding that animal tests are regulated), absent clear evidence to the contrary, the assumption is that toxicity observed in an animal species is potentially predictive of a similar response in at least some humans.[22]

However, some obvious disadvantages are associated with such studies. The main one is that despite enormous genetic similarities between different species of animals and humans, there remain significant and relevant differences (such as metabolic rates) that must be accounted for.[23] A second concern is that researchers must account for the dosage typically administered to animals that is higher than the concentration in the environment to which humans would be exposed; thus, dose–response or the dose at which there is no effect needs to be examined in the context of actual human exposure.[24] A classic case involved a series of studies on rats exposed to formaldehyde. The rats developed squamous cell carcinoma of nasal tissues. However, this could not be extrapolated to humans because (a) the rats were exposed to levels of formaldehyde that humans would never stand for, and (b) rats, unlike humans, are obligate nose breathers and thus are more susceptible to respiratory pathogens.

After using the data from these test methods to determine which substances are actually hazardous to human health, the next step is to determine at what level of exposure a biological response will manifest.[25] An evaluation of dose–response

---

19. *See id.*
20. *See id.*
21. *See id.* at 58.
22. *See id.*
23. *See id.* at 59.
24. Green, *supra* note 16, at 346.
25. NRC, *Science and Judgment, supra* note 2, at 60.

involves identifying the level at which there is no observed effect level or the lowest level at which there is an adverse effect. For example, in a recent study on rats that were exposed to carbon nanotube dust, it was determined that no effects were observed at levels of 0.1 mg/m^3, but that adverse effects were detected in the lungs at higher levels of 0.5 mg/m^3 and of 2.5 mg/m^3.[26]

1. Evaluating Toxicity of Nanomaterials Given the variety and complexity of nanomaterials (e.g., ranging from carbon fullerenes to single-walled carbon nanotubes to multiwalled carbon nanotubes to quantum dots), a substantial amount of research has yet to be undertaken to determine which nanomaterials pose a hazard and at what levels of exposure biological responses will be generated. The preliminary research using animal bioassay studies, however, shows that nanoscale materials do not share the same characteristics or properties as their conventionally sized counterparts; therefore, the adverse effects of nanoparticles cannot be predicted or inferred from larger-sized materials.[27]

These animal tests indicate the same factors that make nanoparticles technologically interesting could mean they have the potential for increased toxicity and/or an increased ability to evade the body's natural defense mechanisms and thereby cause more extensive damage.[28] Researchers have noted that the toxicity of nanoparticles may be influenced by a laundry list of factors (including size, surface area, mass, chemical composition, crystal structure,[29] shape, purity, and inclusion of a surfactant).[30] Moreover, a particular focus has been placed on the relationship between total surface area of a nanoparticle and its toxicity. Because

---

26. Letter from BASF to United States Environmental Protection Agency (dated July 8, 2008) (on file *at* http://www.epa.gov/oppt/tsca8e/pubs/8emonthlyreports/2008/8eaug2008.htm). In addition to the BASF submission, a number of companies have submitted toxicological data on nanomaterials under Section 8(e) of TSCA. See Chapter 10 on TSCA. A summary of some of that data is found in Richard Denison's blog on nanotechnology. See R. Denison, "Shining a (Partly Shaded) Light on Nanomaterials that Present 'Substantial Risk,'" (October 31, 2008) at www.blog/edf.org.

27. *See* NNI, *Environmental, Health and Safety Research Needs for Engineered Nanoscale Materials* (September 2006) at 20.

28. *See, e.g.*, ALLIANZ, ET AL., SMALL SIZES THAT MATTER: OPPORTUNITIES AND RISKS OF NANO-TECHNOLOGIES 30 (2005).

29. U.S. Environmental Protection Agency, Nanotechnology White Paper, EPA 100/B-07/001 (February 2007) at 54. [hereinafter *EPA White Paper*].

30. National Nanotechnology Initiative, *Environmental, Health and Safety Research Needs for Engineered Nanoscale Materials* (September 2006) at 20. When discussing nanomaterial research, it must be recognized that the primary focus has been on certain types of nanomaterials. Specifically, the research has been conducted with metal oxides (e.g., titanium oxide and zinc oxide), fluorescent crystalline semiconductors, fullerenes, and carbon nanotubes. *See id.* at 21. *See also* G. Oberdorster et al., *Principles for Characterizing the Potential Human Health Effects from Exposure to Nanomaterials: Elements of a Screening Strategy*, 2 PARTICLE AND FIBRE TOXICOLOGY 9 (2005), *available at* http://www.particleandfibretoxicology.com/content/2/1/8.

of their size, nanoparticles have a larger surface area per unit of mass as compared to larger particles.[31] As a result of an increased surface-to-volume ratio, nanoparticles may be more biologically reactive, which in turn may also lead to an increase in toxicity.[32]

This raises the issue of the appropriateness of the existing dose–response model in evaluating the potential toxicity of nanomaterials. The model focuses on accumulated mass and mass concentration to determine at what level there are no observed effects or what is the lowest level at which there is an adverse effect. However, if as noted above, total surface area and surface reactivity are more appropriate indicators of toxicity, then new methodologies and new standards may be necessary to ensure adequate protection of human health.[33]

As an illustration of the potential toxicological issues confronting nanoparticles, a 2004 study of juvenile largemouth bass that had been exposed to aqueous *uncoated* fullerenes (buckyballs) for a 48-hour period demonstrated a 17-fold increase in a form of cellular damage in the brain.[34] As tons of fullerenes are produced each year, the results of the study (which were widely quoted in the popular press)[35] had particular salience. However, fullerenes can be *coated* to make them nontoxic.[36] Thus, the question remains: Would a fullerene lose its coating under real world conditions? In the case of coated quantum dots that are initially nontoxic, when exposed to air or ultraviolet radiation they lose their coating and become cytotoxic.[37] And, what would happen if under real world conditions the coating was removed due to interaction with the environment?

---

31. Oberdorster, *supra* note 30, at 9.

32. *See id.* Linda-Jo Schierow, *CRS Report for Congress: Engineered Nanoscale Materials and Derivative Products: Regulatory Challenges* (Jan. 22, 2008) at CRS-7; International Center for Technology Assessment, *Petition for Rulemaking Requesting EPA Regulate Nano-Silver as Pesticides* (February 2008) at 61.

33. David Rejeski, Testimony to the U.S. House of Representatives, Committee on Science, *Environmental and Safety Impacts of Nanotechnology; What Research is Needed?* (Nov. 17, 2005) at 5 (quoting from *Principles for Characterizing the Potential Human Health Effects from Exposure to Nanomaterials: Elements of a Screening Strategy*, 2 PARTICLE AND FIBRE TOXICOLOGY 9 (2005), *available at* http://www.particleandfibretoxicology.com/content/2/1/8.

34. E. Oberdorster, *Manufactured Nanomaterials (Fullerenes, C_{60}) Induce Oxidative Stress in the Brain of Juvenile Largemouth Bass*, 112 ENVIRONMENTAL HEALTH PERSPECTIVES 1058–1062 (July 2004).

35. *See, e.g.*, Kenneth Chang, *Tiny is Beautiful: Translating "Nano" into Practical*, N.Y. TIMES, Feb. 22, 2005 (stating that "a study last year reported that exotic soccer-ball shaped carbon molecules known as buckyballs in water caused brain damage in fish"); B. Holmes, *Buckyballs Cause Brain Damage in Fish*, 13 NEW SCIENTIST 31 (Mar. 29, 2004).

36. Oberdorster, *supra* note 34, at 1058.

37. *See id.* (citing to A. Derfus et al., *Probing the Cytotoxicity of Semiconductor Quantum Dots*, 4 NANO LETTERS 11–18); *see also EPA White Paper*, *supra* note 28, at 37 (citing to Hardman, 2006).

2. Issues Involved in Evaluating Toxicity with Genetically Modified Organisms
Unlike the examination of toxicity associated with chemicals or nanoparticles, the starting point for analyzing whether a genetically modified organism is toxic begins with understanding the degree to which there is substantial equivalence in the genotypic and phenotypic characteristics of the genetically modified organism and a well-understood nonmodified organism.[38] The purpose of a comparative analysis is to assist in determining what additional testing will be necessary to establish the safety level of the genetically modified plant. There are two important issues in the comparative analysis: will the genetically modified plant or food cause (a) antibiotic resistance or (b) an allergic reaction in humans?

The issue of antibiotic resistance generally arises due to the use of a "selectable marker gene" in the genetic engineering process. Developers recognize the gene they seek to integrate into the plant will actually be taken up, integrated, and expressed in only a small portion of plant cells; they also know these few cells are not easily distinguishable from the vast majority of plant cells that do not to take up the gene.[39] As a result, to increase their chances of finding only those plants that successfully have taken up the desired gene, developers couple the desired gene with an easily distinguishable gene, referred to as the selectable marker gene (e.g., kan[r] gene).[40] Thus, when a plant takes up the desired gene, it also takes up the selectable marker. Now, certain selectable markers encode enzymes that are resistant to antibiotics. Therefore, when a specific antibiotic is applied (e.g., kanamycin), the antibiotic will kill the plants that do not have the selectable marker (and therefore the desired gene).[41] The developer is then able to cultivate those plants that have the desired gene.

This raises the possibility that when the genetically modified plant is processed or used as food, the selectable marker will, along with the desired gene, be processed into the food. The question then arises as what to impact this antibiotic resistance enzyme will have on humans who use the antibiotic or on the bacteria that exist in the human digestive tract that may develop a resistance to that antibiotic or to other antibiotics.[42]

---

38. The concept of *substantial equivalence* was adopted in 1983 by the Organisation for Economic Co-operation and Development (OCED) and has subsequently been adopted by the World Trade Organization (WTO) and FDA. *See* 1992 rule discussed in Chapter 6.

39. *See* Food & Agricultural Organization of the United Nations, *The State of Food and Agriculture: 2003–2004*, (2004), Section 5 [hereinafter *State of Food and Agriculture*].

40. *See* 59 Fed. Reg. 26700, 26702 (May 23, 1994); *see generally* FDA/Center for Food Safety and Applied Nutrition (CFSAN), *FDA's Policy for Foods Developed by Biotechnology* (1995), *available at* http://vm.cfsan.fda.gov/~lrd/biopolcy.html [hereinafter *FDA/CFSAN Biotech Foods Policy*].

41. Council for Biotechnology Information, *The Use of Antibiotic Resistance Markers to Develop Biotech Crops*, (May 2001) at 1.

42. *FDA/CFSAN Biotech Foods Policy*, *supra* note 40. The importance of this issue may diminish with time as new genetic transformation methods rely less on antibiotic-resistant

As an illustration of the way antibiotic resistance may be examined by a developer, the following is a brief summary from the FLAVR SAVR™ tomato submission.[43] The kan^r gene was used as the selectable marker. This selectable marker encoded a enzyme referred to as APH(3')II, which resists two antibiotics, kanamycin and neomycin. As a result, the company had to address the potential impacts of APH(3')II on the human body. Specifically, the company showed that under simulated gastric and intestinal conditions, most of the biological activity associated with APH(3')II was destroyed. The company asserted (and FDA accepted) that use of the selective marker enzyme was unlikely to result in the development of antibiotic-resistant bacteria in the digestive tract. However, the company also had to examine whether the enzyme would significantly affect the potency of orally administered antibiotics. It concluded that even if gastric acids did not successfully deactivate the enzyme, it was unlikely that the potency of the antibiotic would be compromised. Specifically, the antibiotic is administered before a person undergoes an operation and thus, it is unlikely that a person would be ingesting FLAVR SAVR™ tomato in and around the time the antibiotics would be used.[44]

A second issue central to determining toxicity is whether use of the genetically modified plant or organism will result in an allergic reaction. Genetic engineering can cause an increase or decrease of an already existing protein in a food or can transfer an allergen from one plant to another that previously did not have such an allergen.[45] For example, one case study demonstrated that a protein transferred from a Brazil nut to a soybean caused an allergic reaction in individuals who consumed the soybean and were sensitive to Brazil nuts.[46] Thus, to determine the potential allergenic reaction, studies must be performed to compare the structural properties or characteristics of the genetically modified organism with those of known allergens to determine if such characteristics are shared.[47] Comparisons can be made with the information on known allergens contained in various national and international databases (e.g., the NIH genetic sequence database referred to as "GenBank," the European Molecular

---

marker genes, either by using other types of markers that are not antibiotic resistant or by eliminating the use of markers completely. U.S. Department of Agriculture, *Introduction of Genetically Engineered Organisms*, Draft Programmatic Environmental Impact Statement (July 2007) at 85 [hereinafter *USDA Introduction of GEOs*], *available at* http://www.aphis.usda.gov/biotechnology/EIS/EIS_index.shtml.

43. *See* Chapter 6 for a discussion of the FLAVR SAVR™ tomato.

44. 21 Fed. Reg. 26700, 26703 (May 23, 1994).

45. 66 Fed. Reg. 4706, 4709 (Jan. 18, 2001).

46. U. S. Food and Drug Administration, Center for Food Safety and Applied Nutrition, *Safety Assurance of Foods Derived by Modern Biotechnology in the United States*, *available at* http://www.cfsan.fda.gov/~lrd/biojap96.html.

47. 59 Fed. Reg. 26700, 26702 (May 23, 1994). Among the characteristics examined are: proteolytic stability, glycosylation and heat stability.

Biotechnology Laboratory nucleotide sequence database, and the UniProt KB/ Swiss-Prot database).[48]

B. The Basics of Exposure-Dynamics

The purpose of the *exposure assessment* is to determine the extent to which a population is exposed to a material. Exposure assessment ideally involves information from the time that the substance is released into the environment to the point that it is absorbed into the human body.[49] Among other things, the assessment should examine the contaminant and its concentration, the source of the substance, the fate and transport of the substance in the various environmental media (e.g., air, soil, and water), and the routes of entry into the body.[50]

This section will focus on two of the key components: the fate and transport of a substance through the environment and the routes of entry into the body. Once a substance is released into the environment, depending on its nature the type of environmental media, and the biology within the media, the substance may: (a) persist in the environment for a certain period, (b) degrade, and/or (c) alter its chemical and physical characteristics.[51] As chemical and physical alterations are specific to the particular substance, the discussion below will focus on persistence and degradation as both of these can be discussed in general terms. Once the

---

48. *See id.* A debate exists whether the testing and modeling protocols can appropriately determine similarities. The concern about allergens became a more widely appreciated after the Star Link™ incident (see discussion in Chapter 5). In 1998, EPA registered Star Link™ (which was a type of corn that had been genetically modified using a strain of the naturally occurring bacterium *Bacillus thuringiensis* (Bt)) for use as animal feed with the condition that Star Link™ was to be kept out of human food supply. EPA limited the registration to animal feed because the data concerning the "digestibility of the protein was insufficient to make a complete assessment on the potential for the protein to be a potential food allergen." *See* Testimony of Stephen Johnson, Assistant Administrator, Before Committee of Agriculture, Subcommittee on Conservation, Credit and Research (June 17, 2003), *available at* http://usinfo.state.gov. Despite the restrictions on use, the media reported in September 2000 that Star Link™ had through unknown mechanisms found its way into human food. *See id.* The suspected cause was pollen that had drifted to a field of corn that had not been genetically modified with the particular Star Link gene, but now that modified corn was being harvested for use in foods consumed by humans. The result of the Star Link discovery was not only a voluntary recall of various products that had a significant impact on the export of corn, but also a heated debate about whether the mechanisms are adequate for detecting and preventing allergens from entering the food supply without warning.

49. U. S. Environmental Protection Agency, Supplementary Guidance for Conducting Health Risk Assessment of Chemical Mixtures, EPA/630/R-00/002 (August 2000) at 13 [hereinafter *EPA Supplementary Guidance*].

50. *See id.* generally.

51. *See id.* at 17–18.

discussion on fate and transport in the environment is completed, this section will focus on three points of entry into the body—the nose, skin, and mouth.

1. Fate and Transport

(a) Persistence in the Environment Persistence refers to the length of the time a substance will remain in the environment after it has been released. In particular, the longer a substance persists in the environment, the greater its chances are of coming in contact with humans (i.e., a particle that is airborne may be inhaled, a particle that is soluble in water may be ingested, and a particle that remains in the top layers of the soil may be taken up by plants that are then processed into foods for human consumption). The persistence of the substance will depend on multiple factors, including the type of substance and the environment within which it is placed. The following is a brief summary of the different factors that play a role in the various environmental media.

The fate of airborne particles depends on such factors as their aerodynamic characteristics, the degree to which the substance can combine or aggregate with other substances, and the degree to which interaction with other airborne particles will cause degradation or alteration of the particle.[52] In the context of water, the fate of materials is dependent on their aqueous solubility, interactions with natural and anthropogenic chemicals, and biological and abiotic processes.[53] These factors will determine if the material will settle into the sediment, have its physical and chemical components altered, remain or become insoluble, combine with other chemicals or substances, or any or all of these possibilities.[54] Finally, as to persistence in the soil, depending on the type and properties of the soil as well as the nature of the material, the material may be mobile, become fused to the soil and become immobile, or may retain its own chemical composition but remain immobile by being trapped in the soil matrix.[55]

Due to their unique characteristics, the unexpected survival and persistence of escaped or intentionally released bioengineered plants or nanomaterials may become a source for ecological and human harm. However, the diversity and complexity of genetically modified materials and nanomaterials allow only for a generalization to be made about issues that may arise in any particular situation.

As discussed more extensively in this chapter, in the case of bioengineered plants, persistence primarily relates to whether genetically modified genes persist outside designated areas and designated crops, and if so, the long and short-term ecological implications of such persistence (e.g., impacts on nontarget species). The study of persistence of nanomaterials remains in the initial stages.

---

52. M. Holsapple et al., *Research Strategies for Safety Evaluation of Nanomaterials, Part II: Toxicological and Safety Evaluation of Nanomaterials, Current Challenges and Data Needs*, 88 TOXICOLOGICAL SCI. 12–17.

53. *EPA White Paper, supra* note 29, at 34.

54. *See id.* at 34.

55. *EPA Supplementary Guidance, supra* note 49, at 19.

While comparisons have been made to the other micro-sized particles, intentionally produced nanoparticles behave quite differently in some cases from micro-sized particles that may be incidentally released due to things such as combustion.[56] For example, research has demonstrated that waterborne nanoparticles generally settle more slowly than larger particles of the same material; thus, biological organisms could potentially ingest the nanoparticles for a longer period of time than material that settles more quickly.[57] Moreover, once at the bottom, because of their size nanoparticles may be taken up by aquatic and marine filter feeders that may not be able to ingest larger particles.[58]

(b) Degradation or Alteration Once a substance is released into the environment, as it comes into contact with other substances in some instances there will be (a) degradation via photolysis, hydrolysis, or biodegradation (both aerobic and anaerobic); and/or (b) alteration from its original profile.[59] The significance of the degradation and alteration process is that it impacts the routes of exposure for humans.[60]

The degradation and alteration of genetically modified plants can occur in various contexts. For example, after harvest the portions of crops that remain in the field will degrade and be absorbed into the soil. Once in the soil, the microbial and invertebrate communities that reside there will consume these genetically modified crops, including the novel proteins that have been inserted into them. This has the potential to cause adverse impacts to these communities, and in turn, could have a significant impact on agriculture because these communities create appropriate soil conditions to grow crops.[61] A fuller discussion of this issue is provided in Section III(A)(2) below.

Because of the variety of nanomaterials, degradation and alteration must be described in general terms.[62] For organic nanomaterial, there may be physical and chemical breakdowns similar to what occurs with organic chemicals, or there may be changes in the physical structure or surface characteristics of the material that could result in increased toxicity, reactivity, or mobility (e.g., carbon compounds may become more toxic).[63] For inorganic nanomaterials such as those composed of ceramics, metals, and metal oxides, while biodegradation

---

56. *EPA White Paper, supra* note 29, at 33.

57. *See id.* at 36.

58. *Id.* at 36–37.

59. *EPA Supplementary Guidelines, supra* note 49, at 20.

60. *See id.* at 23.

61. The Joint Working Group on Novel Foods and GMOs, *Guidance Document for the Risk Assessment of Genetically Modified Plants and Derived Food and Feed* (Mar. 6, 2003) at 15–16.

62. *EPA White Paper, supra* note 29, at 36.

63. *See id.* at 37.

may not occur, there can be chemical transformation when the materials come in contact with environmental media resulting in increased mobility, toxicity, and reactivity.[64]

2. Routes of Entry into the Human Body The primary routes of entry into the human body for biotech and nanotech products are through the nose, mouth, and skin. Once a material enters the body, it may be metabolized, sequestered in a particular organ, or excreted.[65] This section will generally describe the biological functions involved, then discuss the specific issues relevant for each technology.

For genetically modified organisms, the main route of entry is the mouth when modified crops are ingested after being processed into food. Another mechanism especially relevant for agricultural workers is inhalation of airborne pollen during the milling process. Nanomaterials, on the other hand, may enter the body through different mechanisms. Nanomaterials may be inhaled during the manufacturing process, come in contact with skin due to intentional application of nano-containing products or unintentional occupational exposure, and ingested (if nanomaterials are placed in medicines or food additives).

(a) Inhalation When a nanoparticle or genetically modified pollen is inhaled, that particle has to pass through different regions of the respiratory tract. During its travel, the particle may be cleared through the body's clearance mechanisms, become lodged in a portion of the respiratory tract, become translocated to different regions of the body, and or generate an immunological response from the body. The following is a brief primer on how the respiratory system functions.

First, a particle has to pass through two areas of the respiratory tract—the nasopharyngeal area and then the tracheobronchial region—in order to reach the lungs.[66] During its travel through the respiratory tract, the particle may be deposited in the nasopharyngeal or tracheobronchial region, and as a result it may be cleared through the use of the mucociliary escalator that can send particles to the gastrointestinal tract for disposal.[67] If the particle is able to evade these clearance mechanisms, it will be able to reach the lungs. The lungs are protected by the macrophages (i.e., a type of white blood cell)[68] that perform phagocytosis— that is, these white blood cells engulf and ingest the foreign particle, which can then be cleared through the mucocilliary escalator.[69] If a particle is able to avoid the macrophage clearance mechanism, it may become lodged in pulmonary

---

64. *Id.*

65. D. Balshaw et al., *Research Strategies for Safety Evaluation of Nanomaterials, Part III: Nanoscale Technologies for Assessing Risk and Improving Public Health*, 88 TOXICOLOGICAL SCI. 298 (2005).

66. Oberdörster, *supra* note 30, at 16.

67. *See id.*

68. *Id.* 14–15.

69. *Id.*

interstitial areas, cause an inflammatory reaction, or possibly translocate into other parts of the body including the lymphatic system, bone marrow, spleen, heart, or brain.[70,71]

There is limited scientific literature on the impact of inhaled genetically modified crops or pollen on the respiratory tract (as opposed to respiratory ailments due to ingestion of foods derived from genetically modified crops—see above on the discussion on allergens). Rather, anecdotal evidence has been offered concerning the respiratory reactions of those who handled, cleaned, or picked the genetically modified crop.

Among nanoparticles, researchers have examined the impact of fullerenes and multiwalled carbon nanotubes and have made comparisons and analogies to the well-established data on asbestos,[72] quartz, and ultrafine pollutants. As noted above, a number of factors may be relevant as to how a nanomaterial may react once it has been inhaled or what the consequences will be of this inhalation. These factors include (but are not limited to) the size;[73] shape, composition, total mass of the nanomaterial, and surface area.[74] For example, research has indicated that size and surface area are relevant to where the nanoparticle may be deposited in the respiratory tract and to whether the particle is engulfed and ingested by the macrophages or translocated to other parts of the body.[75]

(b) Dermal The skin also serves as one of the principal routes of entry into the body even though it has a strong protective barrier, the epidermis, which protects the underlying dermis. Typically healthy skin is considered impervious

---

70. Oberdörster, *supra* note 30, at 16. The nasal mucosa and tracheobronchial region are supplied with an effective clearance mechanism consisting of ciliated cells forming a mucociliary escalator in order to clear solid particles from the tracheobronchial region within 24 hours.

71. Rejeski, *supra* note 33, at 4.

72. Recent studies have been published that show that when multiwalled carbon nanotubes are injected into an area of a mice's body known as the peritoneum, it can cause an inflammation and cancers in a manner similar to asbestos. The question is whether inhaled multiwalled carbon nanotubes would result in a similar reaction as those observed with those nanoparticles that were specifically injected. *See* John Balbus, *Are Multi-Walled Carbon Nanotubes More Like Asbestos than We Thought? Part II*, Nanotechnology Notes, *available at* http://environmentaldefenseblogs.org/.

73. For example, research has shown that size and surface area of nanomaterials do matter. One study states that 90 percent of inhaled 1-nm particles are deposited in the nasopharyngeal compartment, only 10 percent in the tracheobronchial region, and essentially none in the alveolar region (lungs). In contrast, 5-nm particles show equal deposition of approximately 30 percent of inhaled particles in all three regions; 20-nm particles have the highest deposition efficiency in the alveolar region (approximately 50 percent). *See* Oberdörster, *supra* note 30, at 26.

74. J. Tsuji et al., *Research Strategies for Safety Evaluation of Nanomaterials, Part IV: Risk Assessment of Nanoparticles*, 89 TOXICOLOGICAL SCI. 42 (2006).

75. *See* Oberdörster, *supra* note 30, at 24–25.

to particle exposure.[76] However, if a particle is able to lodge in the skin though the sweat or sebaceous glands and hair follicles, it is not susceptible to removal by phagocytosis.[77] Therefore, one of the primary research questions is whether the substance can penetrate the epidermal barrier.[78]

Dermal contact can occur unintentionally, such as due to combustion or dispersion of pollen. It can also occur as a result of intentional application, such as application of cosmetics or sunscreen.[79] Although the research is insufficient on dermal interaction with genetically modified crops, dermal contact of nanomaterials has been studied.

Specifically, research has been done on titanium oxide (TiO_2) and zinc oxide (ZnO), which have long been active ingredients in sunscreen but which have now been placed in sunscreen on a nanoscale.[80] On a conventional scale, TiO_2 and ZnO tend to appear as a white film when applied to skin.[81] In contrast, nanoscale TiO_2 and ZnO appear clear when applied to the skin and are claimed to be just as effective in blocking the sun's harmful ultraviolet rays.[82] Therefore, given that most consumers prefer a clear application, it is not surprising a number of companies have manufactured sunscreen containing nanoscale TiO_2 and ZnO.[83] While Chapter 8 will discuss FDA regulation of sunscreen as an over-the-counter drug, this section will examine some of the health issues associated with these two nanoscale materials.

When research was initially conducted on nanoscale TiO_2 and ZnO, the conclusion was that such materials were not likely to penetrate the skin's protective barrier.[84] However, more recent studies indicate that certain nanoscale materials may be able to pass through the layers of the skin or can enter the body through pores or hair follicles.[85] Once under the epidermis, certain nanoscale materials

---

76. National Nanotechnology Initiative, *Environmental, Health and Safety Research Needs for Engineered Nanoscale Materials*, National Science and Technology Council, Committee on Technology, Subcommittee on Nanoscale Science, Engineering, and Technology, Sep. 2006, at 25 [hereinafter *NNI Nanoscale Research Needs*], *available at* www.innovationsgesellschaft.ch/images/fremde_publikationen/NNI_EHS_research_needs.pdf.

77. *See* Oberdörster, *supra* note 30, at 14–15.

78. Holsapple, *supra* note 52, at 13.

79. Another way intentional exposure may occur is through contact with "fibrous materials coated with nanoscale substances for water or stain repellent properties." Tsuji, *supra* note 74, at 44.

80. *Id.*

81. *Id.*

82. *Id.*

83. Woodrow Wilson International Center for Scholars, Project for Emerging Technologies, *New Nanotech Products Hitting the Market at the Rate of 3–4 Per Week* (Apr. 24, 2008), *available at* http://pewnanotechproject.us/news/archive/6697/.

84. Tsuji, *supra* note 74, at 44.

85. *Id.*

can be taken into the lymphatic system via the lymph cells in the dermis and thereby interact with the immune system. The degree to which a nanoscale material may cross the skin barrier depends on a number of factors, including whether the application is to an area where the skin is more porous, thin, or damaged; whether the application is in conjunction with water or oily substances; and what is the number of applications of the nanoscale material containing sunscreen.[86] However, as there does not appear to be any large scale clinical testing of nanoscale TiO_2 and ZnO, caution should be taken in generalizing from only a few studies.

Research also demonstrates that nanoscale TiO_2 and ZnO can cause cellular and DNA damage.[87] At their conventional scale, TiO_2 and ZnO are inert.[88] Nanoscale TiO_2 and ZnO, on the other hand, can be activated by ultraviolet light to create reactive oxygen species (ROS) that can damage skin DNA and cell structures. The solution to this potential problem has been to coat the TiO_2 particle with an inert oxide such as silica, alumina, or zirconium, thus preventing the development of ROS even if the nanoparticle is able to pass through the skin's protective barrier.[89] However, research still needs to done or made publicly available on the biological consequences of such coating along with whether the coating will remain on the titanium oxide and zinc oxide or be removed as the sunscreen interacts with the environment.

(c) Ingestion The primary mechanism for genetically modified organisms for to enter the human body is by ingestion. There has been much debate about the consequences of ingesting genetically modified foods, which contain novel or modified proteins. These proteins are treated by the body's gastrointestinal tract in the same manner as natural proteins.[90] The question then arises whether they have a different impact once ingested and digested. Specifically, the concern is with the impact on the various bacteria that exist in the gastrointestinal tract that perform a variety of beneficial functions for the human body, including repressing pathogenic microbes.

As described in Section II (B)(2) above, developers of transgenic plants have used selectable antibiotic-resistant marker genes that are coupled with the desired gene in order to easily identify those plant varieties that have incorporated the

---

86. *NNI Nanoscale Research Needs, supra* note 76, at 25. *See also* Tsuji, *supra* note 74, at 45.

87. Tsuji, *supra* note 74, at 44.

88. *Id.*

89. *Id.*

90. As with all proteins when ingested, the protein will reach the stomach. Once in the stomach, stomach acids will "straighten and unwind the protein" and concurrently, the acids will activate pepsin (an enzyme) that will break the protein apart into smaller amino acid sequences. These partially broken down proteins, then enter the intestines where finally the body absorbs the amino acids. *See infra* note 74.

desired gene.[91] These genetic marker genes are resistant to certain antibiotics. Two health concerns stem from their use. The first issue is whether the consumption of this marker gene in the course of eating the transgenic plant or the processed food derived from the transgenic plant will have an adverse effect on the efficacy of an antibiotic being administered to the person. The second is whether the ingestion of antibiotic resistance marker gene could lead directly to the growth of human pathogenic bacteria in the gastrointestinal tract because intestinal bacteria have been killed or altered.[92]

Presently, nanomaterials are likely to enter the gastrointestinal tract through an indirect route rather than through direct ingestion. As noted above, once the nanomaterial is cleared through the mucociliary elevator, it is transported and deposited into the gastrointestinal tract.[93] The limited research on nanomaterials indicates that they are quickly excreted from the body.[94] It remains to be seen if these processes are equally applicable once nanomaterials are ingested through pesticidal residues on foods, medicines, and dietary supplement.

IV. ENVIRONMENTAL RISK OF AGRICULTURAL BIOTECHNOLOGY AND NANOTECHNOLOGY

In addition to the potential risks to human health, there has been much debate about the risks agricultural biotechnology and nanotechnology pose to the environment. Environmental impacts may be direct or indirect, and immediate or delayed depending on the organism and the environment into which it has been released.[95]

Given that genetically modified plants comprise most of the releases into the environment, this section will primarily discuss the impacts associated with these releases.[96] Among the concerns are altering the susceptibility of the other

---

91. Council for Biotechnology Information, *The Use of Antibiotic Resistance Markers to Develop Biotech Crops*, (May 2001) at 1, *available at* www.whybiotech.com/html/pdf/The_Use_of_Antibiotic.pdf.

92. *FDA/CFSAN, supra* note 40. Efforts have been taken to use marker genes that are widespread in nature with the corresponding antibiotics either being seldom used or not used at all. *See id.* However, until transgenic plants can be developed without the use of antibiotic-resistant marker genes, it will still be necessary to examine the potential impacts to the human digestive system.

93. Oberdorster, *supra* note 30, at 28.

94. *NNI Nanoscale Research Needs, supra* note 76, at 26.

95. *See USDA Introduction of GEOs, supra* note 42, at 67–68.

96. The issue of hybridization is not limited to plant species. In fact, one of the most hotly debated discussions has been the issue of genetically modified animals. A proposal has been made for the approval of a transgenic salmon. Transgenic salmon are mutated in order that the fish can grow to market weight size 6 to 12 months before their natural

organisms to pathogens or infectious diseases through the acquisition of traits, causing the mortality of nontarget species, generating toxins or allergens in other species, and accumulating and influencing the ecosystem through the flow of genes to wild and weedy relatives.[97] There is only limited information on environmental impacts of released nanoscale materials—even though tons of nanoparticles such as fullerenes are being produced each year. Thus, the discussion on genetically modified organisms could provide a preview of the issues that will eventually involve nanomaterials.

A. Environmental Impacts from Genetically Modified Organisms

1. **Implications of Gene Flow** One concern associated with introducing genetically modified plants into the environment is that genes of the genetically modified plant (i.e., transgenes) will become incorporated in a sexually compatible wild relative.[98] The reason this exchange of genes takes on significance is that unlike crop plants that are often not weedy and have a low propensity to survive without human intervention, wild plants by their very nature have weedy characteristics that assist them in colonizing new areas and that can persist in nature without any human assistance.[99] Therefore, the concern is that a hybrid of a wild/genetically modified plant will both have weedy characteristics and the advantages conferred through the transgene, such as increasing resistance to herbicides or containing its own insecticide.[100] This may turn such a hybrid plant into a "super weed" that could spread and persist in the environment longer than other plants because the transgene could provide it with a competitive advantage (e.g., surviving insects or diseases) and/or give it the ability to invade the territory and occupy other plants.[101]

---

counterparts (18 months versus 24–30 months). As with conventionally farmed salmon, transgenic salmon are initially raised in hatcheries and then placed in ocean pens. With the recognition that escapes of farmed salmon from ocean pens occur, transgenic salmon are treated to be all female and additionally treated to make them sterile. But, as environmental groups and even FDA Deputy Commissioner have noted, sterilization is not 100 percent effective, it is almost inevitable that fish escapes from the net pens would include some females capable of reproduction. As a result, there could be interbreeding with wild Atlantic salmon, hybridization with the closely related brown trout, and disturbance of habitat as a consequence of competition for resources. *See* Speech before American Enterprise Institute, Remarks by Lester M. Crawford, Deputy Commissioner U.S. Food and Drug Administration (June 12, 2003).

97. *See generally USDA Introduction of GEOs, supra* note 42, at 67–68.

98. There is also the converse concern that genes of the wild relative may become incorporated in a domesticated crop. *See id.* at 67–68.

99. *See id.* at 74.

100. *Id.*

101. See discussion in *State of Food and Agriculture, supra* note 39, Section 5; *see also,* Agricultural Biotechnology Risk Analysis Research Task Force, *Agricultural Biotechnology*

Proponents of genetically engineered agricultural products and federal agencies have discounted these concerns as being unlikely. First, they doubt such an exchange of genetic material would in fact occur, as in order to achieve successful hybridization, a number of factors must be present, including simultaneous flowering, sexual compatibility, and sufficient proximity to allow for pollen to be carried via the wind, insects, or other pollinators to the receptor plant.[102] In fact, a number of fruit and vegetable plants do not have sexually compatible wild relatives.[103] Most of the insecticide-tolerant and herbicide-tolerant varieties of genetically modified (GM) crops involve plants that do not have sexually compatible wild relatives in close proximity (e.g., soybean, corn, etc.) as compared to plants that do have wild relatives such as sunflowers.

Second, they discount the idea that the transgene would enhance properties that would allow the hybrid plant to outcompete nonmodified wild relatives. Specifically, the genes inserted into crops that create herbicide resistance or protect the plant by incorporating an insecticide are only relevant in the natural conditions if the herbicide is being applied or if the insects are present.[104] If conditions for which the genetic modification was made are not present, it is unclear that the transgene would make a material difference in the survivability or competitive advantage of the hybrid plant.[105] Conversely, if the favorable factors are present, the hybrid plant would have an advantage and could over time replace nonmodified varieties. However, again the issue arises as to whether there are any sexually compatible wild relatives. Moreover, even if there were such wild relatives, proponents argue that controls such as containment and the creation of buffer zones provide adequate means to avoid the transfer of genes. However, the Star Link™ incident (discussed in Chapter 5) indicates these controls may not always prove to be successful—and the results can be significant to a broader range of players than simply those involved in the cultivation of the genetically modified crop.

2. Impacts on Non-Target Species Another significant issue associated with transgenic plants is their impact on non-target species. While the most recognizable claim of non-target species impact concerns Monarch butterflies, in actuality,

---

Risk Analysis Research in the Federal Government: Cross Agency Cooperation (undated) at 8–9, *available at* www.nsf.gov/pubs/2007/nsf07208/nsf07208.pdf.

102. *USDA Introduction of GEOs, supra* note 42, at 74. Additionally, the regulations governing the cultivation of genetically modified crops requires physically and biological containment (*e.g.*, creation of buffer zones or sterile seeds) to decrease the possibility that pollen will spread and will have adverse impacts on wild varieties.

103. FIFRA Scientific Advisory Panel, *A Set of Scientific Issues Being Considered by the Environmental Protection Agency Regarding: Issues Associated with Deployment of a Type of Plant-Incorporated Protectant (PIP), Specifically Those Based on Plant Viral Coat Proteins (PVCP-PIPS)*, Meeting Minutes (Oct. 13–15, 2004), No. 2004-09 at 16–17.

104. *USDA Introduction of GEOs, supra* note 42, at 75–76.

105. *Id.*

the Monarch Butterfly case may highlight the importance of testing outside the controlled setting of a laboratory to determine the real impacts. In the case of the Monarch butterfly, a laboratory study indicated that larvae exposed to the Bt corn pollen ate less milkweed—the only source of nutrition for the larvae—than those exposed to the conventional pollen, and that nearly half of the larvae feeding on the milkweed dusted with Bt corn pollen died after four days whereas none of the larvae exposed to conventional corn pollen died.

As a result of the firestorm that arose, EPA required manufacturers who sought renewal of their registrations to sell or distribute genetically modified corn to provide field test data.[106] Ultimately, after EPA received updated information on product characterization, human health effects, gene flow, effects on non-target organisms, total exposure, and insect resistance management, the agency concluded there was no unreasonable adverse effect from genetically modified corn.[107] However, due to media coverage of the initial report as well as the public outcry that it generated, Gerber announced it would not use genetically modified ingredients in its baby food, and European opposition to use of genetically modified foods increased.

Although it has not been reported on extensively, an area in which genetically modified plants may actually have an impact is their potential impact on soil microbial and invertebrate communities.[108] Thousands of different types of species exist in the soil with their role involving storing water, mixing the soil, preventing erosion, providing nutrition to plants, and breaking down organic matter.[109] Disturbing the functions of these communities would have significant adverse repercussions for the ecosystem.

The relationship between plants and the soil microbial and invertebrate communities (e.g., bacteria, fungi, and insects) is complex and obviously intertwined. For example, plant roots affect the chemical and biological conditions of that soil and thereby influence the development of microorganisms.[110] Microorganisms decompose organic matter and make nutrients available to the plant.[111] Thus, if the genetic composition of the plant were to be modified, questions arise as to whether it would adversely affect the soil microbial and invertebrate communities. For example, could an antibiotic-resistance marker present in the soil be ingested and incorporated by bacteria, cause them to develop resistance, and then travel up the food chain? To date, because of the complexities involved, it is

---

106. *See* EPA, Bt Plant-Incorporated Protectants October 15, 2001 Biopesticides Registration Action Document, citing Scientific Advisory Panel, at Section V.3.

107. *See id.* at Section V.4.

108. *USDA Introduction of GEOs, supra* note 42, at 82.

109. *See id.* In fact, it is reported that a single gram of soil typically contains several thousand species of bacteria alone. *See id.* at 79.

110. *See id.* at 82.

111. *See id.*

not possible to determine with any certainty whether genetically modified crops have an increased adverse impact on the soil microbial and invertebrate communities than do crops grown using conventional agricultural techniques. However, as different genetic material is introduced into plants in order to allow plants to serve, for example, a pharmaceutical or industrial purpose, the impacts on these communities will need to be examined more closely.

B. Environmental Consequences from Nanomaterials

There has been very little information developed on the risks nanoscale materials pose to the environment, but questions have been raised. For example, during the course of conducting a study on the impact of uncoated fullerenes on large mouth bass (noted above in Section II(B)(1)(a), researchers noticed that the water in the tank that had been dosed with fullerenes was visibly clearer than the water in the control tank.[112] The conclusion was that uncoated fullerenes may act as a bactericide and kill beneficial bacteria normally found in an aquatic environment. If this is confirmed in other studies, and if it is determined that uncoated fullerenes are entering the water supply, this would have significant implications to the regulation of such nanomaterials.[113]

Critics of nanomaterials also point to the research conducted on nanosilver as another example of potential adverse impacts on the environment.[114] Because nanosilver has antimicrobial and antibacterial properties, it is being used in a number of consumer products. For example, it has been used as a coating agent on computer keyboards and other computer accessories so that those surfaces could be antibacterial. In fact, approximately 20 percent of all consumer goods having nanoscale substances incorporated into them have nanosilver.[115] But critics note that nanosilver products (e.g., socks that leach nanosilver when washed) can release nanoparticles into the environment, and that such releases can be harmful to beneficial bacteria and thereby cause an imbalance in the ecosystem.[116] For example, two wastewater utility associations contacted EPA and indicated their concern that nanosilver could enter the sewage system and get discharged into the waterway, killing plankton and other beneficial microbials and thus undermining the food chain.

---

112. *See* Oberdorster, *supra* note 34, at 1059.

113. *See id.* at 1061.

114. *See* International Center for Technology Assessment, *Petition for Rulemaking Requesting EPA Regulate Nano-Silver Products as Pesticides* (May 1, 2008) [hereinafter *ICT Petition*], *available at* http://www.icta.org/namoaction/doc/CTA_nano-silver%20petition_final_5_1_08.pdf.

115. *See* Woodrow Wilson International Center for Scholars, *supra* note 88.

116. Cal Baier-Anderson, *Bacterial Resistance to Silver (Nano or Otherwise)*, Nanotechnology Notes (Apr. 29, 2008) *available at* http://www.envrionmentaldefense.org. *See also*, *ICT Petition*, *supra* note 114, at 67.

V. ECONOMIC AND ETHICAL CONSEQUENCES

A. Economic Consequences of Transgenic Plants

Significant differences of opinion exist as to whether genetic modification of crops has a positive or negative economic impact. The issues under consideration include not only the costs and benefits to farmers and consumers of such crops, but how those costs and benefits are distributed within and amongst societies.

Transgenic plants have been modified primarily to provide herbicide and insecticide functions without the need for conventional application.[117] As such, proponents of genetically modified crops have argued that farmers benefit from (a) increased yields because pests (i.e., weeds and insects) are controlled through genetic modification, and (b) lower costs because of reduced reliance on herbicides and insecticides.[118]

Generally speaking, critics of genetically modified crops note among other things that transgenic plants could (a) inhibit crop rotation resulting in further soil deterioration that in the long-term would mean a decrease in crop yields, (b) harm insects used by organic farmers as a natural mechanism of pest control, and (c) make farmers more dependent on multinational corporations for their seeds instead of using seeds that historically have been available at no cost.[119] On this last point, critics note that only a limited number of multinational companies have been involved in the research and development of transgenic plants, and these companies have developed proprietary technologies and hold patents for the genetically engineered genes.[120] Moreover, if to address the gene flow issues described above, companies create seeds that can only grow for one season, farmer dependence on the seed supplier would increase.

---

117. Proponents point to government figures that indicate that in 2007, an estimated 91 percent of all soybean, 24 percent of all corn, and 28 percent of all upland cotton planted in the United States were herbicide tolerant varieties. *See* USDA Data Sets: Adoption of Genetically Engineered Crops in the US: Soybeans, *available at* http://www.ers.usda.gov/Data/BiotechCrops/. Crops are also being grown to be frost resistant amd/or drought resistant.

118. Letter from Center for Food Safety to EPA (June 13, 2007) re: "Comments on 'Plant-Incorporated Protectants; Potential Revisions to Current Production Requirements.'" Critics have challenged this claim by noting it is based on the false assumption that all or most growers would resort to chemical pesticides when, according to their view, chemical pesticides were not always employed. For example, corn has been genetically modified to control the European corn borer, however, only 5.2 percent of the U.S. corn acreage was sprayed to control the European corn borer prior to the introduction of the GM corn. *See id.* Proponents would counter that the damage caused by the European corn borer was real, but it was simply difficult to address through spay-on pesticides.

119. *USDA Introduction of GEOs, supra* note 42, Appendix G-4.

120. *Id.*

In the future, genetically modified plants may be enhanced to be more nutritious or to be modified so that they can be planted on soil that suffers from drought conditions or salinity. One example is "Golden Rice"—a form of rice still not commercially available[121]—that was developed to address vitamin A deficiency among children in developing countries. Children in developing countries suffer from vitamin A deficiency because they do not have access to fresh fruits and vegetables. This lack of vitamin A can cause blindness as well as compromise the immune system. Golden Rice is a genetically modified rice that contains beta-carotene, which when consumed is converted in the body into vitamin A.[122] Initially, critics of Golden Rice alleged that it was not particularly effective.[123] But even with the development of a Golden Rice variety that contains more beta-carotene, critics noted that fortification programs and reintroduction of vitamin-rich food plants with higher nutritional value are less expensive, more readily available, and more effective mechanisms for addressing the malnutrition problem.[124] Proponents, on the other hand, argue that alternative food sources may not grow in the areas with the greatest need, and that the critics' agenda of generally opposing all genetically modified foods is the real reason for opposing this form of rice rather than any particular safety or health concerns about it.[125]

Field testing of Golden Rice in the Philippines did not commence until April 2008. The test results and the issues that may arise during this field testing stage may determine the fate of this rice. But, regardless of this particular product, as arable land grows more scarce, food prices increase, and populations continue to expand, the demand for creating more nutritious or drought-resistant crops will only increase.

B. Ethical Issues Associated with Transgenic Plants

There are numerous perspectives on whether the production of transgenic animals and plants is ethical. Indeed, various organizations and associations all claiming to represent morality come down differently on this issue. For example, some argue that the development of transgenic plants is a moral imperative to

---

121. The development of Golden Rice began in 1980. However, the rice was not field tested in the United States until 2004, and field testing in a country where there is a high rice consumption and a high vitamin A deficiency (Philippines) began only in April 2008. Proponents argue that product commercialization delays are due to unnecessary, expensive, and time-consuming testing. *See* History of the Golden Rice Programme, *available at* http://www.goldenrice.org/Content1-Who/who2_history.html.

122. *See id.*

123. Greenpeace, *Golden Rice: All Glitter, No Gold* (Mar. 16, 2005), *available at* http://www.greenpeace.org/international/news/failures-of-golden-rice.

124. *See id.*

125. J.E. Mayer, *The Golden Rice Controversy: Useless Science or Unfounded Criticism?*, 55 BIOSCIENCE 726–727 (Sept. 2005). Jorge Mayer is the Golden Rice project manager.

help alleviate hunger and poverty. Others counter that it is morally unacceptable because it is not the natural order, it is a manipulation of God's design,[126] and/or that it causes a redistribution of wealth.

These divisions are reflected, for example, within the Catholic Church. On the one hand, bishops in Brazil and South Africa as well as the U.S. National Catholic Rural Conference oppose the development of genetically modified organisms based, in part, on their concern that use of such crops will disproportionately benefit multinational companies. They fear that as these companies will hold the proprietary technology rights and patents, poor farmers in developing countries will have to depend on these companies for their basic livelihood.[127] On the other hand, in 2004 the Pontifical Academy of Sciences issued a study document that recommended the use of transgenic plants as a means of improving the nutritional consumption of those living in developing countries and the agricultural productivity of arable land.[128] The Academy repeated many of the positions outlined by industry, including that an eighth of the world's population went to bed hungry, that agriculture as currently practiced was not sustainable, that the use of conventional pesticides caused harm to humans and animals, that mixing of genetic material from different organisms has been an important part of the evolutionary process, and that the process for creating transgenic plants was not per se unsafe.[129]

C. Economic and Ethical Issues with Nanomaterials

Nanomaterials have been touted as bringing revolutionary change in a variety of fields. There are already hundreds of products that contain engineered nanoscale materials on the market, including antimicrobial products, stain-resistant clothing, solar cells, sporting equipment, cosmetics, sunscreen, and equipment in computer displays and other electronic devices.[130] According to a publicly

---

126. Raising transgenic animals for human consumption creates ethical concerns distinct from those relating to transgenic plants. Among the criticisms posed is that creating a transgenic animal either is playing with God's design or with nature.

127. W.J. Van der Walt, *Ecological Impact of GM Crops: Time for a Sober Scientific Assessment*, SCIENCE IN AFRICA, (Aug. 2004), *available at* http://www.scienceinafrica.co.za/2004/july/gmo.htm.

128. Pontifical Academy of Sciences, *The Use of "Genetically Modified Food Plants" to Combat Hunger in the World* (Study Document) (Sept. 2004), *available at* http://www.agbioworld.org/biotech-info/religion.cabibbo.html.

129. *See id.*

130. *See EPA White Paper, supra* note 32, at 11; *see also* The Woodrow Wilson International Center for Scholars, Project for Emerging Technologies, *Nanotechnology Consumer Products Inventory, available at* http://www.nanotechproject.org/inventories. Health and fitness products (*e.g.*, cosmetics, clothing, and sunscreen) comprise approximately 60 percent of all nanotechnology-based consumer products. *See* Woodrow Wilson International Center for Scholars, *supra* note 88.

available online inventory that tracks nanotechnology-based consumer products, there was a 185 percent increase in the number of such products from March 2006 to February 2008.[131] In actual dollar terms, approximately $88 billion of nanotechnology-based products were sold in 2007.[132]

The creation of the market is due to the fact that nanotech products are intended to provide superior performance while being more environmentally friendly or simply less expensive. One example is the use of nanoscale particles to increase the efficiency of solar cells and to reduce fabrication costs. If such efforts are successful and there is an increased commercialization of such solar cells, this could have an enormous impact on how energy is produced. The reduction in reliance on fossil fuels would, in turn, have an impact on society's carbon footprint and the amount of global greenhouses gases being produced. Undoubtedly, such advances would have economic benefits for certain companies while extractive industries and those who work in them would presumably suffer from reduced demand.

The ethical debate surrounding nanomaterials has mainly focused on a particular form of nanotechnology—self-replicating nanomachines—and whether there is a potential for such technology to destroy organisms and ecosystems. Some argue there should be a relinquishment of nanotechnology research. This debate began in earnest when Bill Joy, chief scientist at Sun Microsystems, wrote an article that was broadly covered in the popular press in which he hypothesized the potential of self-replicating nanomachines having the capacity to destroy entire ecosystems (he recommended such a relinquishment).[133] However, others such as Richard Smalley, a Nobel-prize laureate, have argued that creating nanomaterials for application is realistic, but that as nanomachines are not realistic, this fear will never be realized.[134] Still others such as Eric Drexler argue that nanomachines are realistic, but that they will not be swarming without control because nanomachines are not living cells and can be digitally controlled. This discussion, which involves the most revolutionary potential for nanotechnology, did not directly address the more practical and mundane applications of the technology. But, it does indicate that as the technology evolves, especially as it becomes used for medical treatment, issues about safety to the collective versus individual benefits will be raised and debated.

---

131. The Project for Emerging Technologies, "Update," (Feb. 22, 2008), *available at* http://www.nanotechproject.org/inventories/consumer/updates/.

132. *See* G.M. Lamb, *How Safe are Nanoparticles?*, THE CHRISTIAN SCI. MONITOR (May 30, 2008), *available at* http://features.csmonitor.com/environment/2008/05/20/how-safe-are-nanoparticles/.

133. B. Joy, *Why the Future Doesn't Need Us*, WIRED (2000), *available at* http://www.wired.com/wired/archive/8.04/joy.html.

134. K. Chang, *Yes, They Can! No They Can't: Charges Fly on Nanobot Debate*, N.Y. TIMES, Dec. 9, 2003.

VI. ASSESSMENT OF RISK UNDER CONDITIONS OF UNCERTAINTY

Uncertainty is a major driving force in the debate over both biotechnology and nanotechnology. Risk assessors and federal regulators recognize that uncertainty pervades all risk assessments because precise scientific knowledge is lacking due to the inability to obtain the information, inadequate analysis, or deliberate ignorance.[135] Generally, uncertainties may arise due to misidentification or the failure to identify hazards, random errors in analytic devices, misclassification of exposure pathways, incorrect accounting or failure to account for aggregation of substances or the synergistic interactions with different substances, or lack of representativeness of the sampling population.[136] Additionally, uncertainty may be due to the way modeling is conducted, such as researchers incorrectly inferring the basis for correlations between substances and biological reactions, excluding one or more relevant variables, or using surrogate variables for ones that cannot be measured.[137] The absence of scientific knowledge, however, does not constitute evidence of an absence of impacts or risks. But, the overall risk estimate cannot be more precise than its most uncertain component.[138]

Uncertainty is a problem not easily resolved. The question is how to factor uncertainty in the calculation of overall risk. If uncertainty is not appropriately analyzed, there may be a failure to appropriately determine the actual risk, incorrect comparisons of two alternatives that may not actually share the same risks, or failure to identify research initiatives that might reduce uncertainty.[139] Moreover, even if uncertainty is accounted for, it is necessary to discriminate between different types of uncertainty and to create a hierarchical view of which uncertainties require further investigation or are significant to decisions regarding the risks posed.[140]

From industry's perspective, because of uncertainty the risks associated with these technologies have been perceived as being greater than what the experimental tests would suggest are the actual risks. This perspective has led to unnecessary restrictions on operations or on the sale of products (see discussion above on Golden Rice). On the other hand, environmental and consumer groups argue that despite inadequate and incomplete tests and the resulting uncertainty about actual impacts, agencies discount or do not focus upon uncertainty because it does not fit into the quantitative risk matrix, which results in the approval of

---

135. See, e.g., NRC, Science and Judgment, supra note 2; Risk Assessment in the Federal Government, supra note 3.

136. NRC, Science and Judgment, supra note 2, at 57.

137. See id.

138. See Executive Order 12866 (Jan. 11, 1996), available at http://www.whitehouse.gov/omb/inforeg/riaguide.html.

139. See id.

140. See id.

or acquiescence to the sale of products that should be restricted or prohibited (see above for a discussion on nanosilver products).

VII. IMPACT OF PUBLIC PERCEPTION ON THE FORMULATION OF REGULATORY POLICY WITH REGARD TO NEW TECHNOLOGY

Public opinion shapes and is shaped by public policy. The basis for public opinion is a complex interaction among a number of factors, including risk calculations that have been made, individual values, and emotions. Public sentiment toward a technology is based on an evaluation of among other things, the risk factors discussed above, as well as the uncertainty and fears associated with the technology, trust in the government's ability to effectively regulate risk, industry's ability to honestly report facts, and media framing of the risks associated with the technology. Public opinion is in constant state of flux, in part because new discoveries are made or new risks are identified, and these factors could galvanize or cause the dissipation of public support. The government, industry, media, and interest groups all respond to events and attempt to shape public opinion. Thus, the following description of opinions toward biotechnology and nanotechnology are snapshots in time.

A. Perceptions of Agricultural Biotechnology

In the United States and Europe, public perceptions toward agricultural biotechnology have had a significant impact on whether the activity is pursued. In the face of public opposition to a new technology, regulators are apt to either prohibit it or to allow a slowing down of the approval process, raising objections and increasing the burdens on those seeking approval of the technology. For example, until very recently, there was no federal guidance pertaining to the manufacture and consumption of genetically modified animals. Advocates for the industry claimed that without regulatory guidance to support such activities and to give the public confidence as to the safety of the product, industry would not be able to develop.[141] Part of the explanation for why regulatory agencies had been reluctant to encourage the industry is the public's perception of the "yuck" factor associated with eating genetically modified animals.

It appears that while the American public has relatively little knowledge or understanding of foods that may actually contain genetically modified organisms, they oppose it by more than 2 to 1.[142] In a survey done from 2001 to 2006, the percentage opposition to genetically modified foods has remained in the high 40s, while support hovered in the mid- to upper 20 percent.[143] Some believe that

---

141. A. Pollack, *Without U.S. Rules, Biotech Food Lacks Investors*, N.Y. TIMES, July 30, 2007.
142. The Mellman Group, *Review of Public Opinion Research* (Nov. 16, 2006) at 2.
143. *Id.* at 3.

support will increase once people have more knowledge about genetic engineering and the benefits it may offer.[144] However, others note that agricultural biotechnology has been stigmatized and that increased knowledge will not alter the opposition.[145] As an FDA 2000 focus group demonstrated, the participants recognized the potential for benefits, but also noted their concerns about unknown long-term health consequences and skepticism as to whether regulators were concerned about the general public and not simply special interest groups.[146]

B. Perceptions of Nanotechnology

The general public has very little knowledge about the risks or benefits associated with nanotechnology. For example, in 2006, in one survey only 10 percent of Americans had heard a lot about nanotechnology, 20 percent had heard some, 27 percent had heard a little, and 42 percent had heard nothing (with 1 percent not sure).[147] A 2007 poll conducted for the Project on Emerging Nanotechnologies indicates that a majority of the respondents are unwilling to make any judgment about anticipated risks and benefits of nanotechnology, although 18 percent say the benefits will outweigh the risks and 6 percent believe risks will exceed benefits.[148]

There are mixed results as to whether education of the public will actually lead to acceptance of the technology as more beneficial than harmful, or whether it will lead to the opposite conclusion. For example, in late 2006, the first large-scale empirical effort was undertaken to (a) compare people's opinions about nanotechnology with other technologies, and (b) analyze risks and benefits of nanotechnology.[149] The results demonstrate that the public is relatively neutral toward nanotechnology with the belief that it is less risky and more beneficial than a number of other technologies such as genetic modification, pesticide or chemical disinfectant application, or human genetic engineering.[150] On the other hand, it was considered more risky than solar power, vaccines, or hydroelectric power. The researchers of this study concluded that if people were given facts about nanotechnology, and if people perceive the information as timely and the reporting transparent, they may not develop a harsh negative attitude toward this

---

144. *USDA Introduction of GEOs, supra* note 46, at Appendix G-3.

145. *See id.*

146. *See* U.S. Food and Drug Administration, Center for Food Safety and Applied Nutrition, Office of Scientific Analysis and Support, Report on Consumer Focus Groups on Biotechnology (Oct. 20, 2000), *available at* http://www.cfsan.fda.gov/-comm/biorpt/html.

147. P. D. Hart, *Report Findings* (Sept. 19, 2006) at 6.

148. Project of Emerging Technologies, *Poll Reveals Public Awareness of Nanotech Stuck at Low Level* (Sept. 23, 2007), *available at* http://www.nanotechproject.org/news/archive/poll_reveals_public_awareness_nanotech.

149. S. Currall et al., *What Drives Public Acceptance of Nanotechnology*, 1 NATURE NANOTECHNOLOGY 153 (December 2006).

150. *See id.* at 154.

new technology.[151] In a contrary result, another study demonstrated that when information is provided on the risks and benefits, those who have no opinion on nanotechnology are more likely to believe that the risks outweigh the benefits.[152]

VIII. THE THEORETICAL COMPONENTS OF THE PRECAUTIONARY PRINCIPLE

The precautionary principle is a theoretical and practical concept about how risk should be regulated in order to protect human health and environment when there is uncertainty. However, there is no single precautionary principle; rather, the precautionary principle exists on a continuum from an active or strong stance or to a weak stance.[153] The weaker position asserts that if the proposed activity carries the possibility of harm—but the evidence does not decisively demonstrate harm—then the lack of absolute certainty is no excuse for failure to regulate the activity or mitigate the harm.[154] On other end of the spectrum is the strong version, which demands that unless proponents can prove that their activity will not cause undue harm to human health or the environment, regulatory agencies should refuse to authorize, severely restrict, or even prohibit the activity. The significance of this strong version of the precautionary principle is that the burden of proof rests with the proponent. This is unlike the risk assessment model that typically allows an activity to proceed unless regulatory officials can demonstrate the risks are of such significance that the activity should not be permitted. In contrast, the burden under the precautionary principle falls upon the proponent of the technology to demonstrate that there is no undue harm resulting from proceeding with the activity.

---

151. *See id.* at 154–155.

152. Rejeski, *supra* note 33, at 2–3.

153. C. R. Sunstein, *The Paralyzing Principle*, Regulation 2002–2003, 32–37, at 32. Professor Sunstein refers to it as a "continuum of understanding," with at one extreme a weak version that no reasonable person would object to and at the other extreme a strong version that would require a fundamental rethinking of regulatory policy. Evidence of the weak version is Principle 15 of the 1992 Rio Statement (on climate change), which states that "in order to protect the environment, the precautionary approach shall be widely applied by States according to their capability. Where there are threats of serious or irreversible damage, lack of full scientific certainty shall not be used as a reason for postponing cost-effective measures to prevent environmental degradation." United Nations Environment Programme [UNEP], 1992.

154. *Id.* at 33. *See also* C. Phoenix & M. Treder, *Applying the Precautionary Principle to Nanotechnology*, CENTER FOR RESPONSIBLE NANOTECHNOLOGY (revised December 2003, January 2004), *available at* http://www.crnano.org/precautionary.htm.

In the debate over agricultural biotechnology and nanotechnology, critics have argued that the products developed using these technologies can persist in the environment and that they are also difficult to control. However, proponents of these respective technologies note that no proof has been offered that products manufactured using these technologies result in undue harm. In the field of nanotechnology, there has been a debate over whether the precautionary principle should be applied. Initially, when focus was on self-replicating nanoparticles, there was a forceful but limited exchange between various major theoreticians of new technologies. This argument was facilitated by Bill Joy's article on the dangers of self-replicating nanobots (as discussed above) and his conclusion that as a precaution, there should be a relinquishment of such technologies. However, as nanotechnology has evolved in ways other than in developing self-replicating nanobots, the precautionary principle (or at least some form of it) has been articulated by environmental and consumer groups that advocate for a range of regulatory responses—from those who want a moratorium until tests demonstrate the "safety" of nanoparticles to those who want increased public information in the form of distinct labels and testing parameters that are commensurate with the unique nature of these particles.

A similar set of arguments has been put forth with respect to genetically modified foods. In fact, during the early years of biotechnology development in the United States, scientists voluntarily agreed not to pursue certain experiments. However, this moratorium was lifted after scientists working with the government were able to establish a set of guidelines for conducting research.

In Europe, acknowledgment of the precautionary principle was specifically noted in a Communication by the European Commission. The Commission noted that an environmental assessment should occur before a crop can be planted or genetically modified food can be placed in the marketplace; based on the results of that study, appropriate controls or limitations should be undertaken to avoid or militate against adverse impacts. However, from 1997 to 2004, countries took the precautionary principle as a justification for prohibiting the growth of genetically modified crops and banning the importation of genetically modified organisms. Since 2004, although it is not used for prohibiting all crops, the precautionary principle still remains a guidepost as to how the Europeans evaluate whether a crop should be approved.

4. CREATION OF A REGULATORY STRUCTURE

The regulatory framework that governs biotechnology today is based on principles set forth in a guidance document published more than 20 years ago entitled the "Coordinated Framework."[1] The position articulated in that document was that the products resulting from modern biotechnology techniques do not pose unique hazards and are not inherently more dangerous than those derived from conventional means of genetic manipulation. As a result, the federal government concluded that the statutes (which were drafted prior to the advent of recombinant genetic modification) and the agencies authorized to enforce them were adequate for the task of controlling this new technology. These laws regulate based on the intended use, the characteristics of, and the risks associated with the product in question (i.e., *product-based*) and not the process that created them (*process-based*).

Thus, no one single statute governs biotechnology. Rather, a manufacturer will need to comply with a mosaic of laws, nomenclatures, standards, and procedures in order to experiment with or commercialize a particular product based on the use of rDNA. These same principles and practical realities are applicable to the regulatory structure governing nanotech research and products as well.

However, it was not preordained that biotechnology would be regulated in this manner. Rather, as history demonstrates, the initial efforts at biotech regulation were based on the view that biotechnology posed unique hazards, and therefore, a prohibition on certain forms of research was appropriate.

This chapter will first examine the historical evolution of the biotechnology regulatory scheme. It will then turn and examine what has occurred with respect to nanotechnology. As this section will demonstrate, regulatory structures are not created or adjusted in a vacuum, but instead are influenced by or based upon the interaction of a number of factors, such as interest group politics, technological advances, bureaucratic turf battles, lobbying efforts, and judicial opinions. Each of these factors is influenced by the overarching political, economic, and social climate of the times. It is necessary to remember the zeitgeist of the era in order to appreciate the regulatory actions that occurred in their appropriate context. Thus, prior to an in-depth analysis of the development of regulatory structures, we begin with a brief historical primer of the meta-trends that were occurring.

---

1. 51 Fed. Reg. 23302 (June 26, 1986).

Modern biotechnology became a regulatory concern in the 1970s when there was increasing public awareness of the negative environmental consequences and adverse health impacts associated with products that had been marketed. As was pointed out in Chapter 3, the reason in part was that new technologies made it possible to find relationships among particular chemicals or other substances and cancer or other chronic health conditions as well as to detect chemicals at lower and lower concentrations that had been missed by earlier detection methods.[2] The public awareness spawned the growth of the environmental and consumer rights movements, which in turn created an organizational framework that advocated for more stringent environmental, health, and safety standards and increased transparency of information. The era saw the birth of a new regulatory agency (i.e., EPA) and the passage of a number of federal environmental statutes (e.g., the Clean Water Act and the Clean Air Act) as well as the development of a skeptical and critical view of the way industries operated and how government regulated health and safety. There was also an increased awareness among scientists about the need to critically examine new technologies and their consequences.

When the Reagan Administration came into office in 1980, it espoused a political philosophy that viewed the federal government as overregulating industries and hindering the operation of the free market. This political ideology had saliency in part because the country was confronting an economic malaise of stagflation. In its first two years, the Reagan Administration sought sweeping reforms of federal environmental laws under the direction of a controversial EPA Administrator and equally controversial Secretary of the Interior.[3] These efforts were met with strong bipartisan congressional opposition, a judiciary that was skeptical about weakening environmental laws, and a coordinated (and increasingly powerful) environmental movement.

The remaining years of the decade were marked by a complicated relationship. On the one hand, distrust of EPA led Congress to impose standards on the agency rather than allowing the agency to exercise discretion. On the other hand, important environmental legislation was enacted such as the amendments to the Resource Conservation and Recovery Act (RCRA) and Title III of the Superfund Amendments and Reauthorization Act of 1986, which is also known as the Emergency Planning and Community Right-to-Know Act. However, one of the

---

2. NATIONAL RESEARCH COUNCIL, ET AL., RISK ASSESSMENT IN THE FEDERAL GOVERNMENT: MANAGING THE PROCESS 10 (1983).

3. The first EPA Administrator President Reagan appointed was Anne Gorsuch, who was forced to resign after being found in contempt of Congress. She was replaced by William Ruckeleshaus, who was the first EPA Administrator under President Nixon, and who had bipartisan support. President Reagan also initially appointed James Watt as Secretary of the Interior, but he had to resign after referring to his own staff in derogatory terms.

most enduring results of the Reagan Administration was its issuance of Executive Order 12291 within a few months after the inauguration. This Executive Order, which was presented to the federal agencies as a fait accompli, required all agencies to prepare a cost–benefit analysis[4] (referred to as *regulatory impact analysis*) for all major final rules and proposed rules, and mandated that the Office of Management and Budget (OMB) review and authorize such rules before the agency could issue them.[5] The Executive Order was controversial because among other things, it allowed the OMB to disallow a rule without judicial oversight or transparency.[6]

In the 1990s, globalization of the marketplace and international competition were viewed both as an opportunity and a threat to America's technological advantages. New technologies were seen as being needed to maintain America's economic position in the world. This was when nanotechnology came to the notice of politicians in Washington as potentially revolutionizing medicine and increasing computing power. On the regulatory front, the Clinton Administration revoked Executive Order 12291, but nonetheless continued the policies of requiring agencies to quantify and monetize the costs and benefits of any proposed regulation as well as performing a quantitative risk assessment (which, as described in Chapter 3, became ingrained in regulatory policy in the 1980s).[7] The Clinton Administration issued an executive order that noted that the economic analysis that agencies had to perform and submit to OMB should indicate that "the potential benefits to society justify the potential costs, but with the recognition that not all benefits and costs can be described in monetary or even in quantitative terms."[8]

The events of September 11, 2001, and the Iraq War steered the national discussion to security issues during most of the first decade of the twenty-first century.

---

4. Executive Order 12291, 46 Fed. Reg. 13193, required that, inter alia, (a) agency action not be taken unless the potential benefits to society for the regulation outweighed the potential costs to society, and (b) in setting regulatory priorities, agencies take into account the condition of particular industries affected by regulations, the conditions of the national economy, and other regulatory actions contemplated in the future.

5. Testimony of Sally Katzen before the House Committee on the Judiciary, Subcommittee on Commercial and Administrative Law (Feb. 13, 2007). Prior to the Reagan Administration, the three successive prior administrations had ad hoc policies toward selectively reviewing significant rulemakings and providing comments.

6. *See id.*

7. *See id.*

8. Executive Order 12866 (Jan. 11, 1996), *available at* http://www.whitehouse.gov/omb/inforeg/riaguide.html. *See also* C. Copeland, *CRS Report for Congress: Changes to the OMB Regulatory Review Process by Executive Order 13422*, Congressional Research Service (Feb. 5, 2007) (noting that the Administration addressed the transparency issue by providing the public with information on (1) substantive changes made to rules between the draft submitted to the OMB for review and the action subsequently announced, and (2) changes made at the suggestion or recommendation of the OMB).

The Bush Administration (which has been seen by many as being opposed to or even outright hostile to new environmental initiatives while favoring the regulated community's position) issued controversial amendments to Executive Order 12866. In 2007, the Bush Administration issued Executive Order 13422; some critics saw this as an expansion of the White House's control over federal agencies, but others noted it as being at least in part an effort toward good government.[9]

One of the major policy shifts in Executive Order 13422 was its requirement that agencies provide a "best estimate of the combined aggregate costs and benefits of all its regulations planned for that calendar year to assist with the identification of priorities."[10] Proponents argue that such aggregate estimates are needed to reveal the cumulative impacts of rulemaking, but critics note that because of the inherent bias of cost–benefit analysis against regulation, such a requirement only raised the bar for issuing regulations. Additionally, the Executive Order expanded the reach of the OMB by requiring agencies to provide advance notice of "significant guidance documents" in addition to significant proposed rules and final rules. This category was broadly defined to include, inter alia, guidance documents that address novel legal and policy issues. Because of the reliance of federal agencies on guidance documents to inform the regulated community and the public as to interpretations of statutes and regulations, OMB review could mean a significant slowing down of the process of issuing guidance documents.[11] It is under these meta regulatory trends that the specific regulations governing biotechnology and nanotechnology need to be understood.

I. DEVELOPMENT OF THE REGULATORY STRUCTURE FOR BIOTECHNOLOGY

A. Early Years of Scientific Self-Regulation: Gordon Conference, National Academy of Sciences, and the Asilomar Conference

Modern genetic engineering began in 1972 when researchers created a novel DNA molecule that had the genetic elements of three different microorganisms.[12] In 1973, when scientists attending the Gordon Conference (an annual meeting of biological and chemical scientists to discuss the frontier research) were informed

---

9. http://www.whitehouse.gov/news/releases/2007/01/20070118.html. *See also* Copeland, *supra* note 8, at CRS-1.

10. *See* Copeland, *supra* note 8, at CRS-8.

11. *Id.* at CRS-12.

12. The Maxine Singer Papers: Risk, Regulation and Scientific Citizenship: The Controversy over Recombinant DNA Research. U.S. National Library of Medicine, 8600 Rockville Pike, Bethesda, MD 20894. *Available at* http://profiles.nlm.nih.gov/DJ/Views/Exhibit/narrative/regulation.html.

of this experiment, their excitement for the scientific prospects was tempered by concerns that a self-replicating genetically modified virus or bacteria for which there were no natural predators would be able to spread.[13] Undoubtedly, these concerns were shaped by the realization of the unintended negative consequences of new technologies with many positive attributes. For example, in 1972, DDT was banned after public outcry that it had decimated bird populations even though for decades it had been effective in reducing human deaths by controlling mosquitoes.

As a result, in an unusual move scientists involved in the Gordon Conference proactively submitted a letter to the National Academy of Sciences and the Institute of Medicine describing the new method, summarizing the concerns raised at the conference, and asking the Academy to establish a committee to assess the possible hazards associated with genetic engineering.[14] The scientists also decided to inform the larger scientific community of their concerns and their request for regulatory oversight by publishing the letter in the journals *Science* and *Proceedings of the National Academy of Sciences.*[15]

In response to the letter, the National Academy of Science (NAS) established a committee headed by Paul Berg, who was part of the team of researchers who had commenced modern genetic engineering. Their recommendations[16] were an application of the precautionary principle (see discussion in Chapter 3). Recognizing that recombinant DNA experiments have the potential for creating novel types of infectious DNA elements whose biological properties cannot be completely predicted in advance, the committee asked the worldwide scientific community to refrain from performing recombinant DNA research relating to the use of bacteria or viruses until potential hazards could be better evaluated and safeguards put into place.[17] This was the first voluntary self-imposed moratorium in the history of science.[18]

The committee also recommended that scientists in the field meet to discuss the implications of this new area of study. The Asilomar Conference (as it is commonly referred to) held in 1975 provided a forum for scientists, regulatory officials, lawyers, and the press to discuss the risks associated with, and the ethical

---

13. M. Singer, *Where the Cloning Discussion Began, Reflections from the Frontiers* (Explorations for the Future: Gordon Research Conferences 1931–2006, GRC's 75th Anniversary, *available at* http://www.frontiersofscience.org/reflections.aspx?category=2&essay=32).

14. *See id.*

15. M. Singer & D. Soll, *Guidelines for DNA Hybrid Molecules*, SCIENCE (Sept. 1973) (commonly referred to as the "Singer-Soll Letter").

16. P. Berg et al., *Potential Biohazards of Recombinant DNA Molecule*, SCIENCE 303 (July 1974).

17. *Id.*

18. Singer, *supra* note 13.

implications of, conducting recombinant DNA research.[19] The conference concluded with the scientists agreeing to lift their voluntary moratorium on the condition that appropriate safeguards be employed (principally biological and physical barriers adequate to contain the newly created organisms).[20] They also agreed that there are certain experiments in which the potential risks are of such a serious nature that they ought not to be done with the then-available containment facilities.[21]

B. National Institutes of Health Recombinant DNA Advisory Committee and the National Institutes of Health Guidelines

The NAS committee had recommended that the National Institutes of Health (NIH) establish an advisory committee that was charged, among other things, with developing procedures to minimize the unintended release of genetically modified molecules and with devising guidelines to be followed by investigators working with potentially hazardous recombinant DNA molecules. In June 1976, the NIH Guidelines (hereinafter "the Guidelines") were issued—and the federal government's regulation of genetic modification research began.

The Guidelines, developed by a NIH committee referred to as the Recombinant DNA Advisory Committee[22] (RAC), are compulsory as to any research conducted at or sponsored by an entity receiving NIH funding for rDNA research or any research being funded through NIH funds.[23] The Guidelines were issued after public hearings, and more importantly, after consultation with many of the same scientists who had been involved in the Asilomar Conference. Thus, while genetic engineering was no longer subject to self-regulation by scientists, the Guidelines continued to give substantial deference to their judgment and reflected the policy prescriptions that these scientists had developed on their own. The Guidelines issued a temporary prohibition on experiments involving highly pathogenic bacteria or genes for toxins, set laboratory standards for experimentation by developing risk-based physical and biological containment procedures,

---

19. P. Berg, *Asilomar and Recombinant DNA*, 1980 Nobel Laureate in Chemistry, Nobelprize.org, (Aug. 26, 2004), *available at* http://nobelprize.org/nobel_prizes/chemistry/articles/berg/index.html.

20. P. Berg et al., *Summary Statement of the Asilomar Conference on Recombinant DNA Molecules*, 188 Science 991–994 (June 6, 1975). Also published in 72 Proc. Nat. Acad. Sci. USA 1981–1984 (June 1975).

21. *Id.* at 991–992.

22. B. Talbot, *Development of the National Institutes of Health Guidelines for Recombinant DNA Research*, 98 Public Health Reports, 361–368 (July–August 1983).

23. The NIH modified the Guidelines to allow private companies to voluntarily submit data in order to receive NIH certifications to conduct research as well as to obtain proprietary protection for the information submitted to NIH.

and prohibited the deliberate release into the environment of genetically modified organisms.[24]

However, within six months of their issuance, it was evident the Guidelines needed revising because many of the fears about experimentation with recombinant DNA had not come to fruition, and it appeared that the probability of such hazards occurring in the future was lower than originally anticipated.[25] A new set of regulations issued in 1978 recognized that the original guidelines were unnecessarily restrictive and therefore relaxed the laboratory containment procedures. The amended guidelines also replaced the absolute ban on deliberate release of organisms containing rDNA into the environment, instead providing for a special review process that permitted the NIH Director, acting on the recommendation of the RAC, to approve testing of genetically modified organisms outside of the enclosed laboratory if there was no significant risk to human health or the environment. However, no standards were established for determining when a release would not constitute a significant risk. The amended guidelines also allowed for revisions as technological changes occurred—a provision that has been subsequently evoked numerous times.

C. Office of Science and Technology Policy's Coordinated Framework

1. **Influences on the Development of the Framework** By the late 1970s, the overall framework for regulating the risks from basic biotechnology research was through the NIH, which allowed the scientific community to self-regulate based on containment procedures and prohibitions outlined in the Guidelines. Despite attempts to impose more stringent regulation, scientists succeeded in their argument that scientific control was appropriate in order to allow for experimentation and innovation with regard to this revolutionary technology. Furthermore, they insisted that they could maintain the necessary level of safety.[26]

However, as the science developed as to genetically modified plants and microorganisms, to determine their commercial potential, it became necessary to test outside the confines of the laboratory. The interaction between genetically modified organisms and their environments gave rise to questions about the potential impacts on the ecosystem (e.g., impacts on wild plants, nontarget arthropods, and soil microbial communities), the adequacy of the oversight to ensure that experiments were properly monitored, and the determination as to which agencies would exercise jurisdiction.

---

24. Talbot, *supra* note 22, at 368.

25. D. S. Fredrickson, *A History of the Recombinant DNA Guidelines in the United States*, 151–156 at 154 (Monograph), RECOMBINANT DNA AND GENETIC EXPERIMENTATION (Joan Morgan & W. J. Whelan eds., 1979), *available at* http://profiles.nlm.nih.gov/FF/Views/Exhibit/documents/rdna.html.

26. N. L. Rave, Jr., *Interagency Conflict and Administrative Accountability: Regulating the Release of Recombinant Organisms*, 77 GEO. L.J. 1787–1792 (1989).

These issues came to the forefront when the University of California, a recipient of NIH funding for genetic engineering research, petitioned the NIH to field test a genetically modified bacterium that prevents ice crystal formation on the surface of plants, thus preventing frost and extending the viable life of crops.[27] After receipt of comments from the RAC, the NIH Director approved the field testing without considering the possibility of environmental effects such as the dispersion of the rDNA organisms. Environmental groups petitioned the courts to enjoin the field testing on the grounds that the National Environmental Policy Act (NEPA), 42 U.S.C. § 4321 et seq.,[28] which had been enacted in 1970, required federal agencies to consider environmental effects of major federal actions. When the courts enjoined the field test, they did so on the basis the NIH's environmental review had been inadequate, and they strongly urged the NIH to reevaluate its approval process.[29] The result was that congressional hearings were held to examine the NIH Guidelines. Additionally, during the same time period, EPA had internal deliberations that led to the assertion of jurisdictional authority over small-scale field tests involving certain genetically engineered microbes.[30] In fact, a private company applied to EPA for an experimental use permit under its Federal Insecticide, Fungicide, and Rodenticide Act (FIFRA), 7 U.S.C. § 136 et seq., authority to test the same genetically modified bacteria about which the University of California had petitioned the NIH. EPA approved the petition, which was then also challenged as failing to adhere to the NEPA requirements.[31] However, in contrast to the earlier ruling, the court ruled against the environmental group petitioners, noting EPA's review process was functionally equivalent to that of NEPA, and therefore it would be legalism carried to an extreme to require formal compliance with NEPA when its purposes and policies were being fulfilled.[32] Thus, EPA was able to develop a regulatory mechanism for the first deliberate release of a genetically modified microorganism into the environment.

As biotechnology research was progressing, two developments in intellectual property law hastened the movement toward commercialization. First, Congress enacted the Bayh–Dole Act that permitted universities to hold the patents for any discoveries or inventions developed with government funding. This made it more attractive for industry to collaborate with academic institutions because

---

27. Foundation on Economic Trends v. Heckler, 756 F.2d 143, 149 (D.C. Cir. 1985).

28. *See* Chapter 5 for a full discussion of NEPA.

29. 756 F.2d at 154.

30. *Id.* at 150; D. L. Uchtmann, *A Case Study of Agricultural Biotechnology Regulation*, 7 DRAKE J. OF AGRIC. L. 159, 169 (2002).

31. *See* Foundation on Economic Trends v. Thomas, 637 F.Supp. 25 (D.D.C. 1986).

32. *Id.* at 28.

universities were able to grant exclusive licenses to industry partners.[33] Second, the Supreme Court's decision in *Diamond v. Chakrabarty*, 447 U.S. 303 (1980), made a significant contribution to the drive toward commercialization of biotechnology. The Court held that while the laws of nature, physical phenomena, and abstract ideas are not patentable, a live, human-made microorganism *is* a patentable subject matter. As the Court noted, the microorganism was a non-naturally occurring manufacture or composition of matter—a product of human ingenuity "having a distinctive name, character and use."[34]

As a result of this increased interest in the commercialization of biotech, industry began lobbying the executive branch (which as previously described was sympathetic to concerns about overregulation of industry) to develop a comprehensive policy that would permit commercialization of biotechnology. In 1984, the Reagan White House responded by forming an interagency working group to examine the issue of whether the regulatory framework for products manufactured and developed by traditional genetic manipulation was adequate to regulate products obtained through modern biotechnology means. The interagency working group issued its proposal for a coordinated policy in 1984, and issued the "comprehensive federal regulatory policy for ensuring the safety of biotechnology" (i.e., "Coordinated Framework") in 1986.[35]

2. The Elements of the Coordinated Framework

(a) The Underlying Principles of the Framework The underlying principle of the federal government in biotechnology regulation was that "genetic engineering processes do not necessarily produce organisms that present risks, nor are non-engineered organisms necessarily safe."[36] Thus, unlike the perspective that animated the scientific conferences and the NIH, the Coordinated Framework set forth a perspective that the process of biotechnology was not conceptually different from traditional methods, and that, therefore, this new process did not per se pose new hazards or risks. Furthermore, the assertion was made that stigmatizing this new technology would create market distortions by favoring the continued use of traditional technologies even though the newer technologies might ultimately be of less risk.[37]

By adopting this fundamental principle, the White House and relevant administrative agencies took the position that the existing statutes—even though they

---

33. D. E. Hoffmann et al., *Future Public Policy and Ethical Issues Facing the Agricultural and Microbial Genomics Sectors of the Biotechnology Industry*, 24 BIOTECHNOLOGY L. R. 10, 13 (2005).

34. Diamond v. Chakrabarty, 447 U.S. 303, 309–310 (1980).

35. 51 Fed. Reg. 23302 (June. 26, 1986) *available at* http://usbiotechreg.nbii.gov/Coordinated_Framework_1986_Federal_Register.html

36. *See id.* at *32.

37. *Id. See also* J. Senker & P. van Zwanenberg, *European Biotechnology Innovation System, EC Policy Overview* (Sept. 2000) at 1212.

were drafted prior to the advent of biotechnology—were, for the most part, sufficient to regulate both research about and the marketing of the product. However, critics asserted that even if biotechnology is not per se risky, it does not necessarily follow that existing laws are adequate to address the risks that it does pose.[38] But a number of practical and political considerations guided how the policy was formulated.

First, every agency had an interest in protecting its regulatory authority over the new technology—leading each to argue that biotechnology products were not dissimilar from products covered by existing regulatory mandates.[39] Second, from the industry's perspective, familiarity with the existing statutes and a degree of certainty with regard to how they might be interpreted and implemented created a vested interest in retaining regulation through these known mechanisms. Finally, from the White House's perspective, not only would the process of developing an alternative statutory approach covering a broad spectrum of genetically engineered products and their many uses have been a time-consuming exercise, but it may have meant a new layer of bureaucracy and potentially caused a reexamination of how other industries were regulated. There was also the powerful argument that because already-existing statutes were broadly written, the agencies already had the flexibility to promulgate rules and provide immediate regulatory oversight.[40]

(b) Regulation under the Coordinated Framework Most existing environmental, health, and safety laws govern the testing, manufacturing, marketing, use, and disposal of a product based on the composition of the product, the intended use of the product (e.g., food, pesticide, cosmetic, etc.), and the risks associated with the product when it performs as intended. As regulation is based on the product, different agencies have jurisdiction over different products and have different requirements to ensure the safety of those products. As noted in the Coordinated Framework, regulation of *product* is through FDA, EPA, OSHA and USDA,[41] and regulation of *research* is through the NIH, National Science Foundation (NSF), EPA, and USDA.[42] Each of these agencies issued a policy

---

38. P. Mostow, *Reassessing The Scope of Federal Biotechnology Oversight*, 10 PACE ENVTL. L. REV. 227, 240 (1992).

39. *See id.* at 242.

40. Coordinated Framework, *supra* note 35, at *3.

41. Coordinated Framework, *supra* note 35, at *6.

42. The FDA has four centers with responsibilities for genetically modified products: the Center of Food Safety and Applied Nutrition (CFSAN), which has responsibility for genetically modified foods; the Center of Veterinary Medicine (CVM), which has responsibility for animal drugs created through biotechnology or genetically modified animals (which are regulated as animal drugs); the Center of Drug Evaluation and Research (CDER), which has responsibility for drugs and certain biologics, and the Center for Biologics Evaluation and Research (CBER), which now has responsibility for biologics. EPA has responsibility for insecticides inserted into plants and for genetically

statement articulating how it would extend existing statutory authority to cover biotechnology. To the extent possible, oversight responsibility for a product would lie with a single agency. However, with the recognition that in some cases, multiple agencies would have jurisdiction, the Coordinated Framework stated that agencies would cooperate with each other[43] and that a lead agency would be established to coordinate the regulatory efforts.

As an example, albeit an extreme one, the regulation of Monsanto's version of Bt corn demonstrates how the multiple jurisdictions come into play.[44] This Bt corn variety was subject to regulation primarily by EPA under FIFRA and the Federal Food, Drug, and Cosmetic Act, 21 U.S.C. § 371 et seq., (FFDCA). It was also subject to regulation by the USDA-run Animal and Plant Health Inspection Service (APHIS) under the Federal Plant Pest Act, 7 U.S.C. § 150aa et seq., and the Plant Quarantine Act, 7 U.S.C. § 151 et seq., (laws that have been subsequently replaced by the Plant Protection Act, 7 U.S.C. § 7701 et seq.,). EPA issued an Experimental Use Permit for large-scale field testing (10 acres or more); and later granted registration for commercial sale and use subject to certain conditions; it also and exempted the pesticidal portion from the requirements of the FFDCA that a pesticide residue limit be set in food. On the pesticide residue issue, EPA obtained input from FDA, which has responsibility for enforcing pesticide residue tolerance limits. APHIS authorized field testing (even field testing at less than 10 acres), then granted a petition that APHIS oversight was not necessary. APHIS conducted a NEPA environmental assessment and made a finding that there were no significant impacts on the environment or issues under the Endangered Species Act, 16 U.S.C. § 1531 et seq., From the company's perspective, data and supporting information had to be developed in order to gain approval for field testing by both EPA and APHIS, to register the product through EPA's FIFRA process, and to obtain nonregulated status under APHIS's statutory authority. Finally, before Bt corn could be used in food or feed, FDA had to be consulted.

3. **Significant Post-Framework Documents** An issue that was not addressed in the Coordinated Framework was how the agencies would exercise their respective discretionary authority provided to them by statute. In the years following

---

altered microorganisms. The USDA, through the Animal and Plant Health Inspection Service (APHIS), has responsibility for regulating potential agricultural plant pests and noxious weeds and for animal biologics.

43. *Guide to U.S. Regulation of Genetically Modified Food and Agricultural Biotechnology Products*, PEW INITIATIVE ON FOOD AND BIOTECHNOLOGY 6 (Sept. 2001).

44. This description of Monsanto's approval process for Bt-Maize referred to as MON810 is based on a summary provided in the Council of Environmental Quality and the Office of Science and Technology Assessment of various case studies involving products derived through the use of genetic engineering. *See* CEQ and OSTP Assessment: Case Studies of Environmental Regulations of Biotechnology: Case Study No. II, Bt-Maize, at 1.

the issuance of the Coordinated Framework, federal agencies sought to develop common principles with regard to how agencies should exercise their individual oversight authority.[45] One important policy statement was the "Exercise of Federal Oversight within the Scope of Statutory Authority," which was issued by the President's Office of Science and Technology Policy (OSTP) in 1992.[46] The starting point of this discussion was adherence to the Coordinated Framework's view that biotechnology was not per se risky; therefore, regulation had to be based on the risks associated with use and characteristics of the particular product.[47] In evaluating how to regulate these risks, the Clinton Administration argued that because agency resources are finite, two steps must be undertaken: (a) prioritization to determine which products pose the greatest risks and (b) cost-benefit analyses to determine whether the risks are greater than the costs associated with regulation. However, oversight should be commensurate with the gravity and the type of risk being addressed, and must achieve the greatest risk reduction benefit at the least cost. As discussed in Chapter 3, critics have noted that there is a bias toward under-regulation because the risks associated with new technologies may be understated and unknown, but the costs may be easily quantifiable.

The Clinton Administration also noted that federal agencies should develop guidelines and interpretative statements to assist industry with compliance, such as providing guidance on what categories of products are of such low risk as to not justify oversight (and conversely what categories of products would require oversight). As will become evident in subsequent chapters, the agencies did develop a number of guidance documents especially during the Clinton Administration, but such efforts ground to a halt under the Bush Administration.

---

45. The President's Council for Competitiveness issued a "Report on National Biotechnology Policy," which recommended four principles that should govern federal regulatory oversight for biotechnology and which reiterated certain aspects of the Coordinated Framework. The first principle is that agencies should focus on "the characteristics and risks of the biotechnology product, and not on the process by which it is created." The second principle is that the regulatory review process for products should be designed such that products likely to pose a lesser risk should be given expedited review and that coordination should occur among agencies to avoid confusion, duplication, and delay. The third principle states that regulatory requirements should be based on performance-based standards (i.e., standards that set goals or ends that need to be achieved, rather than, specifying the means by which such goals could be achieved). Finally, the fourth principle indicates that all regulation in environmental and health areas should use performance-based standards for compliance.

46. 57 Fed. Reg. 6753 (Feb. 27, 1992).

47. The 2000 National Research Council report, *Genetically Modified Pest-Protected Plants: Science and Regulation*, reaffirmed the conclusions of the Coordinated Framework— namely that no unique hazards are associated with the use of biotechnology techniques and that the assessment or risk should be based on the particular organism and the environment into which it is introduced, not on the method by which it was produced.

As described at the beginning of this chapter, the Bush Administration sought to curtail the development of guidance documents as means of regulation.

II. THE FEDERAL GOVERNMENT'S NANOTECHNOLOGY EFFORTS

The federal government's approach to nanotechnology has been to regulate it under the existing product-based statutes that it uses to regulate conventional products. The rationales offered to justify this approach (and to reject calls for new statutes that would specifically regulate the nanotechnology process) are an echo of those employed during discussions of the regulatory approach toward biotechnology. Thus, arguments are made that nanomaterials do not pose unique risks, that stigmatizing nanotechnology would result in the use of older technologies that are themselves more risky or less efficient, that crafting a new statute to cover technology that has impacts in diverse fields would be a time-consuming process, that agency efforts at interpreting statutes and regulations to include nanotechnology would be a better use of finite agency resources, and, that industries have familiarity with—and have developed expertise in—ways to operate within the confines of the existing statutes and regulations. However, for both sides of the debate on the adequacy of existing regulations, the critical missing element is sufficient literature on the risks (or the lack thereof) of this emerging technology.[48] Thus, neither side can fully persuade the other that its approach is the correct response for protecting human health and the environment.

As this section will discuss, though the period the federal government has been interested in nanotechnology now exceeds a decade, and the potential for adverse consequences from such technology was recognized as early as 1986, only recently has there been any concentrated attention on environmental or health risks. At the federal level, the primary focus has been on funding the development of nanotechnology for commercial purposes and understanding its basics. Environmental and health risk management still remain a very low priority, as demonstrated by the allocation of funds. This section will first focus on why funding commercial development of nanotechnology has been the government's priority. Then, the section will examine the federal efforts to develop knowledge about its risks.

A. Funding of Nanotechnology—From IWGN to NNI and Beyond

Federal funding for nanotechnology has been and continues to be shaped by the viewpoint that this new technology will transform society and the economy by revolutionizing manufacturing, enabling the creation of environmentally friendly

---

48. However, both sides of the debate have called for an increase in spending on researching the potential impacts of nanomaterials on the environment and human health.

products, and improving health care.[49] Even though early articles and books on nanotechnology recognized the potential adverse consequences for human health and/or the environment, the concern expressed regarding those risks was vastly outweighed by descriptions of the potential benefits.

Federal recognition of nanotechnology came in the wake of the 1992 U.N. Summit in Rio de Janeiro, Brazil, on global sustainability when the then Senate Subcommittee on Science, Technology, and Space (chaired by then-Senator Al Gore) held a hearing on new technologies that would create a more sustainable future, including molecular nanotechnology.[50] This hearing, and the governmental efforts that followed soon thereafter, occurred in a post-Cold War world in which the federal government was increasingly concerned with American competitiveness in the global marketplace. Significantly, the development of science and mathematics was seen as critical to maintaining American economic power. Proponents of nanotechnology argued that this new technology was being developed in other countries such as Japan, with the government funding basic research. Moreover, they argued that the United States could be placed at an economic disadvantage as those other nations would profit from the development of new products borne of nanotechnology.[51]

The executive branch's response to this challenge was to develop a working group to investigate the potential benefits from nanotechnology. The Interagency Working Group on Nanoscience, Engineering and Technology (IWGN) was established in 1998.[52] The IWGN held workshops and seminars. The result of the IWGN's efforts was the launching of the National Nanotechnology Initiative (NNI) in 2000.[53] The primary aim of the NNI was to develop nanotechnology for commercial and public uses with little focus on the environmental or health issues that might arise. To achieve this objective, the focus has been on funding basic research, not on researching those environmental and health issues. The White House press release announcing the budget for the NNI[54] noted that the initiative would involve five kinds of activities: (a) supporting long-term fundamental nanoscience and engineering research; (b) achieving the "Grand Challenges" (e.g., shrinking the entire contents of the Library of Congress into a device the size of a sugar cube); (c) encouraging the sharing of information by nanotechnology centers; (d) funding metrology, instrumentation, modeling, and

---

49. The literature refers to this position as the Drexlerian view as one of the main advocates was Eric Drexler, author of THE ENGINES OF CREATION (discussed in Chapter 2).

50. *See* W. P. McCray, *Will Small be Beautiful? Making Policies for Our Nanotech Future*, 21 HISTORY AND TECH. 177–203, 183 (June 2005).

51. *See id.* at 184.

52. *See* NNI, Strategic Plan (December 2007), at 35.

53. *See* McCray, *supra* note 50, at 191.

54. While the NNI did not fund research, the federal budget for nanotechnology amongst the various agencies are coordinated through and influenced by the NNI.

simulation facilities to facilitate the commercialization of new discoveries; and (e) studying the ethical, social, and economic impacts while developing a skilled workforce.[55]

As funding for NNI was being debated in Congress in April 2000, Bill Joy, chief scientist at Sun Microsystems, penned the article previously discussed that challenged the optimistic view of nanotechnology.[56] In comparing nanotechnology with genetic engineering, Joy presented the potential horror of self-replicating nanomachines that could destroy entire ecosystems. Joy adopted the precautionary principle, similar to the scientists at early biotechnology conferences who had urged a moratorium on certain lines of research. While Joy was not the first to raise concerns about nanotechnology, his prominence in an industry that was strongly lobbying for NNI funding along with his genuine angst about the research paradigm being followed made his arguments particularly salient. Nevertheless, in fall 2000, Congress passed a NNI budget for fiscal year 2001 of $465 million—$30 million less than the administration's request for $495 million—spread across the six agencies that were formally comprised of the NNI at that time.[57] It should be noted that the funding was directed toward the development of nanomaterials using standard chemistry, not the self-replicating nanomachines feared by Joy.

The NNI is managed within the framework of the President's National Science and Technology Council through its Subcommittee on Nanoscale Science Engineering and Technology (NSET). Currently, 25 federal agencies participate in the NNI, with 13 agencies having nanoscale science and engineering research and development budgets.[58] The NNI budget has grown from $465 million to $1.5 billion in the 2009 federal budget.[59] The budgeting priorities have reflected

---

55. Office of the Press Secretary, The White House, *National Nanotechnology Initiative: Leading to the Next Industrial Revolution*, (Jan. 21, 2000), *available at* http://clinton4.nara. gov/WH/New/html/20000121_4.html. The NNI was funded in the FY 2001 budget, and thus, 2001 is the year referenced with regard to the launching of the program rather than when it was proposed in 2000.

56. B. Joy, *Why the Future Doesn't Need Us*, WIRED (April 2000), *available at* http://www.wired.com/wired/archive/8.04/joy.html; *See* McCray, *supra* note 50, at 190.

57. *Available at* www.nano.gov. Originally, there were six agencies that were members of NNI (such as the Department of Defense, National Science Foundation, and Department of Energy), but these were not the agencies that have primary jurisdiction over environmental, health, and safety (i.e., EPA, FDA and OSHA). In fact, EPA joined the NNI in 2002, and the FDA and then OSHA followed in subsequent years.

58. The other 12 agencies have made nanotechnology relevant to their missions or regulatory roles, but do not have R&D funding associated with nanoscale science. *See* Nanotechnology White Paper, United States Environmental Protection Agency, EPA 100/ B-07/001/Feb. 2007, *available at* http://www.epa.gov/OSA/nanotech.htm, at 15.

59. National Nanotechnology Initiative FY 2009 Budget & Highlights, at 1, *available at* www.nano.gov/NNI_FY09_budget_summary.pdf.

the purposes of the NNI in that thus far they have been geared toward developing a fundamental understanding of how to control and manipulate nanoscale materials in order to allow for eventual commercial development or military use of such materials.

In 2003, the 21st Century Nanotechnology Research and Development Act (Pub. L. No. 108-153) codified the goals of the NNI and set forth the research budget for five agencies (including the NSF and EPA) until FY 2008. As with the NNI, the law notes that the purposes of nanotechnology research include advancing American "productivity and industrial competitiveness through stable, consistent, and coordinated investments in long-term scientific and engineering research in nanotechnology," and "accelerating the deployment and application of nanotechnology research and development in the private sector, including start-up companies." In addition to the importance of nanotechnology for economic competitiveness, the Bush Administration was operating in a post-September 11 world. As such, the Bush Administration's continued support for federal funding of nanotechnology was also directly connected to how this scientific endeavor might assist with national security. For example, for FY 2003, the Bush Administration added a new "Grand Challenge" to the NNI addressing homeland security via detection and protection against chemical, biological, explosive, and radiological threats.[60] Moreover, funding priorities reflected the national security focus, with the actual 2007 budget, the estimated 2008 budget, and the proposed 2009 budget each having the Department of Defense as the single largest recipient of funds for nanotechnology—an amount in excess of $400 million.

B. NEHI and Funding Environmental, Health, and Safety Research

As funding of nanotechnology research increased, a variety of research institutions began scrambling to create "nanocenters" and reword research proposals in order to obtain funding.[61] At the same time, however, concerns about the trajectory of nanotechnology were being raised with greater urgency, and with explicit references being made to the fears and apprehensions that characterized the biotechnology experiences. For example, then-Senator George Allen, a co-sponsor of the 21st Century Nanotechnology Research and Development Act, noted that, "the field of genetically modified organisms [was] a very promising field of science which lost public confidence, especially in Europe, and therefore dramatically lost support and funding. If a large portion of the general public feels that science has overstepped its bounds—whether by misinformation or not— then general enthusiasm and support for further discovery of new technologies

---

60. McCray, *supra* note 50, at 97.

61. *See* R. Monastersky, *The Dark Side of Small: As Nanotechnology Takes Off, Researchers Scramble to Assess Its Risks*, THE CHRONICLE OF HIGHER EDUC. (Sept. 10, 2004).

can whither away very quickly."[62] There was genuine fear that, as had occurred with genetically modified foods, negative public reaction could prevent companies from investing monies in developing nanotechnology-related products.

Recognizing the need for exploration of the environmental and health implications raised by nanotechnology, a specific working group—the Nanotechnology Environment and Health Implications (NEHI)—was informally established by the NSET Subcommittee late in 2003 (and formally chartered in 2005).[63] At least 20 agencies participate in the NEHI, including EPA, FDA, and NIOSH.[64] Broadly speaking, the purpose of the NEHI Working Group is to "facilitate the identification, prioritization and implementation of research and other activities required for the responsible research and development, utilization and oversight of nanotechnology."[65] The NEHI Working Group operates on a consensus basis, with representatives to the working group consulting with experts within their own respective agencies. It has no authority to require compliance or implement even an agreed-upon strategy, and each agency retains its own budget.

One of the NEHI Working Group's initial tasks was to develop a prioritized plan for environmental, health, and safety research. However, when the NEHI Working Group issued its report in 2006, it failed to provide any prioritization, instead offering a list of research topics within five broad research categories.[66] Subsequently, in 2007 the NEHI Working Group issued an interim report that included a refined list of research topics, but still did not prioritize the research. These efforts were roundly criticized by members of Congress, industry representatives, and public policy experts alike for failing to establish a cohesive strategy based on effective leadership, coordination, and communication.[67]

---

62. Remarks Prepared for Delivery, National Nanotechnology Initiative Conference Senator George Allen (Apr. 1, 2004), *available at* http://www.nano.gov/html/res/Sen. AllenNational_Nanotechnology_Initiative_Conference_4.1.04.htm.

63. Statement of Norris E. Alderson, Associate Commissioner for Science, FDA, and Co-Chair of NEHI Working Group, before the Committee on Science, House of Representatives (Sept. 21, 2006).

64. Statement of E. Clayton Teague, Director, National Nanotechnology Coordination Office, before the House Subcommittee on Research and Science Education (Oct. 31, 2007); K. Thomas & P. Sayre, *Research Strategies for Safety Evaluation of Nanomaterials, Part I: Evaluating the Human Health Implications of Exposure to Nanoscale Materials*, 87 Toxicol. Sci. 87, 316–321, 317 (2005).

65. Statement of Norris E. Alderson, *supra note* 63.

66. The five categories are: instrumentation, metrology and analytical methods, nanomaterials and human health, nanomaterials and the environment, human health and environmental exposure assessment, and risk management methods.

67. *See, e.g.*, Statement of Paul D. Ziegler, Chairman, American Chemistry Council Nanotechnology Panel before United States House Committee on Science and Technology (Oct. 31, 2007); Statement of Andrew D. Maynard, Chief Science Advisor, Project of Emerging Nanotechnologies, Woodrow Wilson International Center for Scholars before United States House Committee on Science and Technology (Oct. 31, 2007); Statement of

Finally, in February 2008, the NEHI Working Group published a report that prioritized research. The "highest priority" was given to research related to (a) developing analytical methods that would detect nanomaterials in biological matrices, the environment, and the workplace; (b) understanding how chemical and physical modifications affect the properties of nanomaterials; (c) developing methods for standardizing assessment of nanomaterial's spatio-chemical composition, purity, and heterogeneity; and (d) developing reference materials for toxicological and environmental studies.[68] The NEHI Working Group concluded that precise and accurate measurements at the nanoscale in multiple and complex media is a fundamental requirement for assessing the potential impacts to both human health and the environment.[69] The report also noted that agencies had agreed to coordinate their efforts, and that with respect to the basic categories of research. each of the agencies had agreed to take on one of three roles: "coordinating agencies," "contributing agencies," or "user agencies."[70]

While even critics commended the NEHI Working Group for prioritizing research and assigning supervisory roles for agencies, a number of fundamental issues were not addressed by the report. Critics pointed out that the consensus model for decision making and agency-driven budgets has resulted in the process being slow, in no overall vision being established, and in the agencies protecting their bureaucratic and budgetary turfs to the potential detriment of necessary research.[71] Moreover, virtually every commentator outside of government has indicated the insufficiency of monies directed toward environmental, health, and safety research.

C. The Overall Regulatory Framework Governing Nanomaterials

Both the pace of this research and its results will influence how the regulatory framework will evolve. Currently, nanomaterials, as is the case with genetically modified products, are regulated under the same set of statutes and regulations that apply to conventionally manufactured products. The agencies have issued a few guidance documents and have suggested that manufacturers comply through voluntary reporting mechanisms. As a recent FDA Task Force report noted, while there were concerns about the adequacy of the agency's regulatory authority

---

Vicki L. Colvin, Director, Center of Biological and Environmental Nanotechnology, Rice University, before United States House Committee on Science and Technology (Oct. 31, 2007).

68. NNI, *Strategy for Nanotechnology-Related Environmental, Health and Safety Research* (February 2008) at 12.

69. *Id.*

70. *See id.* at 48.

71. Statement of Andrew D. Maynard, Chief Science Advisor, Project of Emerging Nanotechnologies, Woodrow Wilson International Center for Scholars before United States House Committee on Science and Technology (Oct. 31, 2007).

over such items as cosmetics, they believed that FDA action in amending the requirements should wait until further scientific information is garnered.

The reliance on existing statutes and regulations has been criticized in a number of policy papers as being inadequate. For example, critics note that the current metrics or definitions—such as mass or molecular identity—that are used to determine the potential risks of conventional materials or the applicability of a statute may not be appropriate in the nanomaterial context. Additionally, because nanomaterials can be used in, or made into, a variety of different products, the same nanomaterials may be subject to redundant and inconsistent regulation. For example, some uses of nanomaterials require premarket approval and the submission of scientific data to demonstrate safety and effectiveness, while other uses may require no preapproval and only limited agency review post-commercialization.[72] Moreover, the agencies may differ fundamentally in their approaches. For example, EPA believes that physical characteristics (e.g., size and shape) are not a relevant factor in determining whether a nanomaterial is considered a "new" chemical;[73] on the other hand, the FDA Task Force indicated that it is precisely those same physical characteristics that are relevant in determining how the material should be regulated.[74]

In addition to highlighting such issues in policy papers, environmental and consumer groups are using the same tactics as had been employed in the early years of biotechnology—namely, petitioning the agencies to issue new regulations to cover nanotechnology and contending that any existing or future regulatory programs need to comply with NEPA. For example, environmental and consumer groups filed citizen petitions to FDA in 2006 and to EPA in 2008.[75] The petition to FDA called upon the agency to, among other things, issue a formal

---

72. L. J. Schierow, *Chemical Facility Security*, CRS Report RL31530 (2008) at CRS-12.

73. As discussed in further detail in Chapter 10, EPA's view is that the nanoscale material would be considered a "new" material if its molecular identity is different from an existing chemical.

74. Compare EPA, *TSCA Inventory Status of Nanoscale Substances—General Approach* (Jan. 23, 2008) with FDA, *Nanotechnology: A Report of the U.S. Food and Drug Administration, Nanotechnology Task Force* (July 25, 2007) (hereinafter *FDA Task Force Report*). See Chapter 8 for a discussion on nanoscale titanium oxide and zinc oxide in sunscreen and their regulation as an over-the-counter drug.

75. The International Center for Technology Assessment, "Petition Requesting FDA Amend Its Regulations for Products Composed of Engineered Nanoparticles Generally and Sunscreen Drug Products Composed of Engineered Nanoparticles Specifically," filed with Andrew C. Von Eschenbach, Acting Commissioner, FDA, *available at* http://www.icta.org/doc/Nano%20FDA%20petition%20final.pdf [hereinafter *FDA Petition*]; The International Center for Technology Assessment, "Petition for Rulemaking Requesting EPA Regulate Nano-Silver Products as Pesticides," filed with Stephen L. Johnson Administrator, EPA, *available at* http://www.icta.org/nanoaction/doc/CTA_nano-silver%20petition_final_5_1_08.pdf [hereinafter *EPA Petition*].

opinion clarifying the agency's position on nanoproducts and require sunscreens that contain nanoscale titanium oxide and zinc oxide to be submitted as new drug applications rather than rely on the over-the-counter monograph that covers all sunscreens.[76] The FDA Task Force report noted above addressed some of the petitioners' arguments when it found that physical characteristics of the material were relevant as to how it should be regulated.[77] FDA continues to review the petition and has adopted the positions set forth by the Task Force; thus, the implications for sunscreen or other materials regulated by FDA remain to be determined.

The petition to EPA filed on May 1, 2008 asked the agency, among other things, to regulate all nanoscale silver products under EPA's pesticide regulations (including requiring them to be registered as new pesticides).[78] As noted above, EPA has already stated that physical characteristics (e.g., size and shape) are not a relevant factor in determining whether a nanomaterial is considered a "new" chemical under the Toxic Substances Control Act (TSCA), but rather the molecular identity of the nanoscale material.[79] But, the standards under FIFRA are different from TSCA, and thus, EPA may reach a different conclusion in the context of FIFRA.

---

76. See FDA Petition, supra note 75, at 3.

77. See FDA Task Force Report, supra note 75, at 12.

78. See EPA Petition, supra note 75, at 3–4.

79. It should be noted that since the announcement of its policy on what constitutes "new" in the nanoscale context, EPA has issued a notice informing stakeholders that carbon nanotubes are not equivalent to graphite or other allotropes of carbon, and thus may be subject to regulation as a "new" chemical substance.

5. REGULATION OF TRANSGENIC PLANTS AND ANIMALS

I. OVERVIEW OF THE CURRENT ROLE OF TRANSGENIC PLANTS IN AMERICAN AGRICULTURE

The development of transgenic plants remains in its infancy.[1] Companies began field testing the first generation of transgenic plants in the 1980s, and as data from the USDA indicates, U.S. farmers quickly adopted transgenic plants (such as genetically modified versions of soybeans, corn, and cotton) in the mid-1990s when they became available. The focus of farmers has mainly been the adoption of herbicide-tolerant (HT) and insect resistant varieties of genetically modified crops. HT crops are varieties of crops that are able to survive intensive applications of herbicides that previously would have destroyed them along with the targeted weed.[2] Insect-resistant crops incorporate genes from the soil bacterium *Bacillus thuringiensis* (Bt), which produces a protein toxic to specific insects such as budworm, bollworm, and the European corn borer. Initial field tests in the 1980s involved Bt varieties of tobacco, cotton, and tomatoes[3] with Bt corn first being field tested in the early 1990s.

Unlike genetically modified HT varieties, Bt corn has encountered significant public opposition because of two highly publicized incidents: the Monarch Butterfly and the Star Link™. While the adverse publicity associated with these events has not slowed the adoption by farmers of certain types of already approved genetically modified seeds (e.g., in 2000 an estimated 54 percent of all soybeans planted in the United States was an HT variety, and by 2007, an estimated

---

1. J. Fernandez-Cornejo et al., *The First Decade of Genetically Engineered Crops in the United States*, Economic Research Service (USDA) (April 2006) at 1. The USDA classifies the genetically modified crops (or transgenic plants) into one of three generations. The first generation are transgenic crops with enhanced input traits such as herbicide tolerance, insect resistance, and tolerance to environmental stresses. The second generation are transgenic crops that have added-value output traits such as nutrient enhancements. Finally, the third generation are transgenic crops that have been modified to manufacture pharmaceuticals or industrial chemicals.

2. *See id.* at 8.

3. J. Carpenter, *Case Studies in Benefits and Risk of Agricultural Biotechnology: Roundup Ready® Soybeans and Bt Field Corn* 23 (2001). National Center for Food and Agricultural Policy.

91 percent of all soybeans was an HT variety),[4] it did result in the severe curtailment in the introduction of new varieties of insect-resistant and herbicide-resistant seeds. For example, Monsanto withdrew insect-resistant potatoes and halted development of herbicide-resistant wheat due to fears of losing export sales, and the rice industry did not develop herbicide-resistant varieties of rice.[5] In fact, for the first time since the 1990s, a new type of genetically modified crop was grown in the United States—namely, sugar beets that were planted in the 2008 growing season.[6]

Under the 1986 Coordinated Framework, three separate federal regulatory agencies have authority to regulate the production of transgenic plants and animals.[7] The United States Department of Agriculture's (USDA's) Animal and Plant Health Inspection Service (APHIS) controls the release, importation, or transfer of transgenic plants under its mandate to regulate plants that have the potential to become plant pests. EPA regulates transgenic plants that have been modified to produce their own pesticidal substances (referred to as plant-incorporated protectants, or PIPs). The Food and Drug Administration regulates food and feed products produced from genetically modified crops. EPA and APHIS both review transgenic plants that are modified for pesticidal substances, and each requires information from the developer that complies with its regulatory requirements.[8] Additionally, both FDA and USDA's Food Safety Inspection Service (FSIS) has a role in the regulation of transgenic animals.

The regulation of agricultural biotechnology applies primarily at two distinct points in the development of the product: (a) the cultivation or breeding of the plant or animal, and (b) the products that are derived from the transgenic plant or animal. This chapter will focus on the first aspect by examining the regulations and guidance documents primarily issued by USDA and EPA.

---

4. USDA Data Sets: Adoption of Genetically Engineered Crops in the US: Soybeans. *See* http://www.ers.usda.gov/Data/BiotechCrops/ExtentofAdoptionTable3.htm.

5. Andrew Pollack, *Round 2 for Biotech Beets*, N.Y. Times, Nov. 27, 2007.

6. *Id.* There are certain unique economic aspects to growing sugar beets, which are primarily used as the source for the nation's sugar supply. For example, sugar is a refined product that contains just the chemical sucrose (with no DNA or proteins), and thus foreign genes cannot enter it. As only about 3 percent of American sugar is exported, import restrictions by other countries on genetically modified foods will not have a significant impact on the industry, and sugar processing facilities are owned by farmers who are more accepting of biotech crops. *Id.*

7. Each agency employs already existing product-based statutes and regulations to determine whether a particular genetically modified plant or animal should be allowed for commercial cultivation or breeding and/or should be permitted to enter the consumer marketplace.

8. Testimony of Stephen Johnson, Assistant Administrator, Before Committee of Agriculture, Subcommittee on Conservation, Credit and Research (June 17, 2003), *available at* http://usinfo.state.gov.

Chapter 6 will examine the regulation of genetically modified foods and will primarily focus on FDA.

II. FEDERAL REGULATION OF TRANSGENIC PLANTS

A. The Historical Evolution of Laws

Prior to examining how the USDA, and specifically APHIS, regulates transgenic animals and plants, it is important to consider the historic development of the agency's authority. The current USDA regulatory structure is based on the 1912 Plant Quarantine Act (PQA).[9] The purpose of the PQA statute was to prevent the introduction of nonnative species into the United States. The statute was enacted in response to the devastation caused to an economically significant species, the American chestnut tree, that resulted from the importation in the late 1800s and early 1900s of Chinese and Japanese varieties that inadvertently contained a certain fungus.

Until the mid-1950s, the organizational matrix of the USDA involved various bureaus that had jurisdictional authority over plant health, livestock disease research, animal import regulations, and interstate movement of animals.[10] In the mid-1950s, these bureaus were consolidated under the USDA's Agricultural Research Service (ARS) which divided responsibilities based on whether they entailed research or regulatory functions.[11] In 1972, APHIS was created as an entirely new entity within the USDA and became responsible for the regulatory functions that had previously resided with the ARS.[12] However, ARS continues to function as the research arm of USDA.

In 1985, the USDA designated APHIS[13] as the agency responsible for regulating plants that had been genetically modified (referred to as a "regulated article"

---

9. 7 U.S.C. §§ 151–164(a), 167.

10. *See* APHIS, "History of APHIS," *available at* http://www.aphis.usda.gov/about_aphis/history.shtml.

11. *See id.*

12. The meat and poultry inspection divisions of the Consumer and Marketing Service were added to APHS (putting the *I* into APHIS). In 1977, however, "the Food Safety and Quality Service was established and took responsibility for meat and poultry inspections. That agency is now known as the USDA's Food Safety and Inspection Service." *See id.*

13. Within APHIS, the regulation of genetically modified plants is divided amongst two branches: the Biotechnology Regulatory Services (BRS) and the Compliance and Inspection Branch (CIB). Within the overall APHIS structure, it is the BRS's role to issue permits and acknowledge notifications for transgenic plants. The CIB is charged with assuring compliance with all relevant provisions of the regulations, including authorizations under the permit and notification process. *See id.*

in the regulations).[14] Two years later, the agency promulgated rules requiring the issuance of a permit prior to the release into the environment (i.e., use of a regulated article outside the constraints of physical containment, such as in a laboratory, confined greenhouse or fermenter) of a genetically modified plant. In 1993, the regulations were amended to create a *notification* process, which was offered as streamlined alternative to the permit process, and a *petition process for nonregulated status* for certain genetically modified crops. The original notification process was limited to six plants that had a low risk of having genetic material transferred to wild relatives (i.e., corn, cotton, potato, soybean, tobacco, and tomatoes).[15] However, the notification requirements were amended in 1997 to include all plants except those federally listed as noxious weeds, as well as other plants designated as weeds by APHIS or a state agency.

In 2000, as part of the Agriculture Risk Protection Act, 7 U.S.C. § 1501 et seq., Congress passed the Plant Protection Act, 7 U.S.C. § 7701 et seq.,[16] This act repealed, consolidated, and expanded several laws that regulated plant pests, including the Plant Quarantine Act of 1912, the Federal Plant Pest Act of 1957,[17] and certain portions of the Federal Noxious Weed Act of 1974.[18] The Act provides the Secretary of Agriculture with broad regulatory authority to undertake operations or measures to detect, control, eradicate, suppress, prevent, or retard the spread of plant pests or noxious weeds. Congress provided the USDA with such broad authority because new plant pests or noxious weeds constitute a threat to a major component to the U.S. economy—namely, agriculture.

The USDA, specifically the APHIS, has authority to regulate any genetically modified or transgenic plant if that plant has the potential to become a *plant pest*, a *noxious weed*, or *biological control agent*. These three terms are defined by the

---

14. The implementing regulations, titled as "Introduction of Organisms and Products Altered or Produced through Genetic Engineering Which are Plant Pests or Which There is Reason to Believe are Plant Pests" are set forth in 7 C.F.R. § 340, specifically define a *regulated article* as "any organism which has been altered or produced through genetic engineering, (a) if the donor organism, recipient organism, or vector or vector agent belongs to any genera or taxa designated in the list enumerated in 40 C.F.R. § 304.2 and meets the definition of plant pest or (b) for any unclassified or unknown organism that the USDA determines is a plant pest or has reason to believe is a plant pesticide." 7 C.F.R. § 340.1.

15. Genetically Engineered Organisms and Products; Notification Procedures for the Introduction of Certain Regulated Articles; Petition for Nonregulated Status, 58 Fed. Reg. 17044, 17045 (Mar. 31, 1993). Between 1987 and March 2, 1993, APHIS granted 365 permits for field testing and 1,301 interstate transport permits for transgenic plants. Of that total, 19 percent were for corn, 10 percent for cotton, 20 percent for potato, 18 percent for soybean, 5 percent for tobacco, and 13 percent for tomato. *See id.*

16. Pub. L. No. 106-224, 7 U.S.C. § 7701–7772.

17. 7 U.S.C. § 150aa *et seq.*, 7 U.S.C. § 147a.

18. Repealed 7 U.S.C. §§ 2802–2813; but did not repeal 7 U.S.C. § 2801 and 7 U.S.C. § 2814.

statute as follows: A *plant pest* is broadly defined as "[A]ny living stage of any of the following that can directly or indirectly injure, cause damage to, or cause disease in any plant or plant product: (A) A protozoan. (B) A nonhuman animal. (C) A parasitic plant. (D) A bacterium. (E) A fungus. (F) A virus or viroid. (G) An infectious agent or other pathogen. (H) Any article similar to or allied with any of the articles specified in the preceding subparagraphs."[19] As discussed in this chapter, because most genetic modifications have involved making plants herbicide- or insect-resistant, APHIS has used its authority under this definition of plant pests to regulate genetically modified crops. However, if the modifications to the plant are for pharmaceutical or industrial purposes, then the APHIS's authority under the broadly defined terms *noxious weed* and *biological control organism* may be more appropriate. *Noxious weed* is defined as "any plant or plant product that can directly or indirectly injure or cause damage to crops . . . livestock, poultry, other interests of agriculture, irrigation, navigation, the natural resources of the United States, the public health, or the environment."[20] A *biological control organism* has an equally expansive definition as "any enemy, antagonist, or competitor used to control a plant pest or noxious weed."[21]

The Plant Protection Act, however, contains a savings clause that provides that any regulations promulgated under the preexisting laws (such as the regulations governing genetically modified plants) would remain in effect until APHIS issued new rules under the authority given the Plant Protection Act. While APHIS is now considering issuing regulations relating to new generations of genetically modified plants, no regulations have been issued to date. The APHIS regulatory scheme for transgenic plants is as follows: In order for a company to release into the environment (i.e., field test), import, or transport a transgenic plant, it is necessary to (1) obtain a permit approval, (2) submit a notification and receive acknowledgement of it, or (3) obtain a nonregulated status based on an approved petition. To commercially sell the plant, either the company must obtain a permit or the plant must have nonregulated status.[22]

B. Current USDA Regulations

1. **Introduction** As noted in the preceding section, in 1987 under the 1912 Plant Quarantine Act, Federal Plant Pest Act, 7 U.S.C. § 147a, and other statutes preceding the 2000 Plant Protection Act, APHIS issued regulations that presumptively classified most genetically modified plants as "plant pests" and prohibited their introduction without a permit. Subsequently, APHIS amended

---

19. 7 U.S.C. § 7702 (14).

20. 7 U.S.C. § 7702(10).

21. 7 U.S.C. § 7702(2).

22. The notification process is not available as to plants that have been modified for industrial or pharmaceutical purposes, and deregulation is not available unless the plant poses no greater plant-pest risk than nongenetically modified versions.

those regulations to create the notification and nonregulated status processes.[23] When Congress enacted the Plant Protection Act, the statute provided that unless the Secretary of Agriculture deregulates a particular plant pest based on a petition, "No person shall import, enter, export, or move in interstate commerce any plant pest, unless the importation, entry, exportation, or movement is authorized under general or specific permit and is in accordance with such regulations as the Secretary may issue to prevent the introduction of plant pests into the United States or the dissemination of plant pests within the United States."[24] Thus, the statute codified the regulatory system that had previously been developed by noting that plant pests could be deregulated by petition and that authorization may be acquired through a specific permit or even a general permit (a concept akin to notification in other environmental fields, such as the Clean Water Act).

2. Permitting An applicant seeks a permit in order to release into the environment (i.e., field test), import, transport, or commercially sell a genetically modified plant.[25] The permit process begins with the submission of an application that contains 14 different aspects.[26] Among the detailed descriptions that will need to be provided are: (a) the anticipated or actual expression of the altered genetic material in the regulated article and how it differs from the expression in the nonmodified parental organism; (b) the molecular biology of the system which is or will be used to produce the regulated article; (c) the purpose of the introduction of the regulated article including a detailed description of the proposed experimental and/or production design; (d) the processes, procedures,

---

23. 58 Fed. Reg. 17044. When APHIS initially proposed amending the permit process, it had even considered notification being made on the same day as of the introduction of the genetically modified plant to the environment. As APHIS noted, this procedure was opposed by a coalition of parties as a premature step toward deregulation and/or self-regulation, and thus it was rejected. 58 Fed. Reg. 17050. As of June 20, 2007, a combined 13,833 notifications were submitted to and permits issued by APHIS. However, a closer examination of that number indicates that there were 12,438 notifications and only 1395 permits—thus, almost 90 percent of submissions to APHIS involved notifications and only 10 percentage involved permit applications. Moreover, as of 2007, almost two-thirds of all permits and notifications were issued to just five plants: corn, soybean, cotton, potato, and tomato.

24. 7 U.S.C. § 7711(a).

25. APHIS can also issue *limited* permits for the purpose of interstate movement and importation of a regulated article. These permits are valid for one year from the date of issuance. A single limited permit may be used for interstate movement of multiple articles in lieu of requiring a permit for each individual interstate movement, or it may be used for multiple interstate movements between contained facilities. 7 C.F.R. § 340.4(c)(1).

26. 7 C.F.R. § 340.4(b). If there are portions of the application deemed to contain trade secrets or confidential business information, each page of the application containing such information must be clearly marked as *CBI Copy*. Application forms are *available at* http://www.aphis.usda.gov/brs/pdf/2000.pdf.

and safeguards that have been used or will be used in the country of origin to prevent contamination, release, and dissemination; and (e) the intended destination, uses, and/or distribution of the regulated article.[27] For field testing, a key component an applicant must demonstrate is that effective measures have been undertaken to ensure plantings are confined to a specific area and that it is not likely the genetic material will spread outside the testing area.

The application must be submitted at least 120 days in advance of the proposed release into the environment. However, the 120-day clock for APHIS review does not start running until APHIS receives what it considers to be a complete application. The clock will also not start running if an environmental impact statement, in addition to an environmental assessment, is necessary to comply with the requirements of the National Environmental Protection Act (NEPA)(which is discussed in a subsequent section of this chapter). Thus, when the application is deemed to be complete and the overall NEPA process is also completed, the applicant is notified as to the start of the 120-day clock.[28] The permit may contain limitations or restrictions on planting or transportation or require the adoption of mitigation measures.

3. Notification As noted above, notification has been by far the most prevalent method by which companies seek the environmental release of a genetically modified plant. The notification process allows for field testing, importation, and interstate transport of genetically engineered plants.[29] However, the notification process only applies to plants, so genetically modified microorganisms, insects, or other biological entities do not qualify for the notification process.[30]

---

27. 7 C.F.R. § 340.4(b)(5), (6), (8), (10), (11). In addition to obtaining a permit from the BRS, APHIS's Plant Protection and Quarantine (PPQ) division may require for a container variance. Although BRS is responsible for overall management of the program and directs which field test sites should be inspected to determine compliance with the containment, PPQ has responsibility for actual inspections of genetically engineered plant field test site. An internal audit by the USDA, however, found that BRS does not have a formal risk-based process for selecting individual sites for inspection, and that PPQ does not complete all of the inspections BRS requests. *See supra* note 13.

28. Once APHIS receives a completed application and has completed its initial review, it shall forward this documentation to the appropriate regulatory agencies for the state in which the release is to occur. 7 C.F.R. § 340.4(b).

29. It should be noted that multiple interstate movements and/or multiple environmental release (e.g., field trial locations) may be described in the same notification. Moreover, notification of interstate movement and environmental release may be combined in the same document. Multiple importations may also be described in a single notification, provided all the importations must be shipped from the same origin and to the same destination. Notification of importation may not be combined with the notification of interstate movement or release. *See* USDA-APHIS Biotechnology Research Service, *User's Guide: Notification, Chapter 6*, (Feb. 5, 2008) at 16.

30. Genetically engineered insects, nematodes, bacteria, viruses, and other regulated organisms do not qualify for notification. *See id.*

To qualify for the notification process, six specific eligibility criteria and six performance standards must be met.[31] The purpose of these criteria and standards is to ensure "confidence that the regulated article will not be released beyond the proposed introduction (both in time and space)."[32]

The eligibility criteria are as follows:

1. Recipient organisms are not listed as a noxious weed (as enumerated in 7 C.F.R. Part 360) and are not considered by APHIS to be a weed in the area of release.[33]

2. There is stable integration of genetic material into the host genome (e.g., does not mobilize or replicate naturally)

3. The applicant has to demonstrate sufficient knowledge about the function of the genetic material to assert that it does not cause plant disease.

4. The introduced genetically modified plant (a) does not produce an infectious entity, such as a plant virus, animal virus, human virus, etc.; (b) is not toxic to organisms that are living or feeding on the plant, except for being toxic to the target organism or nontarget organism that is not likely to be living or feeding on the plant; and (c) must not express compounds intended for pharmaceutical[34] or industrial use.[35]

5. It does not pose a significant risk of creating new plant viruses because either (a) it is a noncoding sequence, or (b) the sequences are (i) from viruses prevalent and endemic in the plant species in the area of the proposed introduction, and (ii) do not encode a functional cell-to-cell movement protein.[36]

---

31. 7 C.F.R. § 340.3(a).

32. *See* USDA, *supra* note 29, at 6.

33. The "considered to be a weed" clause applies only to notifications of release into the environment and not notifications of importation or interstate movement. Given that historically the objective of this regulatory system was to prevent the spread of weeds that either on their own or through cross-pollination would degrade the environment by spreading in an uncontrolled manner, it is not surprising that the notification process specifically excludes such plants.

34. A plant is considered as intended for pharmaceutical use if commercialization of the compound would require the approval of the following agencies: FDA's Center for Biologics Evaluation and Research (human biologics); FDA's Center for Drug Evaluation and Research (drugs); FDA's Center for Veterinary Medicine (animal drugs); or USDA's Center for Veterinary Biologics (animal biologics). *See id.* at 9.

35. The following criteria are used to determine if the plant has been produced for industrial use: (a) the plants are engineered to produce compounds that are new to the plant; (b) the new compound has not been commonly used in food or feed; and (c) the new compound is being expressed for nonfood/nonfeed industrial uses (e.g., detergent manufacturing, paper production). *See id.*

36. *See id.* 29, at 9 explaining 7 C.F.R. § 340.3(b)(5).

6. It does not contain any sequences from any human or animal pathogens as enumerated in the regulation.[37]

In addition to the eligibility criteria, notification can only be used if the six performance standards are all met. If any of the six performance standards is not met, the applicant has to seek a permit. The applicant has to develop protocols to demonstrate that it can meet the performance standards listed below. Although the applicant has flexibility in developing the appropriate protocols, in submitting its notification to APHIS the applicant certifies that regardless of the methods selected, the standards can be met.[38] The key element of these standards is that they provide a mechanism to introduce a regulated article in such a manner that it or its offspring are unlikely to persist in the environment.

1. If plants or plant materials are shipped, they must be shipped in such a manner as to ensure that it is unlikely that there is a release from the shipping container in transit.[39] This performance standard can be met by complying with the requirements described in 7 C.F.R. 340.8 or any alternative method that is sufficient to meet this standard.[40]

2. The regulated article must be planted in such a manner that it is not inadvertently mixed with nonregulated plant materials of any species.[41] As APHIS notes, this standard can be met by planting the regulated article in a defined area demarcated by a buffer zone. The buffer zone should be sufficient to allow for the movement of planting and other farm equipment in such a way that transgenic seed do not get deposited outside the designated area of planting.[42] Moreover, individuals should take care when they enter into and use equipment in the designated planting area to prevent inadvertent carrying of transgenic seeds.[43]

3. The plants and plant products must be maintained in such a manner that the identity of all material is known while it is in use, and the plant parts must be contained or devitalized when no longer in use.[44] The disposition

---

37. 7 C.F.R. § 340.3(b)(6).

38. *See* USDA, *supra* note 29, at 9–11.

39. 7 C.F.R. § 340.3(c)(1).

40. *See* USDA, *supra* note 29, at 11. While the regulations also require that the applicant certify the shipped plants will be maintained in facility in such a manner as there will be no release, APHIS does not evaluate the adequacy of such of research and storage facilities. Rather, APHIS encourages applicants to ensure that the destination facility meets the containment guidelines issued by the NIH as provided in the "Guidelines for Research Involving Recombinant DNA Molecules (NIH), http://www4.od.nih.gov/oba/rac/guidelines/guidelines.html.

41. 7 C.F.R. § 340.3(c)(2).

42. *See* USDA, *supra* note 29, at 12.

43. *See id.*

44. 7 C.F.R. § 340.3(c)(3).

may occur through treatment with an herbicide, incineration, or above-ground composting.[45]

4. "There must be no viable vector agent associated with the regulated article."[46]

5. The field test must be conducted in such a manner that (a) the regulated article will not persist in the environment, and (b) no offspring can be produced that could persist in the environment.[47] As APHIS notes, this standard will most often determine whether the regulated article will be introduced under the notification process or permit.[48]

6. Upon termination of the field test, no viable material shall remain that is likely to "volunteer" in subsequent sessions, or volunteers shall be managed to prevent their persistence.[49]

If the applicant can meet the performance and eligibility criteria, then a notification can be submitted for interstate movement, importation, or environmental release. APHIS will provide acknowledgment within 10 days for an interstate notification, 30 days for an importation notification, and 30 days for an environmental release. As to the environmental release, the acknowledgement will apply to field testing for a one-year period from the date of introduction, but can be renewed annually by the applicant submitting further notification to APHIS.[50] If APHIS determines that the proposed action does not qualify for notification, this will not prejudice the applicant's submission for a permit.

Following field testing, if the applicant wants to sell the plant, the typical route is to seek a deregulated status by filing such a petition with APHIS. The reason is that nonregulated status allows for planting and transport without restrictions. Alternatively, an applicant could submit a permit application to commercialize the crop.

---

45. *See* USDA, *supra* note 29, at 12–13.

46. 7 C.F.R. § 340.3(c)(4).

47. 7 C.F.R. § 340.3(c)(5).

48. As APHIS notes, plant species that have *weedy* characteristics are not likely to meet performance standards. Weedy characteristics may include: abundant seed production and dispersal, seed dormancy or long-term seed viability in the soil, reproduction via vegetative structures, rapid population establishment, adoption for long-distance dispersal, existence of feral populations of nontransgenic plants, and plants found in disturbed areas. *See* USDA, *supra* note 29, at 13.

APHIS notes that growers should review the isolation distances published by the Association of Official Seed Certifying Agencies (AOSCA) in its *Yellow Book* as the minimum acceptable distance between the regulated plants and any sexually compatible species. The seed isolation distances based upon AOSCA standards for the most common crops are published in 7 C.F.R. § 201.76. *See id.*

49. 7 C.F.R. § 340.3(c)(6).

50. 7 C.F.R. § 340.3(e)(2–4).

4. Reporting Obligations under Permits and Notifications Applicants who receive an acknowledgement for a notification submission or who obtain a permit have certain reporting requirements. Specifically, within six months of completing field tests, data results from the field test must be submitted.[51] A field test report shall include the APHIS reference number, methods of observation, resulting data, and analysis regarding all deleterious effects on plants, nontarget organisms, or the environment.[52] Additionally, depending on the nature of the genetically modified plant and conditions set forth in the permit and notification, APHIS may also require approved applicants to submit other progress reports during the field test.

Under the APHIS review process, submitted data is evaluated for potential risks, including whether the transgenic plant might "(1) expose other plants to pathogens; (2) harm other organisms, including agriculturally beneficial organisms, threatened or endangered species, and in the case of plants that produce pesticides, organisms that are not the intended target of the pesticide (i.e., non-target organisms); (3) increase the weediness in another species with which it may cross; (4) have an adverse effect on the handling, processing or storage of commodities; or (5) threaten biodiversity."[53] If these issues arise, it is unlikely that an applicant will be able to obtain deregulation status or a permit without significant limitations.

Permit or notification applicants are required to notify APHIS of any "unusual occurrences." In the event that there is an accidental or unauthorized release of the plant pests, the applicant must orally notify APHIS immediately and in writing within 24 hours. As to what constitutes an unusual occurrence, APHIS provides the following examples: the "potential dispersal of plant material outside the approved area of introduction by high winds or flooding; accidental planting of the regulated article in the wrong location; planting a variety with an unauthorized construct; damaged packaging materials; and materials lost in shipping."[54] If the transgenic plant is found to have characteristics substantially different from those listed in the permit application or suffers any unusual occurrence (e.g., excessive plant death or unexpected impacts on non-target species), it must be reported to EPA within five working days.[55]

5. Inspections APHIS regulations also require permit or notification applicants to provide access to and allow inspections of facilities and/or the field test site and any records to demonstrate compliance with requirements necessary to obtain a permit or receive a notification acknowledgement (e.g., compliance with

---

51. 7 C.F.R. § 340(f)(9).

52. *See id.*

53. Pew Initiative on Food and Biotechnology, *Issues in the Regulation of Genetically Engineered Plants and Animals* (April 2004) at 34.

54. *See* USDA, *supra* note 29, at 24.

55. 7 C.F.R. § 340.4(f)(10).

the six eligibility criteria and the six performance criteria.) Personnel from the APHIS Biotechnology Regulatory Services or Plant Protection Quarantine divisions—or state personnel—are supposed to conduct such inspections with the agencies selecting a site for investigation based on factors such as the type of regulated article, the volume shipped, the acreage, and the compliance history of the applicant. The way APHIS conducts inspections has come under some internal criticism. The Office of the Inspector General for the USDA (Southwest Region) criticized BRS and PPQ by noting: "Inspection requirements are vague and there is a lack of coordination between the two APHIS units responsible for the inspection program."[56]

6. Enforcement Tools The Plant Protection Act provides the Secretary with extensive enforcement tools as outlined in sections 414 and 415 of the Act. The Secretary has the power to quarantine or take other remedial measures, or in fact can take extraordinary emergency measures to prevent the interstate movement of a plant pest or noxious weed. Section 414(a) provides the Secretary with a wide range of remedial tools to prevent the dissemination of plant pests or noxious weeds that are either new or not known to be widely distributed in the United States, including the power to hold, seize, quarantine, treat, apply other remedial measures to, destroy, or dispose of any plant, plant pest (including noxious weeds), biological control organisms, plant products, or their progeny.[57] The USDA can apply these remedial measures based on the broad standard of "if there is reason to believe" that these items are pests, infested with pests, or violate any pest-related regulations.[58] The USDA can require an owner of such items or the owner's agent to take actions without cost to the federal government, but the law requires that the USDA take the least drastic action that is feasible to adequately address the dissemination of the pest.[59] If an owner fails to undertake the action ordered by the USDA, the USDA may take the action and recover the associated costs.[60]

Additionally, the Plant Protection Act authorizes the USDA to declare an extraordinary emergency when the presence of a plant pest or noxious weed that is new or not known to be widely distributed in the United States threatens plants or plant products in the United States.[61] In addition to the remedial actions discussed in section 414(a), USDA may quarantine, treat, or apply other remedial actions to premises; quarantine any state or portion of a state; and/or prohibit or

---

56. U.S. Department of Agriculture Office of Inspector General Southwest Region, Audit Report, *Animal and Plant Health Inspection Service Controls Over Issuance of Genetically Engineered Organism Release Permits*, BIOTECHNOLOGY LAW REPORT (April 2006).

57. 7 U.S.C. § 7714(a).

58. 7 U.S.C. § 7714(a)(1).

59. 7 U.S.C. §§ 7714(b)(1),(d).

60. 7 U.S.C. § 7714(b)(2).

61. 7 U.S.C. § 7715(a).

restrict the movement of any plant, plant product, biological control organism, article, or means of conveyance within a state if necessary to prevent the dissemination of or to eradicate the plant pest or noxious weed.[62] Thus, APHIS has the power to regulate within a state, and is not limited to just interstate commerce. However, these actions can only be undertaken after the USDA has reviewed the actions undertaken by the affected states, consulted the officials of the affected states, and made a finding that the state's actions have been inadequate to eradicate the pest.[63] As with section 414, the law requires the USDA take the least drastic action that is feasible to adequately address the dissemination of the pest.[64] The USDA has discretion to compensate for economic losses incurred by a person due to the agency's undertaking action pursuant to its authority in this section, with the amount of such compensation being final and not subject to judicial review.[65]

7. Determination of Nonregulated Status During the 1993 amendment of APHIS's regulations, a petition process was created for nonregulated status.[66] The impetus for the petition process was the ad hoc review of the FLAVR SAVR™ tomato (see discussion in Chapter 6) and its determination that the crop should not be subject to regulation by APHIS because it posed no plant pest risk.[67] APHIS concluded it should formalize its procedures through the issuance of a final rule rather than through guidance documents. Although APHIS noted that in its proposed rules it had used the word *microorganism*, the final rule clarified this position by noting that the nonregulated status was only available to plants. As of July 2007, there have been more than 70 genetically modified organisms that have been deregulated, which covers not only the original genotype, but also any progeny of the genotype.[68]

Once a plant has been afforded nonregulated status, it is completely exempt from all APHIS requirements. Thus, for developers of genetically modified plants, obtaining nonregulated status allows for cultivation of the crop without the restrictions of buffer zones or other mechanisms of ensuring that the genes from the plant did not impact the larger environment.

To obtain nonregulated status, the petitioner must first complete field testing under a permit or notification.[69] A petitioner must then submit a full statement

---

62. 7 U.S.C. § 7715(a)(1)–(4).

63. 7 U.S.C. § 7715(b).

64. 7 U.S.C. § 7715(d).

65. 7 U.S.C. § 7715(e).

66. 58 Fed. Reg. 17044 (Mar. 31, 1993).

67. *See id.* at 17051.

68. USDA, Introduction of Genetically Engineered Organisms, Draft Programmatic Environmental Impact Statement (July 2007), at 29, *available at* www.aphis.usda.gov/brs/pdf/complete_eis.pdf.

69. 7 C.F.R. § 340.6(c)(5).

explaining the grounds for why the organism should not be regulated and include copies of scientific literature, unpublished studies, and data from tests performed. Specifically, to determine whether the genetically modified plant is significantly different from nongenetically modified plants, the APHIS uses these documents to evaluate a variety of issues, including the potential for plant pest risk, disease and pest susceptibilities, changes to plant metabolism, weediness and impact on sexually compatible plants, effects on non-target species, and potential for gene transfer, A petitioner must be prepared to provide the following information:

1. description of the biology of the nongenetically modified plant and information necessary to identify the recipient plant in the narrowest taxonomic grouping applicable;
2. relevant experimental data and publications;
3. detailed description of the differences in the genotype between the genetically modified organism and the nongenetically modified organism (e.g., the donor organism(s), the nature of the transformation; the inserted genetic material and its product); and
4. detailed description of the phenotype of the genetically modified organism and why it is unlikely to "pose a greater plant pest risk than the unmodified organism from which it was derived," including, but not limited to: plant pest risk characteristics, disease or susceptibilities, expression of the gene product, change of plant metabolism, weediness of the regulated article, and the impact on the weediness of any other plant with which it may interbreed.[70] APHIS has specifically noted that the scientific studies need not be "peer reviewed scientific studies" (i.e., data that is published in scientific literature);
5. field test reports for all trials conducted under permit or notification procedures that were submitted prior to the submission of a petition. The reports are to include the APHIS reference number; methods of observation; resulting data; and analysis regarding all deleterious effects on plants, non-target species, or the environment.[71]

Upon submission of a petition, APHIS will inform the petitioner of the number assigned to the file and whether the petition is deficient in any way. If the petition is not deficient or the deficiencies have been addressed, APHIS will publish a notice in the Federal Register and provide a 60-day period for public comments.[72] Depending on the comments received, the process may extend beyond the 180-day response time period contained in the regulations for notifying the

---

70. 7 C.F.R. § 340.6(c)(1–4).

71. *Id.*

72. 7 C.F.R. § 340.6(d). As noted in Section III, APHIS may wait until the environmental assessment as required under NEPA is prepared, then submit both the petition and environmental assessment to public comment.

petitioner if the petition has been approved in whole or in part or if it has been denied. Any person whose petition has been denied may appeal that decision by informing the USDA Administrator in writing within 10 days from receipt of the written notification of denial. The appeal must state all of the facts and reasons upon which the petitioner relies (including any new information) to demonstrate that the petition was wrongfully denied. The Administrator may then either deny or grant the appeal. If there is dispute about a material fact, an informal hearing will be held.

Significantly, a petitioner may also ask for nonregulated status to be extended to organisms that are similar to one that has previously been given unregulated status. APHIS will announce in the Federal Register all preliminary decisions to extend nonregulated status to a similar organism 30 days before the decision become final and effective. If the request for extension of the status is denied, the petitioner may submit a modified request or a separate petition for a determination of nonregulated status as to that particular plant.

III. THE NATIONAL ENVIRONMENTAL POLICY ACT

This section will briefly describe the elements of the National Environmental Policy Act of 1969 (NEPA)[73] and the implementing regulations.[74] NEPA is a broadly worded statute that has been subject to extensive litigation, and thus, this section provides only an outline of the elements that are of particular relevance to applicants for federal permits or approvals. Any applicant that is subject to NEPA must carefully consider its scope and the documents that must be provided in order to comply with the requirements it has spawned.

NEPA is a procedural statute that has significant substantive impacts. It has two major elements: (a) federal agencies have the responsibility to consider the environmental effects of major federal actions significantly affecting the environment, and (b) the public has the right to review that consideration.[75] Specifically, the statute provides that all federal agencies "to the fullest extent possible" include in "every recommendation or report on . . . major Federal actions significantly affecting the human environment," a detailed statement on, and the alternatives to, "the environmental impact" of the proposed activity.[76] In order to translate these general principles into an actual working program, the Council of Environmental Quality, an entity created by NEPA, promulgated a set of regulations and guidance documents that outline the procedural steps that must be

---

73. 42 U.S.C. §§ 4321–4347.

74. See 40 C.F.R. §§ 1500 et seq.

75. Baltimore Gas & Electric Co. v. Natural Resources Defense Council, Inc., 462 U.S. 87, 103 S.Ct. 2246, 76 L.Ed.2d 437 (1983).

76. 42 U.S.C. 4332(C).

followed in order to comply with NEPA. The first step in the process is to deter-mine if NEPA is applicable. Agencies can establish by regulation categories of actions that "do not individually or cumulatively have a significant effect on the human environment." These categories are referred to as "categorical exclusions." If the major federal action does not fall within an agency-established categorical exclusion,[77] and the action has the possibility of a significant impact on the envi-ronment, then the agency must begin the process by reviewing an *environmental assessment* (EA). An EA is a concise and factual, rather than argumentative, docu-ment. It generally includes a brief description of the project, the environmental impacts (if any), a description of the alternatives, and a list of agencies and per-sons consulted in the determination of whether there will be significant impacts on the environment.[78] Each of these elements can be the subject of extensive discussion and debate. For example, in certain cases, the issue of whether the project is being improperly segmented, and thus, there are cumulative impacts needs to be explored and resolved.

If the agency concludes there are either no significant environmental impacts or any impacts can be adequately mitigated, then it may issue a finding or no significant impacts (FONSI). Alternatively, if the agency determines that there are significant impacts, then the *environmental impact statement* (EIS) process begins. Public comment on the EA or FONSI is determined by agency regula-tions. For instance, APHIS provides for a 30-day public review and comment period for EA and FONSI.

Completing and obtaining approval for a project based on an EIS can be a long, arduous, and costly process for an applicant, even though procedurally the steps are straightforward. Once an agency decides to proceed with an EIS, it issues a Notice of Intent in the Federal Register outlining the reasons for the EIS. Then, the extent of the EIS is considered through a scoping process that involves a public meeting. Based on the results of the scoping sessions, a draft EIS is made available for public comment. After the agency reviews, considers, and responds to public comments, a final EIS is developed on which agency action is taken.

The actual contents of an EIS focus on multiple areas with the analysis being project specific. However, generally an EIS will contain an analysis of: (a) the direct and indirect environmental impacts of the proposed action; (b) any adverse environmental impacts that cannot be avoided should the proposed action be implemented; (c) the reasonable alternatives to the proposed action; (d) the rela-

---

77. 40 C.F.R. § 1508.4. NEPA allows each federal agency to establish *categorical exclu-sions* which are deemed to be actions that "do not individually or cumulatively have a sig-nificant effect on the human environment" and which do have not been found to have had such effect in past instances.

78. 40 C.F.R. §§ 1501.4(a)–(b), 1508.9.

tionship between local short-term uses of the environment and the maintenance and enhancement of long-term productivity; and (e) any irreversible and irretrievable commitments of resources that would be involved in the proposed action should it be implemented.[79] Once a final EIS is issued, this could be the start of litigation by parties who participated in the process and are dissatisfied with the legal or factual conclusions the agency reached. It should be noted that in the NEPA process, the agency designated as the lead agency for jurisdictional purpose shepherds the process; however, other agencies will provide comments on areas over which they have jurisdictional authority.

Thus, for example with APHIS's regulation of genetically modified plants, the initial question is whether the applicant's activity is subject to NEPA or is categorically excluded. APHIS's NEPA-implementing regulations, which are at 7 C.F.R. Part 372, have categorically excluded confined field tests from the NEPA process on the basis that adverse environmental impacts may be avoided or minimized through the confinement and containment actions that are built into the way the action will be undertaken.[80] However, the exclusion does not apply when the test involves the field release of genetically modified organisms or new species, organisms, or novel modifications that raise new issues.[81]

Where there is no categorically exclusion, such as for nonconfined field testing of genetically modified plants or more generally for nonregulated status petitions, the agency will require the applicant to provide an EA and potentially an EIS. Of particular significance in the APHIS NEPA analysis will be an examination of impacts on non-target species, especially threatened and endangered species. This would probably implicate the Department of Interior's Fish and Wildlife Services and EPA. From a public participation view, APHIS affords a period of public comment with respect to the EA or FONSI.[82] Additionally, APHIS regulations provide for a 60-day comment period for nonregulated status petitions.[83] Thus, the agency may request comments on the petition and the EA concurrently.

As a final point, NEPA does not apply to most of EPA's actions. The rationale for this exception is that EPA's actions already occur within a substantive regulatory framework that focuses on environmental quality and a procedural framework that provides "full opportunity for thorough consideration of the environ-mental issues and for ample judicial review."[84] EPA review process is

---

79. 42 U.S.C. 4332(C).

80. 40 C.F.R. § 372.5(c).

81. 7 C.F.R. § 372.5(d)(4).

82. 40 C.F.R. § 372.8(b)(2)–(3).

83. 7 C.F.R. § 340.6(d).

84. Foundations on Economic Trends. v. Heckler, 756 F.2d 143, 146–147 (D.C.Cir. 1985) (citing EDF v. EPA, 489 F.2d 1247, 1256 (D.C.Cir.1973)).

considered functionally equivalent to NEPA. Thus, the issue of compliance with NEPA arises in the context of APHIS or other USDA entity, FDA, or other federal agency action.

IV. USDA REGULATION OF INDUSTRIAL AND PHARMACEUTICAL TRANSGENIC PLANTS

The regulatory process for genetically engineered plants that contain pharmaceuticals or industrial chemicals differs from those applicable to transgenic plants. While these regulations remain in transition, the APHIS has outlined some of their parameters. First, unlike transgenic plants for foods, the notification process is not available. When the notification process was initially proposed in 1993, "the types of genetically engineered plants that had industrial uses were typically those in which nutritional components, such as oil contents."[85] As a result, because APHIS had familiarity with regulating such plants, it allowed introductions through the notification process.[86] However, because it did not have regulatory experience or scientific familiarity with pharmaceutical uses, and because industrial uses broadened beyond food and feed uses, APHIS issued an interim rule in 2003, and then a final rule in 2005, that withdrew the notification option for these uses. *Industrial uses* is defined as meeting the following criteria: "(1) The plants are engineered to produce compounds that are new to the plant; (2) the new compound has not been commonly used in food or feed; and (3) the new compound is being expressed for non-food, non-feed industrial uses."[87] Thus, in order to obtain permission to test or market pharmaceutical or industrial plants, it is necessary to obtain a permit. Additionally, APHIS has indicated that industrial and pharmaceutical crops are not categorically excluded from NEPA, but rather are subject to the EA and EIS process described above.

The concerns about cross-contamination through seed mixing and pollen dispersal are only heightened in the case of industrial and pharmaceutical plants. Moreover, the impacts to non-target species and endangered species, adequate containment for the plants, and disposal of the plants are all issues that have heightened interest for these types of transgenic plants. Thus, permit conditions are likely to be more stringent, such as (a) increased isolation of areas where these plants are grown, (b) larger buffer zones, (c) disposal mechanisms that completely destroy the crops, (d) avoidance of sexually compatible plants, and (e) segregated equipment.

---

85. 70 Fed. Reg. 23009 (May 4, 2005).
86. *See id.*
87. *See id.*

V. FEDERAL REGULATION OF PLANT INCORPORATED PROTECTANTS

In 1947,[88] Congress enacted the original FIFRA[89] as legislation to protect farmers from manufacturers and distributors of adulterated or misbranded pesticides. The USDA, which originally administered the statute, required among other things both registration and labeling of pesticides.[90] In 1972, Congress amended the statute by changing its focus from primarily an economic statute to a safety one. Two years prior, President Nixon had transferred pesticide regulation and enforcement to the then-newly created EPA by executive order. The 1972 amendment not only provided the congressional imprimatur for that action, but also established many of the provisions that are part of the current FIFRA regulatory scheme, such as the cost-benefit analysis of the registration process, experimental and special local use permits, and reasonable compensation by tag-on registrants.[91] The statute was subsequently amended in 1975, 1978, 1980, 1988, 1990, and 1991 with each amendment contributing to the current scheme.

In 1996, Congress passed the Food Quality Protection Act (FQPA),[92] which amended the Federal Food, Drug, and Cosmetic Act as well as FIFRA. Among other things the FQPA changed were FIFRA's registration program for pesticides. Specifically, the FQPA amendment established a procedure for pesticide registration review and a goal that every pesticide registration should be reviewed every 15 years.[93] Additionally, the FQPA amendment provided that, at a minimum, prior to reregistration of a pesticide, EPA had to reassess each existing pesticide residue tolerance (or the exemption from the tolerance requirement) as well as determine if any new tolerances or exemptions needed to be placed.

A. Regulatory Scope of the Federal Insecticide, Fungicide, and Rodenticide Act

FIFRA defines *pesticide* as (a) any substance or mixture of substances *intended* for preventing, destroying, repelling, or mitigating any *pest*; (b) any substance or mixture of substances *intended* for use as a plant regulator, defoliant, or desiccant; and (c) any nitrogen stabilizer.[94] A *pest* includes (a) any insect, rodent,

---

88. In 1910, Congress enacted the Federal Insecticide Act, which was an economics-based statute intended to protect farmers from substandard and fraudulent pesticides. After World War II, with the recognition that the number of new products coming into the marketplace had increased, chemical manufacturers and farmers worked with Congress in crafting the legislation, which repealed the Insecticide Act.

89. Pub. L. No. 80-104, 7 U.S.C. §§ 135 *et seq.*

90. In 1964, the statue was amended to require that registrations were made mandatory as well as to require "signal words" on labels.

91. Federal Environmental Pesticide Control Act, Pub. L. No. 92-516; D. STEVER, LAW OF CHEMICAL REGULATION AND HAZARDOUS WASTE 3–6 (2008).

92. Pub. L. No. 104-170 (1996).

93. 7 U.S.C. § 136(d).

94. 7 U.S.C. § 2(u).

nematode, fungus, or weed, or (b) any other form of terrestrial or aquatic plant or animal life or virus, bacteria, or other microorganism (except viruses, bacteria, or other microorganisms on or in a living man or animal) that is injurious to health or the environment.[95]

Under EPA's regulations, there are three factors for determining *intent*: (a) the person who distributes or sells the substance explicitly or implicitly claims (by labeling or otherwise) that the substance (by itself or in combination) can or should be used as a pesticide or that the substance contains an active ingredient that can be used to manufacture a pesticide; (b) the substance has no commercial value other than to be distributed or sold for a pesticidal purpose or for use in the manufacture of a pesticide; or (c) the person knows, or even has constructive knowledge that the substance will be used or is intended to be used as pesticide.[96] As the Third Circuit noted, in determining intent it is relevant to examine labels, industry representations, and advertising as well as general public knowledge which can make a product pesticidal notwithstanding the lack of express pesticidal claims by the producer.[97]

To limit the otherwise broad reach of FIFRA, Congress created a limited number of statutory exemptions that have been more fully explained through EPA's implementing regulations. The statute specifically excludes new animal drugs or animal feed.[98] Additionally, section 25(b) of FIFRA gives the agency authority to exempt a pesticide from regulation under two separate prongs: if EPA determines that the pesticide is adequately regulated by another federal agency *or* that pesticide has a "character which is unnecessary" of regulation.[99]

EPA has exercised its authority by providing exemptions for certain products under specified conditions.[100] One such exemption is for *biological control agents*[101]

---

95. 7 U.S.C. § 2(t), 7 U.S.C. § 136w(c)(1).

96. 40 C.F.R. § 152.15

97. N. Jonas & Co. v. EPA, 666 F.2d 829, 833 (3rd Cir. 1981).

98. 7 U.S.C. § 2(u).

99. 7 U.S.C. § 136w(b).

100. *See, e.g.,* Liquid chemical sterilants that are regulated by FDA, nitrogen stabilizers, vitamin hormone product (e.g., plant nutrients and plant hormones), and products intended to aid in the growth of a plant are excluded provided they meet the applicable criteria under the regulations. 40 C.F.R. § 152.6. EPA also excludes entire classes of pesticides they have determined to not require regulation, including treated articles (e.g., paint treated with pesticide to protect the paint coating, or wood products treated to protect the wood against insect or fungus infestation), arthropod pheromones and pheromone traps, biological preservatives (e.g., embalming fluids), foods, natural cedar, and certain food oils, as well as other oils. 40 C.F.R. § 152.25.

101. 40 C.F.R. § 152.20(a)(4)). As an illustration of why PIPs are not biological control agents, EPA notes the example of chrysanthemums. Chrysanthemums produce pyrethrum, which has insecticidal properties. The use of living chrysanthemum as a means of controlling insects is exempt from FIFRA as a biological control agent. On the other hand, extracted pyrethrum that is dusted on crops is subject to FIFRA as a traditional chemical

on the basis that they are adequately regulated by another federal agency.[102] However, genetic manipulation of plants wherein pesticidal properties (e.g., insect resistance or plant virus tolerance) are produced or used by the living plants to typically protect the plant from insects is not considered to be a biological control agent. Instead, these are considered plant-incorporated protectants" or PIPs[103] that are subject to the full panoply of FIFRA regulations except as noted in 40 C.F.R. Part 174. However, it should be made clear EPA does not regulate the seed or the whole plant under these FIFRA regulations, but only the pesticidal materials expressed by the plants. Further, plants that are genetically altered to contain herbicides are not considered PIPs because herbicides are regulated as conventional pesticides.

B. Specific Regulations or Guidance on Plant Incorporated Protectants

EPA initially proposed to regulate PIPs in 1994, but it did not issue its final rules on the subject until 2001.[104] The only PIPs that are exempt are those derived using conventional breeding from sexually compatible plants.[105] These PIPs are not the result of recombinant DNA or other techniques (such as microinjection, cell fusion, or micro-encapsulation), but rather are the result of "100 years of scientific breeding among sexually compatible plant populations using Mendelian genetics."[106] The agency's rationale for this distinction between conventionally bred PIPs versus PIPs derived through modern biotechnology is that conventionally bred PIPs have only a low probability of producing adverse environmental

---

pesticide because significantly greater number of organisms would be exposed than if it were simply retained within a chrysanthemum. Similarly, if corn was genetically modified to include pyrethrum, such corn would not be considered a biological control agent because of the number of species exposed to that corn. 66 Fed. Reg. 37792–93.

102. 40 C.F.R. § 152.20(a)(4)).

103. 40 C.F.R. § 174.21 provides the exemptions from the FIFRA requirements. Specifically, they are applicable to PIPs where (a) the genetic material that encodes the pesticidal substance or leads to the production of pesticidal substance is from a plant that is sexually compatible with the recipient plant, and (b) the genetic material has never been derived from a source that is not sexually compatible with the recipient plant. In such cases, the manufacturer has only to report to EPA the adverse effects on human health and the environment. 40 C.F.R. 174.71.

104. The 1994 initial rules exempted several categories of products. In the interim, there were congressional and public hearings, interagency negotiations, and scientific advisory council meetings. The ultimate final rule, however, was more modest in scope by exempting only PIPs derived through conventional breeding. Thus, the final rule draws a distinction between conventionally generated PIP and PIPs developed through genetic engineering on the basis of *process* rather than the *product*. *See* Mary Jane Angelo, *Regulating Evolution for Sale: An Evolutionary Biology Model for Regulating the Unnatural Selection of Genetically Modified Organisms*, 42 WAKE FOREST L. REV. 93, 124–126 (2007).

105. 66. Fed. Reg. 37772, 37794–95 (July 19, 2001).

106. 66 Fed. Reg. 37795 (quoting from 59 Fed. Reg. 60524 (1994) (proposed rule).

impacts or of having novel exposure issues while providing farmers with higher yields and consumers with lower prices. The only requirement imposed on those who produce, distribute, or sell such PIPs is that if an actual adverse effect to human health or the environment is known, there is an obligation to notify EPA within 30 days of obtaining the information using the format outlined in the regulations.

C. Experimental Use Permits

Prior to registration of a pesticidal product, EPA allows the manufacturer to experiment (i.e., field test) under an experimental use permit (EUP) to accumulate data necessary for a registration application under section 3 of FIFRA, provided that the experiment will not cause unreasonable risks to human health or the environment. EUPs are required for testing unregistered PIPs or an unregistered use of a PIP when field testing occurs on a cumulative total of 10 or more acres of U.S. land. If the experimentation occurs on a smaller scale or in the laboratory or a greenhouse, the research is exempt from the EUP requirements unless there is a threat of inadequate containment.[107]

The acreage threshold has been controversial. EPA's perspective is that it does not want to unduly burden pesticide research. It believes that when a small amount of pesticide can be effectively controlled and has a low risk of adverse impacts to the environment, it would not be justified in burdening the manufacturer with submitting and obtaining pre-experimentation approval.[108] Critics, however, note that the acreage threshold is inappropriate because genetically modified plants can cause adverse impacts on the environment through gene transfer and impacts on non-target species. Yet small-scale field studies are still subject to regulation by the USDA under the conditions outlined in the preceding section. Moreover, even in such small-scale field tests EPA would still have a role if it is probable the PIP will end up in food or feed—in which case a temporary tolerance or tolerance exemption must be obtained from EPA. As noted in the next section, EPA has granted tolerance exemptions for various Bt crops.

Prior to the submission of an EUP application, the researcher is encouraged to hold a meeting with EPA. The meeting may be in person, via telephone, or a combination of both. Due to the novelty of PIP EUPs, this pre-submission meeting is used to custom tailor the application submission information to the particular experiment, lessen the disclosure burden if possible, and avoid resubmission of applications for failure to provide the necessary information. There is a significant

---

107. One of the criticisms of EPA is that the 10-acre threshold is not appropriate in the genetically modified plant concept as there can be significant adverse impacts to the environment from even a small acreage of genetically modified plants due to the pollination issues and cross-breeding issues discussed in prior chapters.

108. See 58 Fed. Reg. 5878, 5878–79 (Jan. 22, 1993) (citing 39 Fed. Reg. 11306 (Mar. 27, 1974)).

monetary fee associated with the submission of an EUP, and the applicant must forfeit 25 percent of it if the applicant ultimately decides not to pursue the application.[109] The agency strongly advises that the applicant participate in such pre-submission meeting in order to avoid such a future withdrawal.

For PIP EUP, the general application requirements as set forth in 40 C.F.R. 172.4 govern. Additionally, the agency provides a brief sheet outlining the PIP EUP submission requirements: "Plant-Incorporated Protectant EUPs—Preliminary Guidance."[110] Aside from certain basic information about the applicant, the product, and those participating in the experiment, the applicant has to provide information about:

- the purpose and objectives of the proposed testing and information on the testing parameter (e.g., acreage, duration, amount of PIP proposed for use, study protocols, etc.);[111]
- an appropriate description of the prior testing conducted on the impacts/effects on the targeted organisms as well as non-targeted organisms;
- containment procedures to ensure that PIP pollen does not outcross with surrounding crops and that the harvested crop does not enter commerce;
- the proposed method for storage and disposal methods for any unused seed and its containers;

---

109. The submission fees for PIP EUPs range in the hundreds of thousand of dollars depending on the complexity of required agency review. Updated fee tables may be found in the Pesticide Registration Improvement Renewal Act—PRIA II effective October 1, 2007, *available at* http://www.epa.gov/pesticides/fees/.

110. Plant-Incorporated Protectant EUPs—Preliminary Guidance, *available at* http://www.epa.gov/oppo00001/biopesticides/news/pip-eup-prelim-guid.htm.

111. EPA's guidance document indicates the agency would like information on, among other things, the maximum acreage per state (sum of all studies); maximum acreage per study (sum of all states); estimated acreage per state for each study; estimated number of locations per state for each study; if a year-over-year acreage increase is requested for one or more studies, a justification for such increase; and a clear explanation of acreage calculation, including a breakdown of EUP acreage designated for PIP test plants, registered PIPs, non-PIP plants, and border rows included within experimental blocks. *See* Tips for Plant-Incorporated Protectant (PIP) Experimental Use Permit (EUP) Program Submission," www.epa.gov/oppbppd1/biopesticides/pips/pip_hints.htm. In order to calculate the affected acreage, EPA states that it should include the area of "PIP test plants, including plants containing registered PIPs; non-PIP plants used for breeding purposes; non-PIP plants that are not intentional recipients of a PIP; and associated border row contained within the test blocks, including border rows which outline the perimeter of the test blocks." *Id.* This information is submitted in Section G of the EUP application and is arguably the most scrutinized part of the submission.

- percentage ranges by dry weight of PIP and PIP-inert ingredients as they are distributed throughout each part of the plant;[112]
- any "other additional pertinent information as the [EPA] Administrator may require";[113] and
- where PIP seeds and/or crops will be transported, a proposed EUP label listing percent composition of all PIP and PIP-inert parts. The name of PIP may be omitted.[114]

Additionally, if the EUP is to be used in such a manner that any residue can reasonably be expected to result in food or feed, the applicant has to either (a) submit evidence that a pesticide residue tolerance or an exemption from the tolerance requirement has been established; (b) submit a petition to EPA proposing establishment of a tolerance or an exemption from the tolerance requirement; or (c) certify that the food or feed derived from the experimental program will be destroyed or fed only to experimental animals for testing purposes or otherwise disposed of in a manner that will not endanger human health or the environment.[115] Section VI(A) of this chapter contains a fuller discussion of the petitioning for a tolerance or seeking an exemption.

EPA's PIP EUP application review is a transparent process in which all PIP EUP submissions and EPA's ultimate decision are published in the Federal Register.[116] Public comments are sought and considered by EPA.[117] Depending on the content of the comments received, or based on its own evaluation of whether there is "sufficient interest to warrant" one, EPA may hold a legislative-type hearing on the application.[118] Therefore, it should be of no surprise that even with provisions for confidential submission, in most cases EPA will not grant confidential treatment of the PIP's genetic formula as the concern for public safety outweighs the proprietary interest in the formula. It may be possible, however, for a persuasive applicant to seek confidential treatment of the PIP manufacturing method as confidential business information.[119]

EPA reviews a non-food/feed PIP EUP application where no Scientific Advisory Panel (SAP) review is required and where the experimental crop will be destroyed within six months after receipt of the application and all supporting data.[120] The review period extends to 15 months for PIP EUP applications for

---

112. *See* Plant-Incorporated Protectant EUPs—Preliminary Guidance, *supra* note 111.

113. 40 C.F.R. § 172.4(b).

114. *See* Plant-Incorporated, *supra* note 111.

115. 40 C.F.R. § 172.4(b)(2).

116. *See* Plant-Incorporated, *supra* note 111.

117. *Id.*

118. 40 C.F.R. § 172.11.

119. 40 C.F.R. Part 174.9.

120. Upon approval of the state plan by EPA, a state agency designated by the state may issue a EUP. 7 U.S.C. § 5(f). The delegation criteria is published is 40 C.F.R. Part 172, subpart B.

food/feed use and for those in which SAP review is required. Once a EUP has been granted, the holder needs to obtain any necessary state(s) permits.[121]

Use of a pesticide under a EUP is under the supervision of EPA and is subject to the terms and conditions as prescribed in the permit.[122] Within 180 days after the expiration of the permit (unless extended based upon a request to EPA), the applicant will provide EPA with a report on the data gathered, including information on amount of the product applied; the crops or sites treated; any observed adverse effects; any adverse weather conditions that may have inhibited the program; the goals achieved; and the disposition of containers, unused pesticide material, and affected food/feed commodities.[123] Additionally, EPA requires the immediate reporting of any adverse effects from the use of, or exposure to, the pesticide.[124]

EPA may refuse to grant or may revoke any EUP at any time if it is determined that an EUP is not justified, that the issuance of the permit would cause unreasonable adverse effects on the environment, that the terms or conditions of the permit are being violated, or that the terms or conditions are inadequate to avoid unreasonable adverse effects on the environment.[125]

If EPA decides to revoke or refuses to grant a EUP, the potential/actual registrant has an "opportunity to confer" with EPA, provided that an application for such a meeting is made within 20 days of the notice. Within 20 days of the meeting, EPA will issue its final decision.[126] Additionally, interested third parties may challenge EPA's decision to grant a EUP. The judicial review provisions of the statute and the Administrative Procedures Act govern such challenges.[127] To succeed, the plaintiff must overcome the presumption of the validity of the agency's action. The court will affirm the agency's decision if the agency had a rational basis for it. The review consists of a two-step process in which the court looks at whether "(1) there was procedural defect in the agency EUP process, or (2) that the agency's substantive decision to issue the EUP was arbitrary, capricious or an abuse of discretion."[128]

EPA may reject a deficient EUP application that is not cured within the initial review period of 21 days. During the time, the agency reviews the EUP application for completeness and determines if any additional information is needed for the full-fledged evaluation. Notably, if the EUP application is rejected for insufficient supporting data, EPA will retain 25 percent of the application fee. This penalty for deficient submission is supposed to serve as a deterrent against incomplete submissions.

121. *See* Plant-Incorporated, *supra* note 111.

122. 7 U.S.C. § 136c(c).

123. 40 C.F.R. § 172.8.

124. 7 U.S.C. 136c(d). 40 C.F.R. § 172.8(a).

125. 40 C.F.R. §§ 172.10(a),(b).

126. 40 C.F.R. § 172.10(c).

127. *See* Foundation on Economic Trends v. Thomas, 637 F. Supp. 25, 28 (D.D.C. 1986).

128. *Id.*

D. Registration Process

Upon completion of field testing under a EUP, if a developer wants to distribute or sell the pesticide in the United States, then except under limited circumstances, the pesticide must be registered.[129] A registration will be given, and hence, the use or marketing of the pesticide will be permitted, if the pesticide when used in accordance with widespread and commonly recognized practices does not generally cause unreasonable adverse effects on the environment.[130]

FIFRA's elaborate registration process is outlined in Section 3 of the Act,[131] which requires a prospective manufacturer to submit a registration application, a proposed label, directions on the use of pesticide, warnings and precautionary statements, restrictions on use, and information regarding the formula and description of the tests that provide the basis for the manufacturer's claims. Currently, in addition to standard registration (also referred to as a generic registration), EPA may issue a conditional registration. After discussing the testing data requirements that are applicable to either a generic registration or a conditional registration, this section will focus on conditional registrations as most PIPs are registered under the conditional registration program.

1. **Testing and Data Requirements for PIPs** Specifically with respect to the testing data, EPA thus far, has not promulgated data requirements specifically designed for PIPs. But, as Stephen Johnson, at the time Assistant Administrator, Office of Prevention, Pesticides and Toxic Substances noted:

> However, because the PIPs that EPA has reviewed are protein based, EPA has required digestibility information, where the amount of time it takes for a

---

129. In addition, manufacturers of conventional pesticides or the active ingredients used in pesticides, including PIPs, are required to register their establishments prior to production. EPA will issue an establishment number. Within 30 days of registration, and annually thereafter, the establishment must provide information on the types and amounts of pesticides (and if applicable, active ingredients used in producing pesticides) that it is currently producing, produced in the past year, and sold or distributed during the past year.

130. 7 U.S.C. § 136a(c)(5). In addition, its composition must meet its claims and its labeling and other materials must be submitted. FIFRA defines *unreasonable adverse effect on the environment* as "(1) any unreasonable risk to man or the environment, taking into account the economic, social, and environmental costs and benefits of the use of any pesticide, or (2) a human dietary risk from residues that result from a use of a pesticide in or on any food inconsistent" with the standard under section 408 of the Federal Food, Drug, and Cosmetic Act. 7 U.S.C. § 136bb. The fact that the statute requires a balancing between environmental and health impacts and the benefits associated with the pesticide is a significant issue when rejecting or canceling a registration.

131. Under the statute, generic registrations (i.e., not conditional registrations) are for a five-year period and automatically expire unless they are renewed. In addition, if EPA decides to deny an application, it must comply with the same procedures as outlined below for canceling a registration.

protein to break down in gastric and intestinal fluids is determined. This information is relevant to a determination of the potential of the protein of the protein to be toxic or an allergen. Also, EPA focuses on allergenicity information by examining digestibility test information, tests for heat stability and a comparison of the structure of the protein to the structures of known food allergens.[132]

In addition to allergenicity issues, EPA needs information about the potential adverse effects and environmental fate of the pesticide with respect to flora and fauna, especially non-target plants and non-target species, persistence in the environment, and genetic material transferability.[133] Before undertaking tests, EPA encourages each applicant (especially those who are applying for the first time) to consult with the agency.

As an alternative to undertaking its own tests, and in order avoid duplication of expensive scientific and technical work, if there are products that have previously been approved that are substantially similar or identical to the applicant's product, then a submitter may rely upon previously submitted data. The applicant can either comply through the *cite-all* method or through the *selective citation* method.

(a) Cite-All Method Under the cite-all method, an applicant must submit an acknowledgment that the application relies upon (a) all data submitted with or specifically cited in the application, as well as (b) each document in EPA's files that concerns the properties or effects of the applicant's product, an identical or substantially similar product, or any active ingredient in the applicant's product, and that the information in EPA's files is the type of data the agency would currently require for such a proposed product. The cite-all method, however, requires a subsequent registrant attempting to rely on information already in EPA's files (what is known as a *follow-on registrant*) to obtain authorization from the original submitter prior to relying upon any exclusive use data. Thus, the original submitter may delay or deny the registration by not agreeing to provide the authorization.

(b) Selective Citation Method Under the selective citation methodology, it is possible to demonstrate compliance through one or more of the following mechanisms: citing to a previously submitted study or studies, submitting a new valid study, requesting a waiver from the data requirements, citing to publicly available documents or studies generated at public expense, and/or noting that no other entity has submitted data that would satisfy the data requirements in question (i.e., demonstrating the existence of a data gap).[134]

---

132. Johnson Testimony, *supra* note 8.
133. 40 C.F.R. § 158.202.
134. 40 C.F.R. §§ 152.90–152.96.

If the applicant is relying on another party's prior data, and that data is covered under a period of exclusive use, the applicant has to notify the original party and pay compensation (either based on a negotiated sum or one determined through an arbitration process).[135,136] Given that a negotiated payment is involved, it is unlikely that the original submitter will object to the follow-on registrant. However, there may be occasions when an original submitter may petition the agency to deny or cancel a registration of a follow-on registrant. This situation typically arises if there is a dispute about the monetary compensation. That is, the original submitter may seek cancellation or denial based on the fact that the applicant has failed to participate in agreed-upon procedures for reaching agreement on the amount, failed to comply with the terms of an agreement, or failed to either participate in an arbitration proceeding or comply with the terms of an arbitration decision.[137]

2. Conditional Registrations Three types of conditional registrations may be granted,[138] but the relevant one for registrants of PIPs is found in section 3(c)(7)(C), which provides that conditional registration may be granted for a new active ingredient even when certain data are lacking, on condition that such data are received by the end of the conditional registration period (provided that the pesticide does not meet or exceed the risk criteria for serious acute injury to humans or domestic animals; that use of the pesticide during the conditional registration period will not cause unreasonable adverse effects; and that use of the pesticide is in the public interest).[139] In granting a conditional registration, EPA will have already considered the available data on the risks and benefits associated with the proposed use (e.g., whether the protein to make the PIP has a minimal risk of being a food allergen, whether the PIP would be a toxin to non-target species, whether the PIP would have an impact on endangered species, etc.) and made some basic health and safety determinations that during the period of conditional registrations, the PIP will not cause any unreasonable adverse effect on the environment, and that use of the pesticides is in the public interest.

---

135. Under Section 3(c)(1)(F)(i), for a 10-year period after a study is submitted in support of a first registration of a new pesticide, a new combination of a pesticide, or a new use of an existing pesticide, the study will be considered under the exclusive use of the original data submitter.

136. Subsequent applicants may seek an exclusive use of the data, and the rights of the original data submitter may be transferred to subsequent users. 40 C.F.R. § 152.98.

137. 40 C.F.R. § 152.99.

138. EPA may grant a conditional registration when the pesticide and proposed use are identical to an already-registered pesticide. FIFRA Section 3(c)(7)(A). EPA may "conditionally amend the registration of pesticide" to allow additional uses. FIFRA Section 3(c)(7)(B).

139. *See* 7 U.S.C. § 136a(c)(7)(c).

EPA can place numerous conditions on such a registration, including specific relevance for a PIP, an automatic expiration date for the registration, requirements regarding the types of data the registrant will collect and submit to EPA for review, restrictions on where the registrant can plant, the creation of a non-genetically modified planting refuge in conjunction with the genetically modified planting area, the creation and development of a program to educate and promote farmers (referred to in EPA documents as "growers") on following insect resistance management requirements, and the requirement that the manufacturer and grower enter into a grower agreement. Of particular relevance to manufacturers of PIPs are insect resistance management (IRM) requirements and grower agreements. Both items are linked.

3. **Insect Resistance Management** Manufacturers have to develop an IRM plan, but growers have to agree to abide by the conditions set forth in that plan and to cooperate with the manufacturers in implementing it. The purpose of the IRM plan is to create a program that will detect the emergence of insect resistance, evaluate statistically significant and biologically relevant changes in the susceptibility of the target species, and create a remedial plan in the event that any insect resistance is detected. To accomplish these purposes, an IRM plan typically contains refuge requirements as well as sampling and detection methodologies. All of these elements must be approved by EPA as part of the registration process. Once completed, the results of the sampling and the analysis of those samples have to be provided to EPA.

Refuge requirements refer to areas where non-PIPs are planted that are adjacent to, near, and/or embedded within the area where the PIP is planted. For example, one refuge requirement indicated that at least 5 acres of non-Bt corn is planted for every 95 acres of specific BT corn, and that the area must be at least 150 but preferably 300 feet wide. Additionally, EPA required that non-Bt corn must be a comparable variety and managed in the same manner as the Bt corn area.[140] In another refuge requirement, EPA noted among other things that the refuge must represent at least 20 percent of the grower's corn acres.[141]

---

140. *See* EPA Fact Sheet, "Bacillus thuringiensis var. aizawai CryiF and the genetic material (from the insert plasmid pGMA281) necessary for its production in cotton and Bacillus thuringiensis var. kurstaki CryiAc and the genetic material (from the insert of plasmid pMYC3006) necessary for its production in cotton" (September 2005), *available at* www.epa.gov/pesticides/biopesticides/ingredients/factsheets/factsheet_006512–6513.

141. EPA, "Bacillus thuringiensis Cry3Bbi Protein and the Genetic Material Necessary for its Production (Vector ZMIR13L) in Event MON 863 Corn & Bacillus thuringiensis CryiAb Delta-Endotoxin and the Genetic Material Necessary for its Production in Corn" (May 2005), *available at* www.epa.gov/pesticides/biopesticides/ingredients/factsheets/factsheet_006430-006484.

As part of the IRM plan, the manufacturer will outline how it will educate the grower on compliance with the terms of the IRM program. The education program can involve transmittal of written documents, face-to-face meetings, and radio and television commercials. EPA may require this material to be given to it for its records. Manufacturers will also be responsible for designing and implementing an IRM compliance assurance program that will evaluate to what extent a grower is in compliance with the plan, including specific steps such as conducting a survey of a representative sample of growers. The compliance assurance program should also outline the mechanism that a manufacturer would use to address instances of noncompliance by growers. Manufacturers may also be required to provide an annual report to EPA outlining their compliance with the terms of the IRM plan.

4. Grower Agreements Grower agreements arise in the context of PIP sales because of the way EPA views PIPs. Specifically, EPA does not view the seeds of plants that have pesticide incorporated into them as pesticides. Thus, when a grower receives a bag of seeds, failure to comply with the use restrictions on the label does not constitute a violation of FIFRA for the grower. As a result, instead of seeking enforcement against growers for failing to comply with a label, EPA requires that anyone who wants to grow PIPs to enter into an agreement that contractually obligates the grower to comply with the IRM plan. The manufacturer is required to submit the grower agreement to EPA, and it cannot put terms in the grower agreement that deviate from the IRM plan. Additionally, grower agreements must be maintained for a specified period of time and be available for review by state agencies.[142]

There does not appear to be any regulatory requirements that specify the criteria for a grower agreement. However, standard provisions will include requirements for restricting the use of the seed (e.g., that it cannot be used for researching, breeding, or replanting), requiring compliance with the IRM plan, assisting the manufacturer in complying with the IRM plan, and agreeing to indemnify the manufacturer for violations caused by the grower's noncompliance with the IRM plan.

E. Special Local Needs Registration

Although it is not of particular importance to PIP registrations thus far, it should be noted that a state may register an EPA-registered pesticide for additional

---

142. *Id.* As critics of grower agreements note, "Companies with a financial interest in selling their seeds are in effect appointed as officers to enforce restrictions—such as [refuge] rules—which often make their seeds less desirable. This clearly puts them in a conflict of interest situation, in which their interest in selling seeds conflicts with their delegated duty to ensure that sometimes burdensome rules are enforced." Letter from Center for Food Safety to EPA (June 13, 2007) re: "Comments on 'Plant-Incorporated Protectants; Potential Revisions to Current Production Requirements.'"

uses[143] to meet "special local needs" provided that the use has not been previously denied, disapproved, or canceled by EPA and that the use is covered by necessary tolerances or other clearances under the federal Food, Drug, and Cosmetic Act.[144] EPA defines a *special local need* as "an existing or imminent pest problem within a State for which the State lead agency, based on satisfactory supporting information, has determined that an appropriate federally registered pesticide is not sufficiently available."[145] EPA is given veto power over all state registrations if the agency determines the registration is inconsistent with the FFDCA if the use is determined to be an imminent hazard. If EPA believes that the state has not exercised adequate control, the agency may suspend a state's registration authority until such time it believes that the situation has been corrected.[146]

F. Reporting Obligations

Aside from the reporting requirements outlined in a registration or conditional registration, all registrants have an ongoing obligation to EPA to provide any additional factual information regarding unreasonable adverse effects from the pesticide on the environment or human health.[147] If the registrant possesses or receives information (including conclusions and opinions), and that information is relevant to the assessment of the risks and benefits of one or more of the specific pesticide registrations currently or formerly held by the registrant, then the registrant has a reporting obligation.[148] Specifically, unless stated otherwise a registrant has 30 days[149] to report after it first possesses or knows of the following

---

143. Under EPA regulations, a state may register any new use of a federally registered pesticide; any use of a federally registered pesticide as to which some other uses have been denied, disapproved, suspended, or cancelled (provided that there is consultation with EPA); or any use of a federally registered pesticide for which registration some or all of uses have been voluntarily cancelled (again, provided that there is consultation with EPA).

144. 7 U.S.C. § 136v(c)(1).

145. 40 C.F.R. § 162.151(i).

146. *See* 7 U.S.C. § 136v(c) (3) and (4). A state registration that is not vetoed by EPA is considered to be a FIFRA-registered product and thus subject to all the requirements of sections 3 and 6 of FIFRA. 40 C.F.R. § 162.155 provides a list of the basis for suspending a state's authority, thereby elaborating on the term *adequate control*. Additionally, EPA provides a set of procedural rights for the state as well as internal appeals within the agency leading to an adjudicatory hearing and then appeal to the Administrator. Despite silence on the part of the statute, EPA regulations provide that a state may challenge EPA's decision by seeking judicial review after the state has exhausted its administrative appeals.

147. 40 C.F.R. § 159.152.

148. 40 C.F.R. § 159.158.

149. Depending on the severity of the adverse effects incident, a registrant has a different reporting time. For incidents involving human fatalities, a registrant must submit information within 15 days. Information on significant or prolonged human illness, property damage that is alleged to have occurred in a manner that could have caused direct human injury such as fire or explosion, fatalities to wildlife, plant damage over 45 percent

types of information with respect to a pesticide on which it holds a registration, provided that the information, if true, would be relevant (either alone or in conjunction with other information) to EPA's determination of risks and benefits: scientific studies (e.g., toxicological, ecological, epidemiological, and exposure studies), information about studies terminated prior to completion, detection of a pesticide above appropriate levels, failure of performance incident reports, and dietary and environmental pesticide residue incident reports.[150] Additionally, there is a "catch-all" reporting obligation "if the registrant knows, or reasonably should know" that if the information proves to be correct, EPA might regard the information (alone or in conjunction with other information) as raising questions about the continued registration of a product or about the appropriate terms and conditions of registration.[151] The failure to provide timely information could lead to an enforcement action and/or cancellation or suspension of the registration.

G. Cancellation and Suspension

The FIFRA requirements for permanently canceling or temporarily suspending a registration are elaborate and complex. This discussion will briefly discuss the mechanisms for (a) canceling a registration for cause, (b) canceling a conditional registration, and (c) suspending a registration.

　　1. **Canceling for Cause**[152]　According to the statute, EPA may cancel a registration for cause if either (a) it appears that the labeling or other submitted material does not comply with FIFRA, or (b) "when used in accordance with widespread and commonly recognized practices," the pesticide has an unreasonable adverse effect on the environment.[153] In implementing these regulations, EPA has interpreted the broadly worded second prong by focusing on six risk factors, specifically that continuing the registration in effect (a) poses a risk of serious acute injury to humans or domestic animals; (b) poses a risk of having an oncogenic, heritable genetic, teratogenic, fetotoxic, reproductive effect, or a chronic or delayed toxic effect; (c) may result in residues on non-target organisms at levels that are acutely or chronically toxic; (d) may pose a risk to the continued existence of a species protected under the Endangered Species Act; (e) may result in the

---

of the acreage exposed, and water contamination over a certain limit may be accumulated for 30 days and then reported within 30 days after each accumulation period. For all other incidents, information may be accumulated by registrants for 90 days and submitted within 60 days after the end of each 90-day accumulation period.

　　150. 40 C.F.R. § 159.155(a).

　　151. 40 C.F.R. § 159.195. However, if the registrant determines that an analysis is based on data that were erroneously generated, recorded, or transmitted or computational errors, the registrant does not have to inform EPA about that analysis. 40 C.F.R. § 159.158(b)(1).

　　152. It should be noted that a similar analysis must be undertaken if EPA seeks to change a use classification for a pesticide or if it wants to deny a registration.

　　153. 7 U.S.C. § 136d(b)(1)–(2).

destruction or other adverse modification of a critical habitat under the Endangered Species Act; or (f) may pose some other risk to humans or the environment which is of significant magnitude to merit a review of the benefits to justify initial or continued registration.[154]

In determining whether it should cancel a registration, the agency undertakes a *Special Review*.[155] This Special Review is conducted if based on a validated test or other evidence, EPA determines[156] that one or more of the uses of pesticide exceeds the six risk factors set forth above. At that point, a presumption against the registration arises, and the registrant has the burden of persuading the agency that this initial determination was "erroneous, that the risks can be reduced to acceptable levels without formal proceedings, or that the benefits of the pesticide outweigh the risks."[157]

Procedurally, once EPA makes that initial determination, a notice must be provided to the affected registrant; thereafter, the initial determination is recorded in the Federal Register.[158] The issuance of the notice triggers a comment period and possibly informal public hearings. During this period, affected parties may submit comments and/or hold meetings with EPA to discuss their factual or substantive positions as well as options that may be undertaken to avoid cancellation of the registration (e.g., changing the labeling or packaging or complying with additional restrictions).[159] EPA has signaled that any registrant that is willing to voluntarily change the composition, packaging, labeling, or other terms of its registration after issuance of the notice should meet with the agency as that would be "most helpful and productive."

Once the comment period is concluded, EPA issues a notice of preliminary determination outlining, among other things, whether the use satisfies the risk criteria; whether changes in composition, packaging or other alterations has satisfied EPA, and whether the use poses a unreasonable adverse risk to human

---

154. 40 C.F.R. § 154.7.

155. 40 C.F.R. § 154.1(a). There has been significant litigation concerning the openness of the Special Review process, but the scope of that litigation is outside the parameters of this book. Suffice to say, the current regulations note that the process is intended to assess risk and benefits "in an open and responsive manner." 40 C.F.R. § 154.1

156. EPA may decide to evaluate a pesticide on its own initiative or at the suggestion of any third party.

157. 40 C.F.R. § 154.1. EPA notes the "burden of persuasion that a pesticide product is entitled to registration or continued registration for any particular use or under any particular set of terms and conditions of registration is always on the proponent(s) of registration." 40 C.F.R. § 154.5.

158. 40 C.F.R. § 154.21, 154.25. EPA may determine after issuing its notice that it does not want to initiate the Special Review process, and thus, it will put in the Federal Register a notice outlining its rationale. 40 C.F.R. § 154.23.

159. Interested parties may also petition EPA for a hearing to present factual information or to respond to another parties' presentation.

health or the environment. If EPA decides either to move forward with canceling the registration or to hold a formal adjudicatory hearing on whether to undertake such action, it must prepare an analysis regarding the impact on the agricultural economy. Specifically, EPA examines the impact of the proposed action on production and prices of agricultural commodities, retail food prices, and other aspects of the agricultural economy, with this examination reflecting FIFRA's historical roots as an economics-based statute aimed at allowing the free flow of appropriate pesticides. EPA must notify the Secretary of Agriculture to provide the USDA with an opportunity to comment on EPA analysis.[160] The agricultural economy analysis is also referred to a Scientific Advisory Panel (SAP), or some other scientific peer review group to provide EPA with comments, evaluations, and recommendations on the "impact on health and the environment of the action."[161] However, the statute does not include any requirement that the SAP or other scientific peer review group has to provide comments within a specified period of time or that EPA decision has to follow its findings. In contrast, the USDA has only a 30-day window to provide comments. If the USDA Secretary does not provide comments within 30 days, or after review of any comments made, then, without any further delay EPA can make a final determination and notify the registrant and the public of its intent to cancel.

Once a notice of cancellation has been issued, it becomes final and effective at the end of 30 days from the receipt of the notice by the registrant or its publication in the Federal Register, whichever occurs later. The effectiveness of the notice may be forestalled if (a) the registrant makes the necessary corrections, if possible; or (b) the person who is adversely affected (either the registrant or similarly interested party) by the notice makes a request for a quasi-adjudicatory hearing.[162] Once a hearing is requested, the registrant and EPA begin working through a convoluted set of legal procedures.[163] At a hearing, the proponent of the cancellation has the burden of going forward to present an affirmative case

---

160. 7 U.S.C. § 136d(b).

161. 7 U.S.C. § 136w(d).

162. 7 U.S.C. § 136d(b).

163. Other parties who may be interested in the outcome are permitted via filing a motion for intervention to join the litigation as a party, with the regulations indicating intervention should be freely granted provided it would not unreasonably broaden the issues already presented and the intervention is relevant. *See* 40 C.F.R. § 164.31. EPA regulations 40 C.F.R. Part 164 outlines the documents that must be submitted and the process to be followed. The regulations provide for a prehearing discovery, including deposition testimony and issuance of subpoenas for witnesses and documents. *See* 40 C.F.R. §§ 164.50–71. At the hearing, all relevant, material, and competent evidence that is not unduly repetitious will be admitted, even if some of this evidence may not be admissible under the Federal Rules of Evidence. The weight given to the evidence will be determined by its reliability and probative value. *See* 40 C.F.R. § 164.81(a).

for the cancellation. In all cases, however, the ultimate burden of persuasion that a pesticide product is entitled to registration or continued registration for *any particular use* or *under any particular set of terms and conditions* is always on the proponent(s) of registration.[164]

2. Cancellation of Conditional Registration EPA may issue a notice of its intent to cancel a conditional registration if it determines that the registrant has (a) failed to initiate or pursue appropriate action toward fulfilling any conditions imposed, or (b) failed to meet any condition imposed by the end of the period provided for its satisfaction.[165] However, the conditional registrant may continue to sell or use the existing stock of conditionally approved product as long as EPA determines that the sale or use would not have an unreasonable adverse effect on the environment.[166]

If the registrant does not request a hearing, EPA's notice to cancel becomes final 30 days after the registrant receives the notice. If a hearing is requested, the only matters for resolution are whether the registrant has initiated or pursued appropriate action to comply with or satisfy the condition(s) within the time period provided or whether EPA's determination regarding the disposition of the existing stock was appropriate.[167]

3. Suspension If EPA determines that action is necessary to prevent an imminent hazard during the time period required to cancel or change the classification of the pesticide, but that there is no emergency, the agency may order a suspension of the pesticide.[168] The registrant may request a hearing within five days of receiving the suspension notice,[169] and a hearing will be held on an expedited basis within five days of the request. If such a hearing is not requested, a suspension order immediately becomes effective and it is not subject to judicial review.

---

164. 40 C.F.R. § 164.80. The ALJ may make either a initial determination within 25 days of the close of the hearing, or at any time, the ALJ has discretion to issue an accelerated decision in favor of respondents if, inter alia, there was a failure to comply with the discovery process or to state a claim. In either event, the party must file an appeal to the Environmental Appeals Board within 20 days; otherwise, the ALJ's decision becomes final. 40 C.F.R. §§ 164.90, 101, 102. Unless otherwise stipulated by the parties, the Environmental Appeals Board must issue its decision within 90 days of the close of the hearing or filing of accelerated decision. 40 C.F.R. § 164.103.

165. 7 U.S.C. § 136d(e)(1).

166. 7 U.S.C. § 136d(e)(1).

167. 7 U.S.C. § 136d(e)(2).

168. 40 C.F.R. § 164.120.

169. Intervention is permitted in a suspension hearing provided that the request is made within five days of the notice. If the suspension hearing is an emergency hearing, no intervention is permissible.

Even though the presiding officer need not be an administrative law judge, the hearing is governed by the same standards as for a cancellation hearing. Thus, despite the shortened period of time for commencing a hearing and for rendering a decision, a suspension hearing can take an extensive period. Within eight days after the conclusion of the presentation of the evidence, the presiding officer of the hearing submits to the parties his proposed recommendation; two days later, the presiding officer will provide its recommendation to the Environmental Appeals Board.[170]

EPA regulations provide that any interested party other than the registrant may petition for a reconsideration of a final order of cancellation or suspension order. For reconsideration to be warranted, EPA would need to find that the petitioner has presented substantial new evidence that may materially affect the prior order and that such evidence could not have been discovered with due diligence by the affected registrant prior to the final order being issued.[171] If the petition is granted, a hearing will held in which the burden of proof rests on the petitioner. The issues of the hearing will be limited to whether substantial new evidence exists and whether such evidence warrants a reversal of or modification to the existing order. The determination of these issues will be made by taking into account the human and environmental risks associated with this pesticide as well as the cumulative effect of all past, present, requested, and reasonably anticipated uses.[172] If the petitioner's request is denied, judicial review is then available.

4. Judicial Review A final order issued after a cancellation, suspension, or emergency suspension hearing process is completed is subject to judicial review.[173] The registrant or other interested parties (with the concurrence with the registrant) may seek judicial review in any federal circuit court wherein the person lives or has a place of business, provided such an appeal is filed within 60 days after the entry of the order. However, in reviewing a suspension order, the court can determine only whether the suspension order was arbitrary, capricious, or an abuse of discretion. This action may be maintained simultaneously with any other administrative proceeding. In other words, a court may be reviewing the suspension order while an administrative law judge (ALJ) may still be hearing the arguments concerning the cancellation order. If the court issues a decision on the suspension favorable to the registrant, it will only stay the suspension order as the agency may continue with its efforts to cancel or reclassify the pesticide.

An administrative decision on cancellation is also subject to judicial review. However, FIFRA does not require that the same court hear the appeal of suspension and cancellation orders. Thus, given the potential that interveners and

---

170. 40 C.F.R. § 164.121.
171. 40 C.F.R. § 164.131.
172. 40 C.F.R. § 164.132.
173. 7 U.S.C. § 136n; 7 U.S.C. § 136d.

applicants may seek appeals, it is possible that two different courts will be reviewing essentially the same underlying evidence on a pesticide.

VI. FEDERAL REGULATION OF PESTICIDAL RESIDUES

Directly linked to the registration of pesticides or PIPs is the issue of pesticidal residues in food and identification of the appropriate level or tolerance (if any) for such residues. As noted in the preceding section, a PIP will not be registered unless there is tolerance or an exemption from the tolerance requirement has been secured (a list of PIP tolerances is noted at 40 C.F.R. Part 180).[174] The regulatory basis for controlling pesticide residue lies with the FFDCA and the amendments thereto in 1954, 1959, and 1996. Specifically, in 1954 Congress amended the FFDCA by adding section 408, which stated that any pesticide residue that is on or in a raw agricultural commodity[175] shall be deemed unsafe unless either a tolerance has been established and the quantity of the pesticide in or on the agricultural commodity is within that limit or the pesticide residue is exempted from the tolerance requirements. The amendments in 1996 (referred to as the Food Quality Protection Act in 1996) established a new safety standard—namely, that EPA must determine that there is "reasonable certainty of that no harm will result from aggregate exposure to the pesticide chemical residue, including all anticipated dietary exposures and all other exposures for which there is reliable information."

The establishment of pesticide residue levels or exemptions from residue levels is handled by EPA, rather than FDA, which retains responsibility for enforcement of tolerances in foods that are either imported or sold across state lines. If a food contains residues in excess of the tolerances, that food is considered adulterated, and as such, its manufacturer or distributor is subject to the enforcement provisions outlined in Chapter 6.[176] Moreover, if the food contains residues in excess of the tolerances, the FIFRA registration for that pesticide can be suspended or canceled. This chapter will focus on how actually pesticide residue tolerances are established by first examining the substantive criteria and then the procedural steps.

---

174. 21 U.S.C. § 346a(1) specifically refers to harmonization between FIFRA registration and tolerances.

175. EPA defines *raw agricultural commodities* broadly to include, among other things, fresh fruits, vegetables in their raw or natural state, grains, nuts, eggs, raw milk, meats, and similar agricultural produce. It does not include foods that have been processed, fabricated, or manufactured by cooking, freezing, dehydrating, or milling. 40 C.F.R. § 180.1. Tolerances apply indirectly to processed foods as well. A processed food will be considered adulterated unless levels of the pesticide residue are in conformity with its tolerance or unless the pesticide residue is subject to an exemption from the tolerance requirement.

176. 21 U.S.C. § 346a(a)(4).

A. Substantive Criteria for Establishing Tolerances

In order to establish or leave in effect a tolerance for a pesticide, EPA has to determine that the tolerance is "safe."[177] The statute defines *safe* as "that there is a reasonable certainty that no harm will result from aggregate exposure to the pesticide chemical residue, including all anticipated dietary exposures and all other exposures for which there is reliable information."[178]

EPA will consider the following nonexclusive set of factors when it is establishing, modifying, leaving in effect, or revoking a tolerance or exemption for a pesticide residue:

(1) the validity, completeness, and reliability of the available data on the pesticide and the residue;[179]

(2) the nature of any toxic effect shown to be caused by the pesticide or residue;

(3) the available information concerning the relationship of the results to human risk, the dietary consumption patterns of consumers (and major identifiable subgroups of consumers),[180] the cumulative effects of such residues and other substances that have a common mechanism of toxicity, the aggregate exposure levels of consumers (and major identifiable subgroups of consumers) to the residue and to other related substances, and the variability of the sensitivities of major identifiable subgroups of consumers;

(4) such information as EPA may require on whether the pesticide chemical may have an effect in humans that is similar to an effect produced by a naturally occurring estrogen or other endocrine effects; and

---

177. EPA tolerances for a residue preempts state and local restrictions on food if the state and local restrictions are based on lower residue levels. States may petition for an exception if EPA-set residue level threatens public health.

178. 21 U.S.C. § 346a(b)(A)(ii).

179. EPA may consider available data and information on the anticipated residue levels of the pesticide chemical in or on food and the actual residue levels of the pesticide chemical that have been measured in food, including residue data collected by FDA, provided that five years after the tolerance is established the developer demonstrates that the tolerance levels are not exceeded—otherwise the tolerance is subject to a modification or revocation. 21 U.S.C. § 346a(b)(2)(E).

180. When assessing chronic dietary risk, it is possible to consider the available information on the percent of food actually treated with the pesticide only if EPA determines that the data are reliable and provide a valid basis to show what percentage of the food derived from such crop is likely to contain such pesticide chemical residue; the exposure estimate does not understate exposure for any significant subpopulation group; if data are available on pesticide use and consumption of food in a particular area, the population in such area is not exposed to residues (in their diet) above EPA's estimates; and there is periodic reevaluation of the estimate of anticipated dietary exposure. 21 U.S.C. 346a(b)(2)(F).

(5) safety factors, which in the opinion of experts qualified by scientific training and experience to evaluate the safety of food additives are generally recognized as appropriate for the use of animal experimentation data.[181]

EPA will not establish or modify a tolerance or an exemption unless there is a practical method for detecting and measuring the levels of residue in or on the food. If a tolerance is to be established, it will not be established at a level lower than the detection limit.[182] EPA will examine if there is an international tolerance by examining the residue limits established by the Codex Alimentarius Commission (Codex).[183] If Codex has established a maximum residue limit but FDA does not propose to adopt the Codex level, EPA shall publish for public comment a notice explaining the reasons for departing from the Codex level.[184]

B. Procedural Criteria for Establishing Tolerances

EPA may issue, establish, modify (but not expand), or revoke a tolerance for a pesticide residue in or on a food in response to a petition (either by the product proponent or an interested party) or based on the agency's own initiative.[185] A petition to establish a tolerance or exempt a residue must be supported by the following data:[186] (a) the name, chemical identity, and composition of the pesticide residue; (b) the amount, frequency, and time of application of the pesticide; (c) full reports of tests and investigations made with respect to the safety of the

---

181. 21 U.S.C. § 346a(b)(2)(D). Additionally, EPA will review the data for the impact of the pesticidal residue on infants and children, including whether the children or infants are likely to have a disproportionately high consumption of foods containing or bearing such residue or they have a special susceptibility to the pesticide chemical.

182. 21 U.S.C. § 346a(b)(3).

183. 21 U.S.C. § 346a(b)(4).

184. See id.

185. EPA may establish, modify, suspend, or revoke a tolerance or exemption from tolerance for a pesticide or pesticide residue. 21 U.S.C. § 346a(e)(1). Additionally, for tolerances that are already issued or exemptions granted, EPA has the authority to request additional information in order to continue the tolerance or the exemption granted. If the additional information is not provided in the specified time, then EPA may modify or revoke the tolerance or exemption in question. If EPA takes such an action, it is subject to the administrative hearing and judicial review provisions set forth in further detail below. 21 U.S.C. § 346a(f).

186. Data and information that are or have been submitted to the Administrator under section 408 are entitled to the same confidential business information and exclusive use and data compensation requirements as set forth in the Federal Insecticide, Fungicide, and Rodenticide Act. The only statutory exemptions are that the data and information may be provided to federal government employees or contractors authorized by the government, or to the Congress of the United States. Furthermore, the information provided by the petitioner will be published when EPA issues its proposed or final regulation or order. 21 U.S.C. § 346a(i).

pesticide; (d) full reports of tests and investigations made with respect to the nature and amount of residue likely to remain in or on the food, including a description of the analytical methods used; (e) practicable methods for detecting and measuring the levels of the residue in or on the food (or for exemptions, a statement why such a method is not needed); (f) proposed tolerances for the pesticide chemical if tolerances are proposed; (g) information on impacts on infants and children and information whether the pesticide may have the same effect as naturally occurring estrogen or other endocrine effects; (h) practical methods for removing any amount of the residue that would exceed any proposed tolerance; and (i) any other information EPA may reasonably require in support of the petition.[187] Additionally, a petitioner must provide an information summary of the data and the arguments in the petition as well as a statement that the petitioner agrees this information summary can be published for public review.[188]

With respect to a petition for either establishing a tolerance or an exemption, once EPA has determined the criteria for a petition has been met, it shall publish a notice within 30 days. The notice will outline (a) the name of the pesticide residue and the commodities for which the tolerance is sought, (b) a description of the analytical methods available to EPA for the detection and measurement of the residue or the reasons the petitioner claims do not require such analytical methods, and (c) a reference to where the informational summary provided by the petitioner is available.[189]

After examining the petition in light of the substantive criteria, FDA has one of three options: (a) issue an order (which may vary from the one that sought by the petition) that establishes, revokes, or modifies a tolerance without further notice and without further period for public comment; (b) issue a notice of proposed rulemaking and provide 60 days for public comment on the proposed regulation,[190] and thereafter issue a final regulation; or (c) deny the petition.[191] If EPA undertakes any of these three actions, it shall take effect upon publication

---

187. 21 U.S.C. § 346a(d)(2)(A)(1)(ii)–(xii), 21 U.S.C. § 346a(d)(2)(A)(1)(ii)–(xii). Additionally, there are some specific requirements, such as if the petition relates to a tolerance for a processed food, reports of investigations conducted using the processing method(s) used to produce that food must be provided, or if any tolerance or exemption has already been granted for the residue, then information regarding exposure to the pesticide residue is needed.

188. 21 U.S.C. § 346a(d)(2)(A)(1)(i). If information or data required is available to EPA, the petitioner may cite the availability of the information or data in lieu of submitting it. the EPA Administrator may require a petition to be accompanied by samples of the pesticide chemical with respect to which the petition is filed.

189. 40 C.F.R. § 180.7(f).

190. Except that the period for comments may be shortened if the Administrator for good cause finds that it would be in the public interest to do so and states the reasons for the finding in the notice of proposed rulemaking.

191. 21 U.S.C. § 346a(d)(4)(A).

unless objections are filed with respect to such regulation or order within 60 days of the publication.[192]

An objection to an order or regulation must specify with particularity the provisions that are deemed objectionable and reasonable grounds for the objection.[193] If such an objection is made, the party can make a request for a public evidentiary hearing. If the objection is properly made, then EPA has to hold a quasi-adjudicatory hearing.[194] EPA can also on its own initiative hold such a hearing. A final determination shall be made in an order setting forth the facts and the conclusions of the law that are made by the EPA Administrator. This evidentiary hearing is a prerequisite for anyone seeking judicial review of the order or regulations. As noted above, EPA on its own initiative can issue a tolerance or exemption, in which case no evidentiary hearing option is available. Thus, once a final order or regulation is issued by EPA, that order or regulation may be immediately subject to judicial review.

Judicial review is available in the federal circuit court where the petitioner resides or has its principal place of business (or the U.S. Court of Appeals for the District of Columbia Circuit), provided that a petition for review is filed within 60 days of publication of the order or regulation.[195] "Upon the filing of such a petition, the court shall have exclusive jurisdiction to affirm or set aside the order or regulation complained of in whole or in part."[196] If the order was issued following a public evidentiary hearing, the findings with respect to questions of fact shall be sustained only if supported by substantial evidence when considered on the record as a whole.[197] The commencement of judicial review will automatically stay the regulation or order unless specifically ordered by the court to the contrary.[198]

The judgment of the court affirming or setting aside (in whole or in part) any regulation or any order and any regulation that had been issued pursuant to such an order shall be final, subject to review by the Supreme Court of the United States.[199]

---

192. 21 U.S.C. § 346a(g)(1). The provisions relating to objections and hearings are the same as if EPA on its own initiative had decided to establish, revoke, suspend, or modify a tolerance or exemption or if EPA required additional data to support the continuation in effect of a tolerance or exemption. If the regulation or order issued in response to a petition (as opposed to EPA's own actions), then a copy of the objection must be served on the petitioner by the EPA Administrator.

193. 21 U.S.C. § 346a(g)(2).

194. 21 U.S.C. § 346a(g)(C). It should be noted that if an objection is made, but there has been no request for a hearing, the EPA Administrator will issue an order outlining the actions taken on each objection.

195. 21 U.S.C. § 346a(h)(1).

196. 21 U.S.C. § 346a(h)(2).

197. Id.

198. 21 U.S.C. § 346a(h)(4).

199. Id.

VII. CASE STUDIES

Two significant events that occurred during the late 1990s serve as important cautionary tales. Both of these events shaped public perception, which in turn continued to create a regulatory atmosphere in which no new GM crops were approved until 2007. Each episode highlighted a particular criticism of the regulatory structure governing genetically modified plants. Specifically, the dispute over whether the monarch butterflies were adversely impacted by pollen from genetically modified corn highlighted for critics of genetically modified products that, among other things, the testing and oversight were inadequate prior to the marketing and wide-spread use of a product.[200] The mistaken placement of a variety of genetically modified corn approved for livestock in food meant for human consumption (commonly referred to as the Star Link™ incident) highlighted for critics the inadequacy of post-approval oversight.[201] Proponents of genetically modified products and the regulators note that in each case the claims were investigated promptly and action was taken in a timely manner to directly address the incident. Nevertheless, the post-hoc responses and attendant negative publicity that accompanied these incidents only increased public skepticism about the adequacy of government controls.

A. Bt Corn and Monarch Butterflies Case Study

Bacillus thuringiensis (Bt) is a naturally occurring bacterium found in the soil and deciduous and coniferous leaves.[202] The bacterium produces certain crystal proteins (Cry proteins) that when ingested by insects (which are most sensitive at the larvae stage) result in gut paralysis, blood poisoning, starvation, and eventual death of the insect.[203]

Since 1961, Bt proteins have been registered as a general use spray-on pesticide.[204] In the 1980s, however, Bt genes were inserted into tobacco, cotton, and tomatoes to produce proteins that were not previously present in these plants. Genetically modified corn seeds started to become commercially available in 1996.[205] The purpose of genetically modified Bt corn was to address

---

200. PEW Initiative on Food and Biotechnology, *Genetically Engineered Corn and the Monarch Butterfly Controversy*, Washington, DC, Pew Initiative on Food and Biotechnology (May 30, 2002) at 17, *available at* http://www.pewagbiotech.org/resources/issuesbriefs/monarch.pdf.

201. J. Kuzma & P. VerHage, *Nanotechnology in Agriculture and Food Production: Anticipated Production*, Project on Emerging Nanotechnologies, Woodrow Wilson International Center (September 2006) at 14.

202. USDA, *supra* note 68, at 99.

203. *Id.*

204. History of BT Cotton. University of California San Diego, *available at* http://www.bt.ucsd.edu/bt_history.html.

205. *See* PEW, *supra* note 200, at 5.

the damage caused by the European corn borer (ECB),[206] a nonnative species that arrived in the early 1900s. Despite the fact that the damage caused by the European corn borer has resulted in economic losses of hundreds of millions of dollars due to a loss of yield, because the pests burrow inside corn stalks where they are difficult to discover and treatment is difficult, timing is complex, and the efficacy of the insecticide is questionable, most farmers were willing to ignore the ECB rather than to treat it with insecticides.[207] Thus, once Bt corn became available, farmers immediately adopted it.

In 1996, EPA approved for the first time the marketing of genetically modified corn seeds that had been modified with Bt Cry proteins; the agency also provided a tolerance exemption in order for it to be approved for human consumption.[208] The protein type was the Cry1Ab protein. Different manufacturers produced different seeds with each being a variation of this protein, which resulted in differences in insect control.[209] EPA's conditional approval of the genetic modified varieties for five years was based on its authority under FIFRA Section 3(c)(7)(B).[210]

EPA had not required ecological field studies when it made its original registration decision. FDA had only a very limited role in the registrations, and it appears that the registrants basically described why their Bt corn did not differ from conventional corn. As an illustrative example, in February 1995, the manufacturer Northrup King met with FDA to discuss the proposed safety and nutritional assessment of corn that had been referred to as Bt 11. Later that year and

---

206. Carpenter, *supra* note 3, at 24.

207. *See* PEW, *supra* note 200, at 6.

The insects that create the most damage for the corn crop in the United States are soil-inhabiting insects such as rootworms, which are the common name for the larval stage of four species of beetles that feed on the roots of corn plants. In 2003, EPA approved Monsanto's genetically modified corn, which is called YieldGard Rootworm, to specifically address this particular insect. *See* J. Gillis, *In Key Test, U.S. Allows Sale of Genetically Engineered Corn*, WASH. POST, Feb. 26, 2003 at p. A01.

208. Although as noted above, EPA was the lead agency in the registration of Bt corn, there was an overlap in jurisdiction with APHIS and thus the material submitted to EPA for registration under FIFRA was also submitted to APHIS. APHIS agreed with EPA's evaluation that there is no significant risk of gene capture and expression of any Bt endotoxin by wild or weedy relatives of corn in the United States. *See* United States Environmental Protection Agency, Bt Plant-Incorporated Protectants (Oct. 15, 2001), Biopesticides Registration Action Document.

209. Carpenter, *supra* note 3, at 24.

210. *See* Bt Plant-Incorporated, *supra* note 211, at VI.I. Specifically, FIFRA registrations were issued in Aug. 1996 (Bt11 Cry1Ab field corn amendment), Dec. 1996 (MON810 Cry1Ab field corn registration), February 1998 (Bt11 Cry1Ab sweet corn registration), and May 2001 (Cry1F field corn registration). EPA also issued a tolerance exemption with respect to Cry1Ab. *See* 60 Fed. Reg. 42446 (Aug. 16, 1995).

in the subsequent year, Northrup King submitted a summary assessment of Bt II corn.[211] FDA concluded:

> Based on the safety and nutritional assessment you have conducted, it is our understanding that Northrup King has concluded that corn grain (kernels), fodder, and silage derived from the new variety, are not materially different in composition, safety, and other relevant parameters from corn grain, fodder and silage currently on the market and that the genetically modified corn does not raise issues that would require premarket review or approval by FDA. Based on the information Northrup King has presented, we have no further questions concerning grain, fodder, and silage from event Bt II at this time. However, as you are aware, it is Northrup King's continued responsibility to ensure that foods marketed by the firm are safe, wholesome and in compliance with all applicable legal and regulatory requirements.[212]

In 1999, however, a short paper in *Nature* magazine caused an international firestorm and debate on the potential dangers of agricultural biotechnology by raising the specter of the highly recognizable and evocative monarch butterfly as the first casualty.[213] Cornell University entomologist John Losey reported laboratory findings that monarch butterfly larvae died after eating milkweed plants dusted with pollen from genetically modified corn.[214] Losey had noticed the large amount of milkweed, which is the only source of nutrition for monarch butterfly larvae, in and around the cornfields. Losey's paper reported that, based on laboratory tests, monarch larvae exposed to the Bt corn pollen ate less milkweed than those exposed to the conventional pollen, and that nearly half of the larvae feeding on the milkweed dusted with Bt corn pollen died after four days whereas none of the larvae exposed to conventional corn pollen died.[215]

As a result of the uproar caused by the publication of this paper—and with the expiration of the registration period pending—EPA requested an opinion from the SAP on whether it should consider "field scouting to supplement acute testing of a few indicator insect species." In 2000, the SAP called upon EPA to include ecological data from field tests when making a determination on the reregistration. EPA reported the SAP's comments as, "'only a limited number of species can be tested in laboratory bioassays, but field studies can be used to detail the impacts on species appropriate for the [PIP] being tested and in a manner that is relevant to determining ecological impacts. It is important that

211. FDA Agency Response Letter, May 22, 1996. ("Biotechnology Consultation").
212. *See id.*
213. *See supra* note 200, at 3.
214. *Id.*
215. *Id.* at 6–8.

the conclusions drawn from the field studies be scientifically sound and not just correlative and that it reflect actual exposure to the [PIP].'"[216]

Subsequently, EPA issued a data call-in to industry to address by April 2001 issues such as, "the distribution of butterflies, milkweed plants and corn; corn pollen release and distribution in the environment; toxicity of Bt corn Cry proteins and Bt corn pollen to lepidopterans; monarch egg laying and feeding behavior; and monarch population monitoring."[217] EPA's deadline was based on the fact that its five-year registration of certain genetically modified corn was about to expire and the agency needed to acquire this information before it decided to reregister the product in time for the 2002 planting season.

Based on studies published in the Proceedings of the National Academy of Sciences at that time and the response of the Agricultural Biotechnology Stewardship Technical Committee to EPA's data call-in, EPA concluded that after receiving updated information on product characterization, human health effects, gene flow, effects on non-target organisms, ecological exposure, insect resistance management, and benefits, there was no unreasonable adverse effect from genetically modified corn.[218] With respect to non-target species, EPA's conclusion was that based on its knowledge and the assumption that butterfly or caterpillar exposure to *Bt* corn in the environment would be low (e.g., exposure would be limited to caterpillars developing on weeds, such milkweeds, within cornfields or very near to cornfields during pollen shed, but the pollen was expected to travel only short distances and cornfields have only a limited concentration of weeds, including milkweeds).[219] As a result, EPA reauthorized the registrations of genetically modified corn for another seven years to expire in October 2008. However, EPA still required additional data with respect to the ecological impacts of certain varieties of genetically modified corn, and also required manufacturers to comply with the Insect Resistance Management program. The latter involved, among other things, mandatory Bt corn refuge requirements and implementation of grower agreements.[220]

Notwithstanding EPA's ultimate conclusion that monarch butterflies were not unreasonably adversely affected by growing Bt corn, public impressions and policy repercussions followed from this incident. By the time Bt corn had to be reregistered, 5 of the 11 products and/or uses that had been originally registered

---

216. *See* Bt Plant-Incorporated Protectants, *supra* note 208, citing Scientific Advisory Panel, at V.3.

217. *See id.* at 30–31.

218. *Id.* at V.4.

219. http://www.ars.usda.gov/sites/monarch/sect1_2.html.

220. The industry response to the monarch butterfly issue was to announce a voluntary implementation of a uniform, industry-wide Bt corn IRM plan for 2000, a year ahead of the schedule EPA required. This plan made it mandatory to plant a non-Bt refuge in each area where Bt corn was planted.

were either voluntarily cancelled or were in the process of being phased out.[221] For example, a genetically modified Bt corn known as Event 176 made by Sygenta Seeds, Inc. (formerly Novartis Seeds Inc.) and Mycogen Seeds (Dow AgroSciences, Inc.) proved capable of harming some larvae because of high levels of toxin in its pollen. Both companies agreed to phase Event 176 Bt corn out of the market. The registration for the Event 176 Bt corn expired in 2001, and all remaining stock was to be used by the end of the 2003 growing season.[222] Moreover, as discussed in Chapter 6, there has been increased opposition to the development of Bt crops in Europe and to the importation of U.S. crops that were Bt varieties.

B. Star Link™ Case Study[223]

Another significant incident regarding genetically modified crops that occurred in the late 1990s involved Star Link™ corn. Specifically, the incident arose out of corn that had been registered under FIFRA for only animal feed purposes—yet was discovered in taco shells. The incident not only highlighted for opponents of genetic engineering the weakness of the regulatory regime because the discovery of the contaminated taco shells was made by an environmental organization, not regulatory officials, but also caused growers to incur significant costs and had implications for the export of grain.

The origins of the incident began in 1997 when the company that manufactured Star Link™ applied to EPA for approval of a registration for food and feed use as well as petitioned for a pesticide residue tolerance exemption in all raw agricultural commodities. However, unlike the previous Bt corn registrations and tolerance exemptions that had been based on the Cry1Ab protein, Star Link™ was based on the Cry9c protein. Because of the long history of Bt usage in a spray fashion, EPA had approved each Cry1Ab protein as being exempt from the tolerance requirement and allowed registrations for use in food and feed. In contrast, because EPA had concerns that Cry9c had characteristics similar to food allergens, the agency exempted Cry9c protein from the tolerance requirements, but that exemption was limited to corn for animal feed and for meat, milk, and eggs derived from animals that were fed with such corn. Thus, effectively, there was a zero tolerance for the presence of Cry9c in foods other than those specified foods. Additionally, the FIFRA registration was also approved for only animal feed purposes. In fact, the company specifically asked that the registration be expanded to human food, but that request was rejected by EPA.

---

221. *See* Bt Plant-Incorporated Protectants, *supra* note 208, at II.1.

222. *See supra* note 200, at 14.

223. The discussion in this section is based on information contained in: EPA, Draft White Paper Regarding Star Link™ Corn Dietary Exposure and Risk (March 28, 2008); Aventis, "Star Link Corn Containment Program," (April 10, 2001); W. Lin et al., "Start Link: Imports on The U.S. Corn Markets and World Trade," Economic Research Service/ USDA (April 2001) 46–54.

However, as noted above, because a tolerance exemption was granted (even though it was limited), there was no FDA oversight over the presence of Star Link™ corn in food products.

In 2000, the Friends of the Earth announced that its testing had detected Cry9c protein in taco shells in supermarkets. A firestorm resulted for the manufacturer and for the regulatory agencies from that discovery. For the manufacturer, Aventis, in addition to the negative publicity it meant having to incur the cost of buying back all of the corn that been genetically modified with the Cry9c protein. Additionally, as the contamination of human food occurred because drifting pollen from the Cry9c protein corn had fertilized non-Cry9c corn, Aventis purchased corn grown within 660 feet of Cry9c protein corn. Moreover, consumers claimed that they had suffered allergic reactions, and Aventis settled their claims even though a study later on indicated that such claims might not have been accurate.

The presence of the Cry9c protein corn had broader repercussions than simply for Aventis, as mills that purchased corn had to institute programs to test the incoming corn while requiring growers to certify that their corn did not contain the protein. Initially, testing was problematic because the government did not have an appropriate detection test mechanism. Moreover, the presence of Cry9c protein corn in the food supply had significant impacts on the export of U.S. corn as other countries had zero tolerance for the protein.

Aventis tried to establish temporary tolerance for 20 ppb, but the EPA SAP rejected the petition, claiming there was insufficient information to justify that level as avoiding allergic reactions to food. However, the company concluded that by the time additional data could be derived, Cry9c protein would be out of the food supply, and it decided that it would produce that type of corn. For opponents of genetic engineering, this episode highlighted the problems with the regulatory surveillance system and the significant impacts there could be in the event of errors.

VIII. FEDERAL REGULATION OF GENETICALLY MODIFIED MICROORGANISMS

In addition to genetic modification of plants, companies have been involved in the genetic modification of microorganisms. Naturally occurring microorganisms have been involved in food production for thousands of years (e.g., yeasts in the making of bread and enzymes in beer brewing), in the manufacture of pharmaceuticals, and in the treatment of sewage and wastes. The genetic modification of microorganism is a more recent development. This section examines the regulations under FIFRA that apply to genetically modified microorganisms that are intended for pesticidal purposes as well the regulations under TSCA that apply to genetically modified microorganisms intended for industrial purposes.

A. Jurisdiction and Role of EPA under the Toxic Substances Control Act

1. **TSCA and Coverage of Microorganisms** In the wake of concerns about toxins in the environment and their impacts on human health, in 1976 Congress enacted the Toxic Control Substances Act, 15 U.S.C. § 2601 et seq.[224] As noted in detail in Chapter 10, TSCA provides EPA with authority to broadly regulate "chemical substances and mixtures" that present an unreasonable risk of injury to health or the environment by allowing the agency to prohibit or limit the manufacture, use, or processing of these substances before they enter the marketplace. This section will address the limited issue of genetically modified microorganisms and their regulation under TSCA. While the regulatory scheme that is applicable to microorganisms incorporates many of the same procedures applicable to traditional chemicals, the section will focus on some significant distinctions. The regulations addressing microorganisms are segregated at 40 C.F.R. Part 725.[225] For a fuller discussion of TSCA, consult Chapter 10.

2. *Chemical Substances* **and Microorganisms** The term *chemical substance* is broadly defined to mean any "organic or inorganic substance of a particular molecular identity, including any combination of such substances occurring in whole or in part as a result of a chemical reaction or occurring in nature and any element or uncombined radical." However, the statute specifically excludes any pesticide as defined under FIFRA,[226] and more broadly, any food, food additive, drug, cosmetic, or device regulated under the FFDCA.[227] Thus, if the intent of the manufacturer is to create a microorganism that will be solely used as a pesticide, drug, food, food additive, cosmetic, or medical device, that microorganism is not subject to the requirements set forth below in this section. On the other hand, if the researcher or manufacturer is unsure of the final use or potential use of the

---

224. Federal legislative efforts to regulate the commerce of chemical substances were originally proposed in 1971 by the President's Council of Environmental Quality. The House and Senate both passed bills in 1972 and 1973 respectively. However, these statutory efforts stalled. It was only after the highly publicized discovery of polychlorinated biphenyls in the Hudson River and polybrominated biphenyls in farm animals in Michigan that the TSCA was enacted. *See* L. J. Schierow, *The Toxic Substances Control Act (TSCA): Implementation and New Challenges*, CONGRESSIONAL RESEARCH SERVICE, Aug. 3, 2007 at 1.

225. Microorganisms that are not subject to specific regulations are those (a) that would be excluded from the definition of chemical substance, (b) that are any microbial mixture as defined in the regulations, or (c) that are microorganisms manufactured or processed solely for export, that are labeled, and that the manufacturer or processor can document that the intention is for export.

226. The FIFRA exclusion does not apply to raw materials, intermediates, or nonpesticidal inert materials.

227. 15 U.S.C. § 2602(2)(A). The FFDCA exemption includes components and inactive ingredients involved in the production of food, food additives, drugs, cosmetics, and devices, but the exemption does not cover by-products.

product, then the microorganism is subject to TSCA if it is being developed for "commercial purposes."

Even if the microorganism is intended for industrial purposes (e.g., bioremediation, biomass conversion, manufacture of specialty chemicals, etc.), the question arises as to EPA's rationale for terming living entities as chemical substances. EPA's position is that microorganisms resulting from deliberate combinations of genetic material from organisms classified in different genera constitute a "new" microorganism which the agency refers to as an "intergeneric microorganism."[228] EPA explained that it decided to classify intergeneric microorganisms as new chemical substances because "of the degree of human intervention involved, the significant likelihood of creating new combination of traits, and the greater uncertainty regarding the effects of such microorganisms on human health and the environment."[229] It should be noted that microorganisms that contain introduced genetic material consisting of well-characterized, noncoding regulatory regions from another genus (e.g., operators, promoters, origins of replication, terminators, and ribosome-binding regions) are not considered intergeneric microorganisms.

3. TSCA Inventory Once a manufacturer has determined that its particular microorganism is not statutorily excluded and does qualify as an intergeneric microorganism, then as with chemical manufacturers, the manufacturer has to determine if the microorganism is already listed on the TSCA Inventory (Inventory). As noted in Chapter 10, under section 8(b) of TSCA, EPA is required to compile and maintain a list of each chemical substance that is manufactured or processed in the United States (except those manufactured or processed in small quantities). If a chemical is not listed on the Inventory, it is considered "new" for regulatory purposes and as such, under section 5(a)(1)(A), a person must notify EPA 90 days before he manufactures or imports such a substance for commercial purposes.

The Inventory has two aspects: a publicly accessible database and a confidential database that can be accessed by EPA at the manufacturer's request if the manufacturer can demonstrate a bona fide business interest. To identify and list microorganisms on the Inventory, EPA uses taxonomic designations and supplemental information (such as phenotypic and genotypic information) to the

---

228. 62 Fed. Reg. 17910, 17913 (Apr. 11, 1997). EPA articulated this position first in its 1986 policy statement, which was published as part of the Federal Coordinated Framework. Subsequently, in 1994 EPA published its proposed rule, "Microbial Products of Biotechnology; Proposed Regulation under the Toxic Substances Control Act." In 1997, EPA adopted the proposed rules with only a few revisions. EPA specifically defines *intergeneric microorganisms* as "a microorganism that is formed by the deliberate combination of genetic material originally isolated from organisms of different taxonomic genera." 40 C.F.R. § 725.3.

229. 62 Fed. Reg. 17913.

extent necessary to accurately and unambiguously identify the microorganism.[230] A manufacturer or importer has to determine whether its microorganism is on the Inventory by examining these factors. If a bona fide intent of manufacturing or importation letter has to be sent to EPA, the applicant must provide the information about the microorganism it would like to manufacture in order for EPA to compare the information to the confidential information in its database. The following information should be contained in a bona fide intent letter:

1. Taxonomic designations, phenotypic information, and genotypic information.
2. A signed statement certifying that the submitter intends to manufacture, import, or process the microorganism for commercial purposes.
3. A description of the research and development activities conducted with the microorganism to date, demonstration of the submitter's ability to produce or obtain the microorganism from a foreign manufacturer, and the purpose for which the person will manufacture, import, or process the microorganism.
4. To the extent known by the submitted, an indication of whether a related microorganism was previously reviewed by EPA.
5. A specific description of the major intended application or use of the microorganism.[231]

If there is no match in the publicly available or confidential databases, the substance is considered "new." It should be noted that all naturally occurring microorganisms (i.e., microorganisms that exist without human intervention) are automatically listed on the Inventory. However, once it has been determined that a microorganism is new, the next question is whether there is a specific regulatory exemption from the premanufacturing requirements. There are exemptions for (a) R&D for noncommercial purposes, (b) R&D but with containment, (c) experimental releases that are approved by EPA, (d) test marketing with EPA approval, and (e) certain low risk microorganisms provided EPA approval is given.

4. R&D Exemption from Pre-Manufacture Notice Requirement Section 5(h)(3) of TSCA exempts any manufacturing and importation of chemical substances "only in small quantities" for scientific experimentation or R&D from premanu-

---

230. 40 C.F.R. § 725.12. Phenotypic information refers to "pertinent traits that result from the interaction of a microorganism's genotype and the environment in which it is intended to be used and may include intentionally added biochemical and physiological traits." Genotypic information refers to "pertinent and distinguishing genotypic characteristics of a microorganism, such as the identity of the introduced genetic material and the methods used to construct the reported microorganism." This also may include information on the vector construct, the cellular location, and the number of copies of the introduced genetic material.

231. 40 C.F.R. § 725.15(b).

facturing notice requirements, provided that the person seeking the exemption notifies those involved in the experimentation or R&D activity of the potential risks associated with this substance.[232] If the activity satisfies the requirements in this section, it is not necessary for the person to seek approval from EPA before taking advantage of this exemption.

Unlike conventional chemical substances in which researchers do not have to concern themselves with the "commercial purpose" criteria because the definition of *small quantity* typically excludes their research, researchers with microorganisms have to determine if they fall under one of the two R&D exemptions: noncommercial purpose or containment.

The noncommercial purpose exemption is only available to researchers at academic institutions. As any research conducted at a company is considered to be for commercial purposes, it therefore is not exempt. For academic researchers, the question centers on whether there is the indicia of commercial purpose. Thus, a researcher must examine whether any of the funding, in whole or in part, for the proposed research comes from a commercial entity (e.g., situations in which the commercial entity contracts directly with the university or researcher, or in which a commercial entity will hold the patents or licensing rights).[233] The researcher also has to consider if there are potential indirect indicators of commercial intent such as whether the research is directed towards developing a commercially viable improvement of a product already on the market or whether the researcher or the university is seeking commercial funding or a patent.[234] The researcher has an ongoing obligation to determine if the research has potential commercial use as the research evolves, and if at any point it develops such a use, the researcher must give proper notification to EPA.

If the research is for commercial purposes, then the next threshold question is whether it involves only small quantities of the substance being studied. EPA has noted that it has adopted a different approach for defining small quantities with respect to R&D involving microorganisms than it did with conventional chemical substances.[235] The traditional definition of small quantities refers to a certain amount necessary to conduct the research. However, the reason that this definition was not adopted for microorganisms is that, unlike traditional conventional chemicals, microorganisms may reproduce and thereby may increase their own volume or amount.[236] Thus, the definition of small quantities refers to

---

232. 15 U.S.C. § 2604(h)(3).

233. 40 C.F.R. § 725.205(b)(1). EPA provides an exemption from section 5 requirements if the researcher is receiving funds from another federal agency that requires compliance with NIH Guidelines. Additionally, any research, whether or not directly funded by an agency, at any institution that adheres to the NIH Guidelines on an institution-wide basis as a condition for receiving federal funding is also exempt.

234. 40 C.F.R. § 725.205(b)(2).

235. 62 Fed. Reg. 17921.

236. *See id.*

the fact that the R&D is conducted within a building or vessel that effectively surrounds and encloses the microorganism and includes features designed to restrict the microorganism from leaving through any mechanism.[237] As such, the exemption does not cover field tests, which are discussed in further detail below.

Although EPA does not require that researchers comply with the NIH Guidelines, the agency bifurcates the obligations that are applicable to those who follow the NIH Guidelines and those who tailor their own containment procedures by using the services of a technically qualified individual who selects the appropriate containment procedures. For instance, there is a requirement that notification of risks must be given to those involved in the research. For those who are following the Guidelines, that means it provides compliance with section IV-B-4-d of the Guidelines, but for all other researchers notification must be made after EPA determines that there is health risk associated with the microorganism based upon the information and data in its possession or control. In addition, there are distinctions in the reporting and record-keeping requirements.

5. TSCA Experimental Release Application For experimentation outside of confined conditions such as field testing (except with respect to small-scale field testing of certain microorganisms listed in section 725.239), EPA approval is required. However, unlike FIFRA EUPs, there are no acreage thresholds to trigger this requirement. To receive approval, an applicant must submit a TSCA Experimental Release Application (TERA) and comply with the terms of approval given by EPA. A TERA and its supporting documentation are an abbreviated version of the information that would be provided under a Microbial Commercial Activity Notice (the microorganism equivalent to the Pre-Manufacture Notices for a chemical). In order to evaluate a TERA, "EPA must have sufficient information to permit a reasoned evaluation of the health and environmental effects of the planned test in the environment."[238] The application will be approved if the proposed activity does not pose an "unreasonable risk of injury to health or the environment," but the burden is on the applicant to demonstrate that there is no such unreasonable risk.

EPA requires, among other things, as set forth in 40 C.F.R. § 725.255, information about the microorganism's identity, a description of the recipient microorganism, the genetic construct of the new microorganism, and its phenotypic and ecological characteristics. The agency also requires information about the proposed experimentation, including:

1. objectives and significance of the activity and a rationale for testing the microorganism in the environment;
2. number of microorganism released;
3. characteristics of the test sites (e.g., location, geographical, physical, chemical and biological characteristics, proximity to human activity);
4. target organism, if applicable, and the anticipated interaction;

---

237. 40 C.F.R. § 725.3.
238. 40 C.F.R. § 725.255(a).

5. duration of the test; and

6. whether state or local authorities have been notified.

The TERA applicant must also provide information on mitigation and emergency measures, confinement procedures, and measures to detect and control potential adverse effects. In addition, the TERA applicant has a general obligation to provide "all available data concerning actual or potential effects on health or the environment of the new microorganism that are in the possession or control of the submitter and a description of other data known to or reasonably ascertainable by the submitter that will permit a reasoned evaluation of the planned test in the environment."[239]

Sixty days prior to the initiation of R&D activity, a person must submit a TERA for review and approval, with or without conditions.[240] EPA has 60 days to review a complete TERA, but the agency may extend that time if it unilaterally determines that it has good cause. Once the TERA is approved, the experimentation has to be conducted in accordance with the terms and conditions of that TERA approval. If, after approval, EPA receives information that raises significant questions about the research, then the applicant has to submit additional information. EPA will then reevaluate whether to continue to approve the activity, and if so, under what conditions. If after approval, EPA receives evidence of unreasonable risk, EPA may issue a notice imposing additional restrictions or requiring the applicant to suspend the activity. The applicant has only 48 hours to implement the provisions of the notice, but then it may submit additional arguments challenging EPA's conclusion. The agency will then evaluate these arguments to make a decision if the activity may be resumed.

6. Test-Marketing Exemption Section 5(h)(1) of TSCA creates a test-marketing exemption (TME). The standard for granting a TME is similar to that for a TERA exemption—namely that the activity cannot present "any unreasonable risk of injury to health or the environment."[241] EPA strongly recommends that the applicant consult with the agency prior to formally submitting an application for a TME exemption to determine if the applicant's activities are eligible for the exemption. Once the issue of eligibility has been decided, the TME application must be submitted at least 45 days before the person intends to commence the test-marketing activity.[242]

The applicant must provide information on (a) the submitter; (b) the taxonomy of the recipient microorganism and the new microorganism, as well as its morphological and physiological features; (c) the genetic construction of the new

---

239. A person need not submit a data that has been previously submitted provided sufficient information is given pursuant to section 725.25(h).

240. 40 C.F.R. § 725.250. *See generally*, 40 C.F.R. §§ 725.270, 725.288 for EPA's review and revocation or modification procedures.

241. 15 U.S.C. § 2604(h)(1).

242. 40 C.F.R. § 725.350(b).

microorganism; and (d) the phenotypic and ecological characteristics of the recipient and new microorganism.[243] Additionally, the applicant must provide information on the maximum quantity of the microorganism that will be manufactured or imported for test marketing, the maximum number of persons who will be provided with or who may be exposed to the microorganism as a result of the test marketing, and a description of how the test marketing activity can be distinguished from commercial marketing or R&D activities.[244]

Evaluation of a TME is similar to that of a TERA in that EPA requires "sufficient information to permit a reasoned evaluation of the health and environmental effects of the planned test marketing activity."[245] Therefore, the applicant has to provide EPA with all test data in the submitter's possession or control and descriptions of other data that is known or reasonably ascertainable by the submitter that concern the health and environmental effects of the microorganism. The regulations provide that EPA must approve (with or without conditions) or reject the application within 45 days of receipt of a complete application.

7. Tier I and Tier II Some chemicals, as well as some microorganisms, have only a low risk when used under certain conditions; thus, such chemicals or microorganisms can be used in certain conditions without full pre-manufacture notice submission. Section 5(h)(4) of TSCA allows EPA by rule to grant an exemption from any and all of the requirements of section 5 if EPA determines that the manufacture, processing, distribution in commerce, use, or disposal of a new chemical substance will not present an unreasonable risk of injury to health or the environment.[246] Two exemptions are typically available to conventional chemicals under section 5(h)(4): the low volume exemption and the low release and exposure exemption (each of which is discussed in further detail in relation to nanotechnology in Chapter 10.

With respect to microorganisms, there are two tiered exemptions: Tier I and Tier II. For a microorganism to be subject to either exemption, EPA must conclude that there is no "unreasonable risk." To reach such a conclusion, EPA primarily examines (a) whether the recipient microorganism is eligible, (b) whether the introduced genetic material contained in the new microorganism meets certain criteria noted below, and (c) whether there are appropriate physical containment and control technologies. Tier I exemptions must comply with all three conditions, but Tier II exemptions need to comply with only the first and second conditions. It is EPA's view that if the recipient is shown to have little or no potential for adverse effect, the introduced genetic material meeting the specified criteria noted below is not likely to significantly increase the potential for adverse effects.

---

243. 40 C.F.R. § 725.355(b),(c),(d).
244. 40 C.F.R. § 725.355(e).
245. 40 C.F.R. § 725.355.
246. 15 U.S.C. § 2604(h)(4).

The first step in examining whether a Tier I or Tier II exemption is available is to determine if the recipient microorganism has to be listed in 40 C.F.R. 725.420. The next step is to decide if the introduced genetic material meets the following criteria as being: (a) limited in size (e.g., reduces risk by excluding extraneous and potentially uncharacterized genetic material), (b) well characterized (e.g., ensures that the functions introduced with the genetic material are sufficiently understood to predict likely behavior of the resulting microorganism), (c) free of certain nucleotide sequences, and (d) poorly mobilizing (e.g., the ability of the introduced genetic material to be transferred and mobilized is inactivated).[247]

It is then necessary to analyze whether the facility in which the new microorganism will be used achieves certain physical containment and control technologies. These requirements include: (a) using a structure that is designed and operated to contain the new microorganism, (b) limiting entry to only those persons whose presence is critical to the reliability or safety of the activity, (c) providing and implementing a written set of personnel safety and hygiene procedures, (d) using inactivation procedures that have been demonstrated to be effective, (e) using mechanisms that are known to be effective in minimizing viable microbial populations in aerosols and exhaust gases, (f) using systems for controlling dissemination of new microorganisms, and (g) establishing emergency clean-up procedures.[248]

The procedures for filing a Tier I exemption include manufacturers or importers filing a certification with EPA that states: (a) their name and address, (b) the date when manufacture or import is to commence, (c) the genus or species of the recipient microorganism listed in 40 C.F.R. 725.420 that will be used to create the new microorganism, (d) a statement on meeting the introduced genetic material criteria and the containment procedures noted above, (e) the site for waste disposal and permits for such a disposal, and (f) a generic statement on the accuracy and validity of the information being submitted.[249] The information outlined above must be submitted at least 10 days prior to commencement of the initial manufacture or import. However, EPA will not review the submission; rather, this serves as a one-time alert to EPA that the submitter is conducting such activities and that the activities are in compliance with the structure outlined by EPA.

The Tier II exemption, on the other hand, is subject to EPA review. The criteria set forth above for "recipient microorganisms" and "introduced genetic material" is equally applicable in this exemption, with the main distinction being that the manufacturer or importer may fashion its own containment and control

---

247. 40 C.F.R. § 725.421; 62 Fed. Reg. 17910, 17916–19 (Apr. 11, 1997).

248. 40 C.F.R. § 725.422.

249. 40 C.F.R. § 725.424(b). It should noted that the generic statement is based on 40 C.F.R. § 725.25(b), but the language concerning the test data certification is not necessary because no data is included in the submission to EPA.

measures, although these must be approved.[250] A submitter for a Tier II exemption should have a meeting with EPA prior to submitting the notice. The notice (which should include information on the identity of the submitter, a more extensive description of the identity of the microorganism than under the Tier I exemption, production volume, and process and containment information) should be submitted at least 45 days before manufacturing or importation commences.[251] With respect to process and containment information, a submitter must describe: (a) the identity and location of manufacturing site(s); (b) process flow diagrams illustrating the production process, identities, and quantities of feedstock; (c) sources and quantities of potential releases to the environment; (d) description of engineering controls or other mechanisms for reducing worker exposure and environmental releases; (e) description of measures that will be undertaken to prevent fugitive emissions; and (f) description of the measures to prevent accidental releases.[252] Additionally, a certification must be provided indicating which elements of the Tier I containment measures are also being followed.[253]

Once a complete notice is submitted, EPA has a 45-day review period. EPA will grant the exemption if it determines the activities will not present an unreasonable risk of injury to health or the environment; it also has the option of granting the exemption with certain conditions. However, EPA may determine that the manufacturer or importer is not eligible for an exemption because the above criteria are not met or there is insufficient information for the agency to make a determination as to risk. If the request is denied, the applicant must submit a Microbial Commercial Activity Notice (MCAN) to manufacture or import this chemical substance. Any microorganism manufactured or imported in violation of these provisions may be seized or activities may be enjoined.

8. Pre-Manufacture Notification System As a general matter, section 5(a) prohibits the manufacture or importation of a new chemical substance that is not on the TSCA Inventory or the manufacture or process of any chemical substance for significant new use unless a pre-manufacture notice (PMN) or a Significant New Use Notice (SNUN), respectively, has been submitted to EPA 90 days prior to the intended date of the activity. Although Chapter 10 will discuss the requirements as applied to chemicals and nanomaterials, this section will describe the particular requirements for genetically modified microorganisms.

Any person who manufactures or imports a new microorganism, or any person who manufactures, imports, or processes a microorganism for a significantly

---

250. 40 C.F.R. § 725.428.

251. 40 C.F.R. § 725.455(a)–(d).

252. 40 C.F.R. § 725.455(d)(1).

253. 40 C.F.R. § 725.455(d)(2). A generic certification similar to a Tier I must also be submitted.

new use, must submit a MCAN[254] 90 days prior to the intended date of the activity. Because the significant new use requirements are currently limited to a specific set of microorganism, this section will primarily address the requirements as applied to new microorganisms.

In submitting the MCAN, the applicant must include the following information:

1. Submitter information (e.g., name, address, and technical contacts);
2. Sufficient information to identify the microorganism on the TSCA Inventory, including data substantiating the taxonomy of the recipient microorganism and the new microorganism, information on the morphological and physiological features of the new microorganism, data on the taxonomy of the donor organisms, detailed description of the genetic construction of the new microorganism, and information on biological interactions and ecological factors;
3. A description of the by-products resulting from the manufacture, processing, use, or disposal of the new microorganism;
4. The estimated maximum amount of the new microorganism intended to be manufactured or imported during the first year of production. and the estimated production for the next 3 years thereafter;
5. A description of different uses and the estimated percent of production volume devoted to each use and the percent of new microorganism in that use;[255]
6. A detailed description of worker exposure and the environmental release. Specifically, the submitter, who controls the operational site, must identify the sites where the microorganism will be manufactured, processed, or used; describe the entire process involved (e.g., provide a diagram of each of the major operations, engineering controls to prevent releases); describe the potential sources of worker exposure (including duration of activity and number of employees involved); provide information regarding the release into the environment; provide a narrative description of means of transportation and method for containment; and describe procedures for the disposal of any articles, wastes, clothing, and other equipment involved in the activity.[256]

---

254. 40 C.F.R. § 725.100. If a person (a) contracts with a manufacturer to produce or process a new microorganism, (b) the manufacturer completes such work exclusively for that person, and (c) that person specifies the identity of the microorganism and controls the total amount produced and the basic technology for the plant process, then that person must submit the MCAN. 40 C.F.R. § 725.105(b). If several individuals are involved in importing the microorganism, the principal importer must submit the MCAN.

255. 40 C.F.R. § 725.155(c)–(g).

256. 40 C.F.R. § 725.155(h)(1). For a site not controlled by the submitter, EPA requires essentially the same information, but allows the submitter to provide estimations and information available in open scientific literature. 40 C.F.R. § 725.155(h)(vi)(2).

As noted above, if there is a significant new use rule issued for a microorganism, and if a manufacturer, importer, or processor is engaging in the new use, then a MCAN has to be submitted. To date, the only microorganisms that are covered by this rule are those identified as the *Burkholderia cepacia* complex. If a company intends to manufacture, import, process, or use *Burkholderia cepacia* complex in the manner of the new use, it should consult with the additional requirements imposed on it through Subpart M of the Part 725 rules. For instance, there are specific requirements for record-keeping applicable to significant new use submitters.

Unless EPA extends the review period for a good cause, an MCAN review must be completed within 90 days of receipt. If EPA does not notify the applicant that it has taken action within the 90-day period, the submitter may begin to manufacture or import. Once a person has commenced manufacturing or importing, within 30 days of the first day, the person must submit a Notice of Commencement (NOC). Submitting an NOC before actual commencement is, in fact, a violation of the statute. However, EPA has the option to take action under sections 5(e), 5(f), or (6)(a) of TSCA. Chapter 10 discusses each of these options in detail.

In brief, section 5(e) allows EPA to issue an order that prohibits or limits certain activities, provided the agency determines there is insufficient information to permit a reasoned evaluation of the health and environmental effects and either (a) in the absence of such information, the manufacture, processing, use, disposal, or distribution in commerce presents an unreasonable risk; or (b) the substance may be produced in substantial quantities (100,000 kg/yr.) and (i) have a *significant exposu*re to humans, or (ii) may reasonably be expected to enter the environment in substantial quantities.[257] However, a consent decree is only binding upon the company that enters into it. The procedural steps EPA undertakes for effectuating a section 5(e) order are outlined in Chapter 10. Alternatively, under section 5(f) EPA may limit the amount of production or impose other restrictions through various mechanisms. In order for EPA to take action under 5(f) and before it can issue a rule under section 6 of TSCA, the agency must have a reasonable basis to conclude the manufacturing, processing, distribution in commerce, or disposal of the chemical substance presents or will present an unreasonable risk. Section 6 provides EPA with the authority to prohibit or limit the amount of chemical substance for a particular use or otherwise restrict the manufacture, import, or processing of the substances if the agency believes the requested activity would present an unreasonable risk of injury.[258] EPA may,

---

257. 15 U.S.C. § 2604(e)(1)(A). EPA has established thresholds for what constitutes a "significant exposure." *See* www.epa.gov/oppt/newchems/expbased.htm. However, those thresholds are based on conventionally sized chemicals, and thus it remains to be seen if they will applied to nanoscale chemicals.

258. 15 U.S.C. § 2605(a).

however, impose only the least burdensome requirements adequate to address the risk.

9. Reporting and Retaining Information The general provisions of the section 8 reporting requirements are equally applicable to microorganisms. This statute provides various mechanisms that can be used by EPA to obtain additional information from companies on toxicity and human exposure. Specifically, section 8(a) provides EPA with the authority to issue rules that can broadly ask for information, section 8(b) requires EPA to compile and maintain the TSCA Inventory database and provides the agency with the authority to request information from manufacturers to update the Inventory, section 8(c) provides EPA with authority to require the reporting of any allegations of significant adverse reactions, and section 8(d) provides EPA with the authority to require the submission of any ongoing and completed unpublished health and safety studies that are known or available. Additionally, section 8 also creates certain obligations on manufacturers, importers, distributors, or processors regardless of whether EPA issues any additional rules or requests. Specifically, section 8(c) obliges EPA to note any allegation of significant adverse reaction that must be maintained internally (regardless of whether the agency makes a request that such information to be reported to it), and section 8(e) creates an obligation to immediately notify EPA if there any new, unpublished information (e.g., preliminary results of animal bioassay studies or epidemiological studies) that reasonably supports there being a substantial risk associated with the chemical.

10. Disclosure of Confidential Information Part 725 of the regulations provide that any person may assert a confidentiality claim when submitting information about microorganisms. The information that will need to be provided to substantiate such a claim will depend on the areas covered by the confidentiality claim, such as whether it encompasses aspects of the specific microorganism's identity, the microorganism's use, and/or the microorganism's health and safety.

Additionally, a company has to answer the following generic questions that are relevant to all claims of confidentiality regardless of the specific nature of the claim: (a) for what period of time the claim is being asserted and why the information should remain confidential until this time; (b) what physical or procedural restrictions exist within the company relating to the use and storage of the information claimed as confidential; (c) whether the information claimed as confidential has been disclosed to a third party, and if so, under what conditions; (d) whether the confidential information been presented in any advertising or promotional materials, material safety data sheets, professional or trade publications or other media, patents, or public filings; (e) whether any federal regulatory agency, federal court, or state has made any confidentiality determination regarding the claim; (f) what harm would occur to the company's competitive position if the information was disclosed and why would such a harm be substantial; and (g) whether a competitor would enter the marketplace if the information were disclosed based on a realistic appraisal of market conditions and barriers to entry.[259]

B. Jurisdiction and Role of EPA under FIFRA

EPA has created a specific set of requirements for specifically genetically engineered microbial pesticides. Even before the issuance of the Coordinated Framework, in 1984 EPA issued an interim policy statement that required notification at least 90 days prior to the small-scale testing of specific genetically modified microbial pesticides. As indicated in Chapter 4, one of the earliest interactions between EPA and a genetically modified organism concerned the regulation of the genetic modification to bacteria. EPA's policy involves small-scale testing at either (a) a facility (greenhouse or laboratory) that does not perform adequate containment or inactivation controls, or (b) intentional environmental release (i.e., field testing). However, a contained facility that has adequate containment or inactivation controls is excluded from these notification requirements.

In 1994, the interim policy statement was codified through issuance of a final rule (59 Fed. Reg. 45600 (Sept. 1, 1994)). Significantly, the policy statement and the subsequent final rule limited the notification requirement to "microbial pesticides whose pesticidal properties have been imparted or enhanced by the introduction of genetic material that has been deliberately modified.[260] As EPA explained, "enhancements" may include giving a microorganism the ability to produce a more potent toxin or increase its survivability.[261] However, while each of these changes may create a more effective microbial pesticide, it may also have adverse consequences for the environment. For example, non-targeted insects may have increased opportunities for contact if the microorganism remains in the environment.[262] On the other hand, microbial pesticides resulting from deletions or rearrangements within a single genome that are brought about by the introduction of genetic material that has been deliberately modified are not subject to the notification requirement.[263] The reason for this difference in treatment is that because no new genetic material was added, no characteristic could be expressed that was not expressed by the parental microorganism or by a natural variant.[264]

With respect to notification, as set forth in 40 C.F.R. § 172.48, EPA requires the submission of sufficient data and information to allow the agency to

---

259. 40 C.F.R. § 725.94(c). There are additional questions if the confidentiality claim is being made to protect the microorganism's identity or if the confidentiality claim is being made to protect the health and safety studies of the microorganism.

260. 58 Fed. Reg. 5878, 5883 (Jan. 22, 1993); 59 Fed. Reg. 45600, 45612 (Sept. 1, 1994).

261. 58 Fed. Reg. 5878, 5883–84.

262. 58 Fed. Reg. 5878, 5884.

263. See id.

264. 58 Fed. Reg. 5878, 5885.

review the proposed test, including information regarding: (a) identity of the microbial pesticide; (b) characterization of its relevant biology and ecology; (c) a description, if applicable, of the way the microbial pesticide has been modified; and (d) a description of the objectives, experimental design, and other relevant parameters of the proposed test. Importantly, EPA will not routinely require the notification to address a broad range of impacts on non-target species, but can request it if the characterization of the microbial pesticides justifies the need for such information.

During the 90-day review period, EPA will make a determination as to whether to: (a) approve the test without requiring an EUP, (b) approve the test without an EUP as long as certain modifications made in the proposed test plan are incorporated, (c) require additional information, (d) require an EUP for the test, or (e) disapprove the test because of the potential for unreasonable adverse effects. EPA has informed submitters that parties should operate under the presumption that no EUP will be necessary.

C. Regulation of Nanopesticides
As a recent petition filed by a group of consumer and environmental groups indicates, nanomaterials, especially nanosilver, have been used in a number of different consumer products for their antimicrobial or antibacterial properties.[265] The issue the petition raises—and one that, in fact, has been raised even before the filing—concerns the way nanomaterials should be regulated under FIFRA. As noted above, the statute defines a pesticide based on intent. Typically EPA examines various factors to determine intent, such as whether the person who distributes or sells the product claims or even implies that the product can be used as a pesticide or whether there is constructive knowledge that people will use the product as a pesticide. Moreover, there is a subset of pesticides referred to as *antimicrobial pesticides* that are pesticides intended to "disinfect, sanitize, reduce, or mitigate growth or development of microbiological organisms." Thus, two questions arise: (a) should products infused with nanomaterials that have antimicrobial properties be considered pesticides; and (b) if so, how should these products be regulated?

Though it appears that EPA's position is still evolving, the agency's actions thus far seem to suggest that it does believe the presence of nanomaterials may result in a product being considered a pesticide under specific circumstances. First, in 2006 two wastewater utility associations (the National Association of Clean Water Agencies and Tri-TAC) transmitted a letter to EPA expressing their concern about the presence of nanosilver (referred to as silver ions) in washing machines because of the possibility that a certain percentage of silver ions would go through the sewage system and be discharged into waterways, killing plankton and thus

---

265. International Center for Technology Assessment (ICTA), *Petition for Rulemaking Requesting EPA Regulate Nano-Silver as Pesticides* (February 2008).

undermining the food chain.[266] The associations requested that silver ions be considered a pesticide and be subject to registration requirements. The washing machines were built by Samsung, which used silver ions to sterilize clothes during the wash cycle without the need for hot water or bleach.[267] From Samsung's perspective, the process was an energy saver for consumers and not environmentally harmful because free silver ions in the wastewater would bind to organic material and become inactive.[268]

Initially, it was reported that EPA had concluded the washing machines would be considered "devices," and as such the silver ions used in them would not need to be registered as a pesticide.[269] Then, it was reported that EPA had reconsidered its decision and concluded that, in fact, the washing machines were not devices and that the agency was planning on announcing that companies using nanosilver and making pesticidal claims had to register their products.[270] However, the ultimate notice issued by EPA did not address nanomaterials but rather "silver ion generating equipment," noting that such equipment would be regulated as a pesticide. Given the expectations that media reports had set for EPA's actions, the actual notice was disappointing to consumer and environmental groups. However, the agency's position was similar to the one it had taken with products derived from biotechnology using its case-by-case approach.

In February 2008, EPA Region 9 entered into a consent decree with a California company whereby the company agreed to pay $208,000 because it sold or distributed four components of a computer (e.g., mouse, keyboard, etc.) that had been nanocoated, and health and pesticidal claims (both implicit and explicit) had been made without the four products being registered as pesticides.[271] The brief consent decree does not lay out in particular what claims were made, what was the nature of the nanocoating, or why the decision was made to classify that nanomaterial as a pesticide. Nonetheless, the consent decree provided precedential support for the arguments made by a coalition of environmental and consumer groups that nanosilver meets the definition of pesticide and that it should be regulated as a "new pesticide," with applicants having to

---

266. P. Phibbs & T. Baltz, Pesticides: EPA Examining Use of Nanosilver in Washing Machines as Possible Pesticide, Daily Environmental Report (May 15, 2006) at A-5–6.

267. See id.

268. See id.

269. J. Kinney, *Pesticides: EPA to Regulate Nanoscale Silver Used in Washing Machines to Kill Bacteria*, Daily Environmental Report (Nov. 21, 2006) at A-3–4. A device is not required to be registered. 40 C.F.R. 152.500(b). A *device* is defined as "any instrument or contrivance . . . intended for trapping, destroying, repelling, or mitigating any pest or any other form of plant or animal life . . . , but not including equipment used for the application of pesticides." 40 C.F.R. § 152.500(a).

270. ICTA, *supra* note 266, at 21–22.

271. In re ATEN Technology Inc., d/b/a IOGEAR, Inc. Docket No. FIFRA 09-2008-0003 (filed Feb. 27, 2008).

provide toxicological, bioaccumulation, non-target species impact data, and other information specifically relating to nanosilver rather than relying on conventional silver data. The lead nonprofit group in this coalition is the International Center for Technology Assessment, the same organization that is the lead nonprofit in filing a petition that FDA reconsider the way it is regulating nanoscale titanium oxide and zinc oxide (see Chapter 8).

The petition outlines the arguments why nanosilver (and more generally any nanoscale product that has antimicrobial and antibacterial properties) should be considered a pesticide. As noted above, it appears EPA seems to recognize that certain products using nanoscale active ingredients should be considered pesticides. However, central to the petitioners' argument is that EPA has focused only on products whose seller or distributor makes explicit claims about antimicrobial properties. They claim the result is that companies remove any explicit labels or marketing information about pesticidal properties, and thus they have a loophole for placing products in the marketplace. The petitioners call upon EPA to clarify that pesticidal intent can be demonstrated through various means, and that removal of explicit labels will not mean that EPA exempts these products from registration requirements.

The second aspect of their argument is the nanosilver should be regulated as a "new pesticide" that is not covered under the bulk silver registrations. This argument is based on distinct risks to human health and environment that nanomaterials may have in comparison to their conventionally sized counterparts (see discussion in Chapter 3). Moreover, because of these risks, the petitioners call upon EPA to employ a new toxicity testing and risk assessment that is specific to nanoscale materials. Interestingly, in addition to using scientific studies to support their argument, the petitioners point to patent filings to argue that the companies themselves view these products as being different from their bulk counterparts. On November 19, 2008, EPA opened a 60-day comment period on this petition.[272]

The agency's experience with PIPs and genetically modified organisms may serve as potential guideposts for what EPA may do. For instance, as with genetically modified organisms, EPA may seek to address the lack of knowledge about human health or environmental effects of particular nano-pesticide applications by asking the parties to submit a notification before undertaking their particular tests (something akin to requiring notification in order to see if an EUP is necessary) so as determine whether the proposed tests would sufficiently address the registration standard that there be no unreasonable adverse effects. Additionally, even if EPA required registration for nanosilver or other nanoscale substances that are intended for pesticidal usage, there are exemptions from FIFRA registration requirements that may be applicable to particular products. For example,

---

272. 73 Fed. Reg. 69644 (Nov. 19, 2008).

products that are "treated" with antimicrobial pesticides are exempt, provided that pesticide is added only to protect the article itself and the pesticide is "registered" for such use. Accordingly, products such as the nanocoated computer accessories discussed above could take advantage of such an exemption. In addition, there is an exemption for products that are intended for use only against microorganisms, bacteria, or fungi living in or on a human or animal and that are labeled accordingly—for example, lotions used to treat athlete's foot.[273] Thus, despite the petitioners attempt to cover a broad range of consumer products, the application of any final rule may be more limited.

IX. FEDERAL REGULATION OF TRANSGENIC ANIMALS

Genetic engineering of animals, such as mice, was achieved in the early 1980s. However, today, more species, especially those that are used for food or food products, can be genetically engineered. Specifically, animals may be genetically altered, among other reasons, to have enhanced quality traits, to produce pharmaceuticals, to express human genetic sequences in order that drugs can be tested, and to produce industrial or consumer products.[274]

Typically, in order to create a genetically engineered or transgenic animal, the gene of interest is injected into a single-cell embryo. A more recent and more efficient technique is to put the gene into a skin cell and create an embryo from that cell by cloning. In both cases, the embryo with the foreign gene is then implanted into the womb of a surrogate mother. After some transgenic animals are born, additional animals can be made by conventional breeding because the foreign gene generally will be passed on to some of the offspring as would any other gene.[275]

The regulatory agency with the primary responsibility for transgenic animals is the Food and Drug Administration, specifically the Center for Veterinary Medicine (CVM). Until late 2008, although FDA had indicated its intention to regulate these animals under the animal drug laws, it had not issued any guidance document or interpretative statement as to how it believed the existing regulations would be applied. However, FDA has now issued a final guidance document.

---

273. EPA, Office of Pesticide Programs, "Chapter 2: What is a Pesticide?" Label Review Manual (Dec. 2006) at 4.

274. FDA, "Guidance for Industry: Regulation of Genetically Engineered Animals Containing Heritable Recombinant DNA Constructs," (January 2009) at 3. While the section will mainly discuss transgenic animals in the context of their being raised to provide foods, transgenic animals can also be used in the medical field. FDA is planning on issuing additional guidance on "biopharm animals."

275. A. Pollack, *Without U.S. Rules, Biotech Food Lacks Investors*, N.Y. Times, July 30, 2007; J. Kuzma & P. VerHage, *Nanotechnology in Agriculture and Food Production: Anticipated Production, Project on Emerging Nanotechnologies*, Woodrow Wilson International Center (Sept. 2006) at 15.

To date, no transgenic animal for human consumption has been approved, but an application has been made to the agency for approval of a transgenic salmon. The facts from that submission provide additional insight as to the way transgenic animals may be regulated.

Finally, it is believed that the USDA's FSIS also has jurisdiction over transgenic animals. Although the USDA issued a 1994 guidance document, the announcement of FDA's draft guidance document has now increased pressure on the USDA to also announce its policy positions that are consistent with the draft FDA guidance.

A. Jurisdiction and Role of FDA under the Food, Drug, and Cosmetic Act

FDA issued a draft guidance document in September 2008 entitled "Regulation of Genetically Engineered Animals Containing Hereditable rDNA Constructs" and called for a 60-day public review and comment period. FDA received 28,000 comments on the draft guidance. However, most of those comments were "form letters" that expressed support or opposition to the use of genetic engineering generally and to genetic engineering of animals in particular.[276] A final guidance document referred by the same name was issued by January 2009. The final guidance document addresses only the regulation of transgenic animals—not cloned animals, which is discussed in Chapter 6. Moreover, it should be noted that while the guidance document only refers to animals with heritable rDNA constructs, animals can also possess non-heritable rDNA constructs (e.g., those modifications intended to be used for gene therapy). FDA may issue a separate guidance for these animals, but it intends to regulate them in a similar manner.

Under the draft guidance, other than a few exceptions, FDA requires that the production of transgenic animals be in compliance with the new animal drug review process.

The Federal Food, Drug, and Cosmetic Act (FFDCA) prohibits the introduction into interstate commerce of new animal drugs that are not the subject of an approved new animal drug application (NADA) or investigational new animal drug (INAD).[277] The term *new animal drug* refers to "any *drug* intended for use for animals other than man," that is not recognized by qualified experts as being safe and effective for use under the conditions prescribed, recommended or suggested in the drug's labelling, and that has not been used to a material extent for a material time.[278] Section 201 of the FFDCA broadly defines a *drug* as any "article" intended "to affect the structure or function of the body of man or animal."

---

276. 74 Fed. Reg. 3057 (Jan. 9, 2009). *See also*, FDA, "Transcript of Media Briefing on FDA's Release of a Final Guidance for Industry on the Regulation of Genetically-Engineered Animals," (Jan. 15, 2009).

277. A copy of the form is *available at* http://www.fda.gov/opacom/morechoices/fdaforms/FDA-356v.pdf.

278. *See* 21 U.S.C. § 321(v).

Because genetically modified animals are typically the creation of insertions of a foreign gene, and because the inserted gene is intended to affect the animal's structure or function, the foreign gene is considered a drug. Thus, the animal is not a drug, but the animal contains a "drug." Moreover, the genetically modified embryo that is inserted into a surrogate mother could also be labeled as a drug.[279]

In order to distribute a new animal drug, its sponsor[280] must obtain approval for a new animal drug application, unless the drug is for only investigational use, in which case, a INAD has to be obtained. Otherwise, the new animal drug will be considered unsafe for the purposes of the adulteration standards.[281] However, according to the final guidance document, generally neither an investigation use permit nor a permit to market is necessary for (a) genetically engineered laboratory animals, or (b) animals of nonfood species that are regulated by other agencies (e.g., genetically engineered insects for pest control). Additionally, on a case-by-case basis FDA may exempt other nonfood species based upon the likelihood of risk, including environmental risks (e.g., glow-in- the-dark zebra fish). In all of these cases, FDA will retain its discretion to take enforcement action if it learns that safety concerns are associated with the transgenic animal. In determining when to exercise this discretion, FDA will take into account environmental risks, such as if the animal were to be released would the potential danger to the environment be greater.

1. **Investigational Use** Before conducting clinical investigation on genetically engineered animals, a sponsor must submit an INAD application[282] and commit to meet FDA's use-testing requirements as set forth in 21 C.F.R. Part 511. As noted in the final guidance document, it is recommended that prior to submitting an INAD, the sponsor should have a pre-submission review and conference with the CVM staff.[283] The purpose of the pre-submission conference is for the CVM to discuss the contents of the INAD. FDA strongly recommends that the sponsor

---

279. *See, supra* note 53, at 112.

280. As discussed in further detail in Chapter 7 in the context of human drug trials, a sponsor is the person (*e.g.*, corporation, individual, partnership) responsible for the investigation, including responsibility for compliance with applicable provisions of the FFDCA and the implementing regulations. A sponsor typically enters into a contractual relationship with an investigator—the person who is actually conducting the tests, in order to ensure that FDA's requirements are complied with.

281. *See* 21 U.S.C. § 360b(a)(1).

282. The form is officially titled as "Notice of Claimed Investigational Exemption for a New Animal Drug" (Form 3458) and is available at www.fda.gov/cvm/Documents/Form_3458_NCIE.pdf.

283. The person or firm distributing or causing the distribution of the new animal drug or animal feed containing the new animal drug shall use due diligence to assure that the new animal drug will be used on animals and not in humans. *See* 21 C.F.R. § 511.1(b)(2). Moreover, the transporter also has an obligation to maintain certain specified records for a two-year period of time. *See* 21 C.F.R. § 511.1(b)(3).

communicate with it early in the process of developing the transgenic animal. Moreover, the applicant should contact FDA after the INAD has been submitted in order to discuss the sponsor's specific regulatory obligations under the INAD.

The INAD submission should contain information about the animal, the introduced gene(s), the intent of the modification, and any gene product that may be produced. FDA is also likely to discuss: (a) the movement of animals during the investigational phase, and the need for labels that specifically note that edible products derived from the investigational phase may not actually be consumed without prior FDA approval; and (b) a plan for disposition of the animals once the investigational phase is completed (e.g., incineration, burial, or composting) to ensure the animal does not mistakenly enter the food supply. A party must also consult with FDA if it wants to introduce an animal into the food or feed supply as well as consult with the USDA's FSIS to deal with any of their concerns regarding the transgenic animal entering the food supply.

While not specifically addressed in the guidance, it appears the standard reporting and record-keeping requirements must be followed unless specific conditions are set out in the permit. Moreover, as with any INAD, FDA may terminate the permit if the sponsor fails to notify the agency of evidence of a significant hazard related to the safety of the drug, attempts to commercially distribute or test market the new animal drug being studied under an INAD, or represents that the drug is safe and effective.[284] Additionally, if FDA determines that continuance of the investigation is unsafe or otherwise contrary to the public interest, or that the drug is being or has been used for purposes other than bona fide scientific investigation, the agency will provide the sponsor with an opportunity to immediately correct the deficiencies—or the sponsor can seek a regulatory hearing on FDA's determination.[285] If the INAD authorization is terminated, the sponsor will have to recall or have destroyed the unused supplies of the new animal drug.

FDA also monitors the activities of the investigator.[286] If FDA has information indicating that an investigator has repeatedly or deliberately failed to comply with the INAD requirements or has submitted false information either to the sponsor or in any report, the CVM will furnish the investigator with a written notice and offer him an opportunity to explain the matter in an informal meeting and/or in writing.[287] If the explanation is not satisfactory to the regulatory

---

284. 21 C.F.R. § 511.1(d).

285. *Id.*

286. The sponsor also has an obligation to assure himself that the investigator is qualified to conduct the study and evaluate the safety and/or effectiveness of the drug, that the investigator will furnish adequate and timely reports about the investigation, and that investigator will maintain complete records of the investigation for two years after completion of the investigation. 21 C.F.R. § 511.1(b)(7).

287. 21 C.F.R. § 511.1(c)(1).

agency, the investigator will have an opportunity for a regulatory hearing. If after the hearing, FDA's initial allegations are upheld, FDA will notify the sponsor that the investigator is not entitled to investigational use of new animal drugs. Once the investigator has been stripped of his ability to receive the drugs, FDA will examine if the investigator submitted unreliable data.[288] If FDA concludes that after the unreliable data is eliminated from consideration, the remaining data is inadequate to support a conclusion that it is reasonably safe to continue the investigation, the sponsor will have an opportunity for a regulatory hearing on whether the INAD authorization should be terminated.[289] But if FDA determines that there is danger to public health, it can terminate the INAD authorization immediately and have a post-termination regulatory hearing on whether the authorization should be reinstated.[290]

Therefore, from the sponsor's perspective, it is critical to use due diligence prior to selecting an investigator and to maintain adequate controls on the research so as to know what is being done. Otherwise, the sponsor risks losing some or all of the investment made in development of this new drug.

2. New Animal Drug Application For a new animal drug to be marketed, it is necessary for the sponsor to obtain approval via a new animal drug application (NADA) and to comply with the requirements of 21 C.F.R. Part 514. In order to evaluate transgenic animals under the existing NADA rules, the agency issued recommendations as to what information will be necessary.[291] However, it should be noted that these recommendations are not binding. Nonetheless, it provides a sponsor with an outline as to what data and information needs to be collected.

The sponsor will need to report information on the "product," including information on: (a) the animal, including its ploidy and zyosity; (b) the name, the number of copies, and intended functions of the rDNA construct; (c) the number and characterizations of the insertion site(s), (d) the intended use of the animal, and (e) the name of the animal line and a description of the animal (e.g., genus and species). FDA will also ask for information that will identify and characterize the rDNA construct that will be introduced as well as the method by which the rDNA

---

288. 21 C.F.R. § 511.1(c)(3).

289. 21 C.F.R. § 511.1(c)(4).

290. *Id.*

291. A NADA requires (a) full reports of investigations that have been made to show whether or not such drug is safe and effective for use; (b) a full list of articles used as components of the drug; (c) a full statement on the composition of the drug; (d) a full description of the methods used in, and the facilities and controls used for, the manufacture, processing, and packing of the new animal drug; (e) a representative sample, if determined to be necessary by FDA, as well as full information regarding each sample's identity and the origin of any new animal drug, and detailed results of all laboratory tests to determine the identity, strength, quality, and purity of the batch represented by the sample; and (f) copies of proposed labels.

construct will be introduced into the initial animal. Data and information will need to be provided on the health and physiological status of the genetically engineered animal. The sponsor would need to conduct a genotypic and phenotypic durability assessment to demonstrate that the rDNA construct is stably inherited and that there is a consistency of the expressed trait over multiple generations. Finally, the sponsor will need to demonstrate food/feed safety (i.e., examining the direct and indirect toxicity risks) as well as undertake an environmental assessment (see above NEPA).

Procedurally, within 180 days of a complete application, FDA has a statutory obligation to: (a) approve it if there is no basis for rejection (in which case the FDA Commissioner shall concurrently place a notice in the Federal Register indicating the approval), (b) deny the petition, or (c) provide a tentative response indicating why the agency has been unable to reach a decision on the petition (e.g., because of the existence of other agency priorities or a need for additional information). The tentative response may also indicate the likely ultimate agency response and may specify when a final response will be furnished.[292] The applicant can petition for reconsideration and ultimately for a judicial review.[293]

Because the statutory clock does not run until the application is deemed complete, the typical approval process for a new animal drug is much longer than 180 days. The more usual situation is for FDA to provide written comments to the applicant noting the deficiencies in the application and requesting that they be addressed.[294] For example, the application for the only genetically modified new animal drug approved, which is for recombinant bovine somatotropin (see discussion in Chapter 6), was under review for nearly six years.[295]

In addition to providing a list of deficiencies regarding the application, FDA may notify the applicant that a site inspection is necessary and/or that samples of the drug, edible tissues, and by-products of animals treated with the drug must be provided.[296] The failure to comply with these requests within the requisite time frames will result in either the application being deemed withdrawn or insufficient.[297]

Once the NADA is approved, the Federal Register notice will contain the name and address of the applicant and the conditions and indications of use of the new animal drug.[298]

---

292. 21 C.F.R. § 10.30.
293. 21 U.S.C. § 360b(c)(1).
294. Pew, *supra* note 53, at 109.
295. *Id.*
296. 21 C.F.R. § 514.100(b).
297. 21 C.F.R. § 514.100(c).
298. 21 U.S.C. § 360b(i).

3. Confidentiality of NADA and INAD Neither the draft nor the final guidance documents discuss how the new animal drug regime's confidentiality provisions operate with respect to transgenic animals—and they are controversial. Under the existing regulations, until FDA has published in the Federal Register its approval or the sponsor has made public or acknowledged the existence of an INAD or NADA, FDA will neither disclose the existence of a NADA or INAD file nor will it make publicly available the data or information in the respective applications.[299] If the sponsor discloses or acknowledges a NADA or INAD file before the publication in the Federal Register notice, the FDA Commissioner may, at his discretion, disclose a summary of the selected portions of the safety and effectiveness data as is appropriate for the public to consider any pending issues.

After the Federal Register notice is published, except for extraordinary circumstances, all safety and effectiveness data will be disclosed to any person who makes a *request* for such information. Thus, only if a party makes a request will the information be disclosed. The party making the request must submit a verified statement that the documents regarding safety and effectiveness will not be used to make, market, or use outside the United States the drug for which the data was submitted. Additionally, if the information is to be shared with other parties, then each party that receives this information must make an identical verified statement.

As far as the general public in concerned, after the Federal Register notice is published, FDA may make publicly available: (a) a summary of the safety and effectiveness data submitted or incorporated by reference in the file (but not full reports); (b) a protocol for a test or study; (c) adverse reaction reports, product experience reports, consumer complaints, and similar documents; (d) a list of all active and inactive ingredients that have been previously disclosed to the public; (e) an assay method or other analytical methods that are not subject to trade secret or confidential commercial information protection; and (f) all correspondence that is similarly exempted from trade secret protection.[300] If at that time there are objections to FDA's decision, a citizen may file a petition requesting reconsideration.

However, these confidentiality provisions come into conflict with the National Environmental Procedures Act (NEPA) requirements, which are applicable to new animal drug permits. The NEPA analysis is part of the INAD and the NADA process. As noted above, FDA will not disclose the contents of a INAD or NADA file until the publication of the Federal Register notice. As a result, the NEPA analysis

---

299. 21 C.F.R. § 514.11(b)–(c).

300. 21 C.F.R. § 514.11(e)(1)–(7). The regulations also note that (a) manufacturing methods and processes (e.g., quality control); (b) production, sales, distribution, and similar data; and (c) quantitative or semi-quantitative formulae are not be disclosed unless this information been previously disclosed to the public or it relates to a product or ingredient that has been abandoned and no longer represents either a trade secret or confidential commercial or financial information. 21 C.F.R. § 514.11(g).

will not be disclosed until after the INAD or NADA approval decision has been made. As discussed above, NEPA is a procedural rather than substantive act, which is intended to ensure that the agency considers environmental impacts prior to granting a permit, providing funding, or taking other regulatory action. But central to the way NEPA works is that the public be involved in evaluating the alternatives, cumulative effects, and related social and economic impacts. This is especially the case with transgenic animals because key concerns surround their impact on the ecological community in which they will exist, including whether they will outcompete other species for prey, whether they will breed with wild relatives or nongenetically modified versions, and whether practical and adequate measures have been taken to prevent adverse impacts (e.g., the disposal practices of such animals). For example, in addressing hybridization, questions would be raised as to whether the transgenic animal can be made sterile or whether the animal could be raised in confined conditions. Moreover, unlike Bt crops in which Bt had a long history of use, neither FDA nor any other agency is likely to have an extensive familiarity with the transgenic animal, and thus information from all sources would be necessary, especially to preserve public confidence in the safety of the transgenic animal. The case study below of the NADA application for transgenic salmon explores these issues further. A key component of NEPA, however, is that the public is made aware of the agency's evaluation and has an opportunity to review and comment on the agency's determination. The analysis the agency performs is whether the application has adequately been considered. However, this analysis is not disclosed to the public until after the decision on the application.[301] FDA has indicated that it is looking into ways to make the process more transparent and may use the advisory committee meetings as a means of making the public aware.

B. Transgenic Salmon Case Study

In 1999, Aqua Bounty Farms, a subsidiary of A/F Protein Corp., began to seek approval for the marketing of a transgenic salmon. As of 2008, the Aqua Bounty is still awaiting FDA approval for its salmon. As noted above, typically, the submission of an INAD or NADA is held confidential. However, because the company disclosed the filing of the INAD, the public has been made aware of the submission. The following is a summary of some of the key elements that have been discussed about that application, and more generally about transgenic fish. It is likely given the increased demand for seafood, the collapse of many commercial stocks, and the presence of industrial-scale aquacultures that transgenic fish will be one of the first vehicles for FDA consideration of a transgenic food animal.

As a general matter, transgenic fish are expected to contain at least one introduced structural gene for growth hormone and one introduced regulatory

---

301. Office of Science and Technology Policy (OSTP), Case Study No. 1, Growth Enhanced Salmon, at 30.

sequence for the control and expression of the introduced structural gene, thereby eliciting the phenotype of enhanced growth rate and feed efficiency.[302] The growth hormones that could be employed are not limited to genes from other fish species and include those from any other animal.

Aqua Bounty proposes to produce an Atlantic salmon that can grow to market weight size 6 to 12 months before their natural counterparts (18 months versus 24 to 30 months). The specific genetic modification involves taking a growth hormone gene from Chinook salmon as well as a second gene from an ocean pout, a distant relative of salmon, that functions as an "on-switch" to keep the first gene constantly producing is hormone.[303] Then, all subsequent generations of the fish will contain these modifications. From the company's perspective, there are certain economic benefits associated with raising transgenic fish, including each fish requiring less feed to grow to the appropriate commercial size. There could also be an increase in the number of cycles at which fish could grow at a particular location as the size requirement can be reached at an earlier date for these fish.

To obtain approval to market the transgenic salmon as a new animal drug, the company will have to demonstrate the safety and effectiveness of the transgenic animals. The safety determination encompasses the safety of the transgenic animal itself to humans who will eat the food and to humans and other animals in the environment. For example, the company would have to demonstrate that the genetic modifications will not trigger an increased production of new allergens or toxins and that these genetic modifications are stable and consistently expressed in the fish.[304] As noted above, a number of studies and investigations are needed to support a NADA, and thus transgenic fish will need to be investigated not only in controlled laboratory settings but in "real life" situations.

In evaluating the safety of the drug, FDA has the authority to examine the environmental impacts, both direct and indirect, that would affect human health or animal health as a result of the use of the product.[305] With respect to the

---

302. See id. at 2.

303. C. K. Yoon, *Redesigning Nature: A Special Report; Altered Salmon Leading Way to Dinner Plates, but Rules Lag*, N. Y. TIMES, May 1, 2000.

304. See id.

305. For example, the National Marine Fisheries Service (NMFS) and the Fish and Wildlife Service have, among other things, "considerable knowledge of reproductive sterilization techniques that might be used to mitigate the interbreeding of escapees with wild stock." OSTP, *supra* note 299, at 31.

It should also be recognized that there is an extensive set of statutes and regulations that govern the construction and operation of an aquaculture facility in the waters of the United States. Specifically, the U.S. Army Corps of Engineers has regulatory authority through Section 10 of the Rivers and Harbors Act and sections 401 and 404 of the Clean Water Act. Additionally, EPA requires a NPDES permit for discharges from such facilities. The NMFS may also evaluate the monitoring conducted by the project sponsor with

production of transgenic salmon, given the similarities in the production process to farming salmon, many of the issues relating generally to aquaculture will be raised. These environmental issues include the pollution of the seabed or other waters with the fecal matter and excess feed from net pens; the spread of bacteria, viruses, and parasites (e.g., sea lice); and the introduction of chemicals to address such diseases.[306] However, in addition the question arises whether production of transgenic fish would exacerbate any of these conditions. On the other hand, advocates of transgenic salmon have noted that due to feed use efficiency, less feed would be thrown into pens and less pollution would be created from wasted feed.

Of significant concern for transgenic salmon is the impact on the wild salmon population if the transgenic fish escape their enclosure. The concern is that due to their size advantage, transgenic salmon will be able to breed more success-fully and thus eliminate the wild population. In an attempt to address this par-ticular concern, Aqua Bounty has stipulated that only sterile, all-female salmon will be introduced into the net pens.[307] It is expected that "brood stocks of such fish would be raised in conventional inland hatcheries, where brood stock would be treated to produce 100% genetically female eggs. The eggs would then be treated to cause reproductive sterility. The reproductive sterile, all-female off-spring would be grown initially in hatcheries and then to maturity in ocean net pens, before being harvested for food."[308] However, opponents of transgenic fish challenge the assertion that any technique has been shown to be 100 percent effective, and that an examination of each individual fish to guarantee certainty is neither economically nor practically viable.

C. Jurisdiction and Role of USDA under Inspection Acts

Protection of the public from unwholesome, adulterated, or misbranded meat and meat products, poultry, and eggs falls within the jurisdiction of the USDA's FSIS. Thus, if a genetic modification is made to an animal that is ultimately intended to be part of the food supply, the FSIS has a role. To date, no genetically modified animals have been approved for the food supply.[309]

---

respect to the adverse impacts of the aquaculture activities, including the escapes and water quality. *See id.* at 32–33.

306. *See id.* at 7.

307. *See id.* at 4.

308. *See id.* at 1.

309. The FSIS may also have a role with regulating meats from cloned animals. As FDA's Center for Veterinary Medicine notes, the FSIS will be consulted on its decision on how to regulate meats from cloned animals if safety becomes an issue. *See* Center for Veterinary Medicine (Food & Drug Administration), *Animal Cloning: Risk Management Plan for Clones and Their Progeny* (Jan. 15, 2008), *available at* www.fda.gov/cvm/CloningRA_RiskMngt.htm.

The FSIS derives its authority from three statutes: the Federal Meat Inspection Act, 21 U.S.C. § 601 et seq., the Poultry Products Inspection Act, 21 U.S.C. § 451 et seq., and the Egg Products Inspection Act, 21 U.S.C. § 1031 et seq. Each of these statutes was enacted in the early 1900s as part of the Progressive Era's attempt to address the highly publicized reports about the conditions in slaughterhouses, the concerns they generated about the safety of foods being consumed, and the economic ramifications if consumers did not trust their food supply.

The mandate of the FSIS it to protect the public from unwholesome, adulterated, misbranded, or otherwise unfit-for-human-consumption products.[310] The primary tools the FSIS uses to ensure compliance with its standards are inspections and labeling. In 1994, the FSIS issued a guidance document indicating that it intended to regulate transgenic animals under the same standards as it has applied to nontransgenic animals. With the issuance of the final guidance from FDA, there may be a need to further elaborate on exactly how transgenic animals will be regulated.

Among the issues that will need to be addressed is the adequacy of measures to ensure the proper segregation and disposal of transgenic animals. The FSIS already has standards for examining animals prior to slaughter, segregating animals and separating animals at slaughter that are diseased from those that are not, conducting postmortem examination of animals in order to label them as being approved or condemned, and determining the establishment's compliance with these requirements.[311] These regulations would need to be altered to ensure that transgenic animals are not comingled with nontransgenic animals.

The issue of labeling meats derived from genetically engineered animals will come to the forefront once a specific food animal has been approved. Under regulations administered by the FSIS, all ingredients used to formulate a meat product must be declared in the ingredients statement on product labeling.[312] Under the statute, a product is misbranded when it contains ingredients that are permitted but not declared on product labeling.[313] As discussed in Chapter 6, under FDA's rules, the mere fact that a food is made from a transgenic plant does not mean that labeling is needed identifying it as such. FDA has reiterated that this position is equally applicable to meat from transgenic animals. Rather, under FDA's construct, labeling is only necessary if there are material differences between the genetically engineered product and its non-genetically engineered counterpart (e.g., the transgenic animal is not nutritionally equivalent or poses a greater threat of an allergic reaction). However, how it remains to be seen how both FDA and the FSIS will have to react to public opinion, which may result in a reconsideration of their currently stated positions.

---

310. 21 U.S.C. § 602.
311. *See* 21 U.S.C. § 603, 605.
312. *See* 21 U.S.C. § 607.
313. *See* 21 U.S.C. § 601(n)(7).

6. REGULATION OF FOOD

The production and manufacture of genetically modified (GM) foods has been a controversial issue both in the United States and abroad, particularly Europe. Advocates on both sides passionately argue about the benefits and risks associated with GM foods. There have been mass demonstrations against GM foods, and political parties in Europe have adopted platforms specifically addressing genetic modification. Thus, the debate about whether genetically modified foods are safe—or even how they are or should be regulated—is not based simply on data and scientific conclusions, but on a mix of political and economic calculations, public opinions, media attention, and a host of other factors. This debate is of significance to manufacturers of nanomaterials, especially those who are involved in placing nanomaterials in food or feed, as well as those who are involved in placing nanomaterials in food packaging that comes in contact with food, because it will affect how such materials may be regulated in the near future.

The legal framework that governs GM foods in the United States and Europe is a manifestation of underlying—and constantly shifting—political, social, and economic dynamics. Moreover, laws may be laggard indicators of where society is headed. Therefore, this chapter will first discuss the issues in the debate between advocates and critics of GM foods, then the applicable laws and guidance documents so that practitioners will know their compliance obligations. Throughout these sections there will be references to the underlying political and social factors that were influencing law or guidance at the time of adoption. Finally, this chapter will close by discussing two emerging technologies: foods derived from cloned animals or their progeny, and foods that have been infused with nanomaterials.

I. THE DEBATE BETWEEN ADVOCATES FOR GM FOOD AND CRITICS OF "FRANKENFOODS"

The debate between those who advocate for GM foods and those who are critical of their introduction is based in part on how each side measures the costs and benefits associated with this technology. In many ways, the term *Frankenfoods*, which was coined by opponents and is popularly used, encapsulates the critics' perspective—namely, their fear that man's hubris will lead to the creation of new species of plants or animals that will disturb the natural order and lead to deadly consequences. In viewing GM foods in this manner, certain critics have advocated the use of the precautionary principle to prohibit their development.

Others have pressed for more detailed and extensive scientific testing and labeling requirements. Proponents, on the other hand, view increased demands for scientific data as a barrier to bringing products to market and as a means of increasing the costs of production (and thus, placing these foods at a competitive cost disadvantage). Furthermore, they view labels as a means of distorting public perceptions in order to create (from their perspective) false distinctions between products that are equally safe.

This debate over GM foods has played out on many fronts—health, ecological, economic, and social. Depending on the time, the product, and the society, different arguments have gained greater or lesser resonance. As these issues have already been discussed in Chapter 3, this section will only briefly summarize them:

On the human health front, proponents note the potential for the development of more nutritious foods (e.g., Golden Rice) that would be of particular benefit to those in poorer societies who have a more limited diet. They also discount the fears about GM foods creating antibiotic resistance among the bacteria in the human digestive system or an allergic response in those who are sensitive, noting the lack of evidence of actual adverse reactions. On the other hand, critics note that testing and monitoring methodologies may not account for, among other things, accumulated impacts and persistence in the environment that may result in long-term exposure. They also point out that there are other means of improving the diet of those in the developing world that are sustainable and cost-effective.

On the ecological front, proponents have argued that growing GM foods is not inherently risky and that safeguards have been instituted to minimize any risks (see Chapter 5 for a discussion of buffer areas under grower agreements). In contrast, critics point out the potential mechanisms by which the ecosystem is being altered—hybridization with wild or weedy relative plants, impacts on the soil microorganisms, and dispersal outside the prescribed growing areas (e.g., using the Star Link™ incident as a real-life counterpoint). Moreover, critics note some impacts may still be unknown, but with further research, these impacts may be discovered.

On the economic and social front, a debate has raged whether the introduction of genetically modified foods is a boon for societies, especially developing ones, by increasing crop yields, reducing the need for pesticides, and potentially improving the nutritional quality of the food. Alternatively, concerns have been raised that farmers will become dependent on a small group of foreign multinational corporations for seeds rather than relying on local sources and that local varieties of a crop will be lost in favor of monoculture agriculture.

As discussed in Chapters 3 and 4, public policy is shaped not simply by scientific determinations of risk, but by public opinion. As the narrative in this chapter will demonstrate, this is no truer than in the context of genetically modified foods (also referred to in the literature and common discussion as *bioengineered foods*).

II. UNITED STATES: JURISDICTION OVER FOODS

Before the chapter focuses on rules and guidance documents relating to genetically modified foods, a brief primer is necessary on the way FDA regulates foods and food additives. FDA derives its regulatory authority over foods and food additives (genetically modified or not) from the Federal Food, Drug, and Cosmetic Act of 1938 (FFDCA).[1] The statute's primary purpose is to prohibit the introduction, delivery, manufacture, or sale of adulterated or misbranded food into interstate commerce.[2] As the date on the statute indicates, the law was originally enacted at a time when GM foods or food additives created through the use of rDNA technology were not even within the realm of possibility. However, the touchstones for regulating GM foods and food additives remain within the adulteration and misbranding standards.[3]

A. Adulteration Standard

The adulteration standard is the most frequently used part of the FFDCA for the regulation of contaminants in foods.[4] A food is deemed to be adulterated if, among other things, it bears or contains any poisonous or deleterious substance that may render it injurious to health or if it contains a *food additive* that is determined to be unsafe.[5]

---

1. The historical antecedent to the 1938 Act was the Federal Food and Drug Act of 1906, a Progressive Era statute that prohibited commerce in adulterated and misbranded food. The law was flawed because it required the government to prove that a substance added to food was dangerous and that the presence of this added substance rendered the food itself dangerous; otherwise, the product remained in the marketplace. The 1906 statute was repealed by the 1938 act (a New Deal Era statute) that expanded federal powers, including shifting the burden on safety from the government to the developer of the food or food additive.

2. The term *food* is defined as: "(1) articles used for food or drink for man or other animals; (2) chewing gum; and (3) articles used for components of any such article." Section 201(f), 21 U.S.C. § 321(f).

3. The statute subjects those who violate these provisions to injunctions, seizures, civil penalties, and potential criminal penalties. Sections 302, 303 and 304; 21 U.S.C. §§ 332, 333, and 334.

4. 57 Fed. Reg. 22984, 22989 (May 29, 1992).

5. Sections 402(a)(1), (a)(2)(C), 21 U.S.C. §§ 342(a)(1), (a)(2)(C). The statute also notes that the following circumstances would lead to a food being considered adulterated: (a) if the food contains or bears an unsafe new animal drug (or conversion product thereof); (b) if it consists in whole or in part of any filthy, putrid, or decomposed substance, or if it is otherwise unfit for food; (c) if it has been prepared, packed, or held under unsanitary conditions whereby it may have become contaminated with filth, or whereby it may have been rendered injurious to health; (d) if it is, in whole or in part, the product of a diseased animal or of an animal which has died otherwise than by slaughter; (e) if its container is composed, in whole or in part, of any poisonous or deleterious substance which may render

The statute defines a *food additive* as any substance that has an intended use that "results or may reasonably be expected to result, directly or indirectly, in its becoming a component or otherwise affecting the characteristics of any food."[6] The statute, however, exempts any substance that is generally recognized as safe (GRAS) based on recognition by experts qualified by scientific training and experience to evaluate its safety under conditions of its intended use.[7] As discussed in further detail below, the fulcrum with respect to food additives turns on whether the substance used as a food additive is GRAS.

As to meeting the adulteration standard:

- Typically, no FDA approval is required prior to marketing a whole food. Such foods can be removed from the marketplace if they can be shown to be "ordinarily injurious to health" or if they have been adulterated to render them injurious to health.[8] As FDA has noted, "most foods derived from plants predate the establishment of national food laws, and the safety of these foods has been accepted based on the extensive use and experience over many years (or even centuries)."[9] As a result, FDA has not found it necessary to conduct routine safety reviews of whole foods derived from plants.
- On the other hand, if a food has been modified using genetic engineering, the food may now be considered a food additive or as GRAS.
- If the substance is considered a food additive, the manufacturer has the statutory obligation to submit a petition to FDA to seek its approval that the food additive is safe prior to marketing.[10] A determination of whether a food additive is safe is based on whether there is technical evidence of safety.[11]
- A GRAS substance is not considered to be a food additive and can be marketed without prior notification to FDA. As outlined below in greater detail, there are two elements involved in deciding whether a substance

---

the contents injurious to health; or (f) if it has been intentionally subjected to radiation, unless the use of the radiation was in conformity with a regulation or an appropriate exemption.

6. Section 201(s), 21 U.S.C. § 321(s). The statute specifically excludes from the definition of *food additive* (a) a pesticide chemical residue in or on a raw agricultural commodity or processed food; (b) a pesticide chemical; (c) a color additive; (d) any substance subject to the Poultry Products Inspection Act (21 U.S.C. § 451 *et seq.*,) or the Meat Inspection Act of March 4, 1907 (34 Stat. 1260), as amended and extended (21 U.S.C. § 71 *et seq.*,); (d) a new animal drug; or (e) an ingredient that is, or intended for use in, a dietary supplement.

7. *See id.* (note that any substance that was used prior to January 1, 1958 will be examined on the basis of its scientific evidence or because of its long history of usage).

8. 21 U.S.C. § 342(a)(1).

9. *See supra* note 4, at 22988.

10. Section 409(b), 21 U.S.C. § 348(b).

11. 62 Fed. Reg. 18938, 18940 (Apr. 17, 1997) (proposed rule).

qualifies as GRAS: (a) technical evidence of safety is generally known and accepted, and (b) common knowledge about the safety of the substance for its intended use.[12]

B. Misbranding and Labeling Provisions

The statute prohibits the introduction into interstate commerce of any food that is "misbranded," and prohibits misbranding of any food already in interstate commerce. Section 403 of the FFDCA states that a food shall be deemed to be misbranded if "its labeling is false or *misleading* in any particular manner."[13] The statute notes that "if an article is alleged to be misbranded because the labeling or advertising is misleading, then in determining whether the labeling or advertising is *misleading* there shall be taken into account, among other things, not only representations made or suggested by statement, word, design or any combination thereof, but also the extent to which the labeling or advertising fails to reveal *material* facts in the light of such representations or *material* with respect to consequences which may result from the use of the article."[14] (emphasis added).

Congress did not provide clear guidance as to what constitutes something material. FDA has noted that differences in performance characteristics (e.g., physical properties, flavor characteristics, functional properties, and shelf life) are material facts under the Act because they bear on the consequences of the use of the product. However, FDA's position is that widespread consumer demand is not material as to whether a label is required. Courts have upheld this position with the distinction that, by itself, consumer demand is not sufficient to require labeling.[15] However, once FDA makes a determination that there are material differences between the product and the product it purports to be, consumer opinion can be used to determine if a label is the appropriate mechanism for disclosing that fact.

Additionally, in the case of food made from two or more ingredients, the Act requires that a label be placed indicating the common or usual name of each ingredient.[16] The statute, however, excludes any "processing aid" or any substance that is an inherent component of food as an ingredient from the labeling requirements. The regulations define a *processing aid* as a substance that is present in finished food at insignificant levels and that has no technical or functional effect in that food. As demonstrated below, the question of what constitutes

---

12. *Id.*

13. 21 U.S.C. § 343(a).

14. 21 U.S.C. § 321(n).

15. Alliance for Bio-Integrity v. Shalala, 116 F.Supp.2d 166, 178–179 (D.D.C. 2000); Stauber v. Shalala, 895 F. Supp. 1178, 1193 (W.D. Wis. 1995).

16. 59 Fed. Reg. 26700, 26709 (May 23, 1994).

a processing aid or inherent component is specifically relevant in the case of genetic engineering.

C. GRAS and Food Additive Petitions

In 1958, Congress enacted the Food Additives Amendment that established, among other things, a procedure for a premarket review and approval by FDA of food additives. The purpose of the amendment was to require producers of food additives to demonstrate with scientific evidence to a reasonable certainty (not an absolute) that no harm would result from the intended use of the additive.[17] Congress, however, recognized that there were some substances whose safety had already been established because of: (a) a long history of use, (b) the very nature of the substance itself, and/or (c) the information generally available to scientists regarding the substance.[18] As noted above, these substances (referred to as "GRAS substances") have been excluded from the definition of food additive, and as such do not have any preclearance requirements. This section will first describe a manufacturer's obligations with respect to GRAS substances, then examine the obligations with respect to food additives.

1. GRAS Requirements and Submissions With respect to GRAS substances, it is not sufficient to demonstrate that the substance is "safe" for its intended use.[19] Rather, a manufacturer must show that there is a consensus—not necessarily uniformity—among experts that the substance is "generally recognized" as safe. To make such a determination, initially a manufacturer should look at FDA's non-exhaustive list of additives that are considered to be GRAS.[20] However, it would be impossible to list of all of the intended uses of a product that are GRAS.[21]

In instances in which the manufacturer has to determine if a particular substance may be classified as GRAS, there are two means of classifying a product: scientific procedure and common knowledge.[22] As to the scientific procedure

---

17. *See supra* note 11, at 18938.

18. *See supra* note 11, at 18939.

19. The FDA defines *safe* as being when "there is a reasonable certainty in the minds of competent scientists that the substance is not harmful under the intended conditions of use. It is impossible in the present state of scientific knowledge to establish with complete certainty the absolute harmlessness of the use of any substance. Safety may be determined by scientific procedures or by general recognition of safety. In determining safety, the following factors shall be considered: (1) the probable consumption of the substance and of any substance formed in or on food because of its use; (2) the cumulative effect of the substance in the diet, taking into account any chemically or pharmacologically related substance or substances in such diet; and (3) safety factors which, in the opinion of experts qualified by scientific training and experience to evaluate the safety of food and food ingredients, are generally recognized as appropriate." 21 C.F.R. § 170.3.

20. The lists can be found at 21 C.F.R. Parts 182, 184, 186, 582, 584 as well as the FDA Web site under the GRAS Notification Program Web page.

21. *See supra* note 11, at 18939.

22. *See supra* note 11, at 18940.

mechanism, the regulations require the same quality and quantity of scientific evidence (e.g., published studies corroborated with unpublished studies) to demonstrate safety as is needed for obtaining a food additive approval (as will be shown later in this chapter).[23] The technical information is usually used to demonstrate that the new substance is substantially equivalent to a substance already in use. The common knowledge mechanism has two conditions that must be met: (a) the data and information for the technical element must be generally available, and (b) there must be a basis to conclude a consensus exists among qualified experts about the safety of the substance for the intended use.[24]

Because the burden falls onto the food manufacturer to determine if the food is indeed GRAS, FDA has traditionally encouraged consultation even though it is not legally required.[25] From FDA's perspective, there is a significant benefit to the developer to engaging in premarket consultation. If the developer makes an independent determination that the substance is GRAS, but FDA subsequently concludes that the substance is not, the agency "can and will take enforcement action to stop distribution of the ingredients and food containing it on the ground that such foods are or contain an unlawful food additive."[26]

When the statute was enacted, manufacturers who determined that a substance was GRAS sought an opinion letter from FDA, who would render an informal opinion as to whether the manufacturer made the appropriate determination.[27] These letters were not binding on the agency and were not available to anyone other than the requestor. This policy was revoked in 1970.[28]

Subsequently, in the 1970s, FDA created a voluntary "affirmation" process whereby a manufacturer could ask for premarket guidance as to whether a substance could be considered GRAS.[29] The GRAS affirmation process involved resource-intensive rulemaking whose benefit to the public was questionable because by the time FDA made a decision on an affirmation petition, the manufacturer had already placed the substance in the marketplace.[30]

In 1997, FDA published a notice that it would not review affirmation petitions, and that it was replacing such petitions with a notification procedure.[31] Though it has not finalized the notification procedure rule, and therefore, any submission under this proposed rule is a voluntary submission, FDA nonetheless has

---

23. 21 C.F.R. § 170.30(b).

24. *See supra* note 11, at 18940.

25. Pew Initiative on Food and Biotechnology, *Guide to U.S. Regulation of Genetically Modified Food and Agricultural Biotechnology Products,* at 20.

26. *See supra* note 4, at 22989.

27. *See supra* note 11, at 18939.

28. *See id.*

29. 21 C.F.R. § 170.35

30. *See supra* note 11, at 18945.

31. *See supra* note 11, at 18938. Any outstanding affirmation petitions were converted into notices. *See id.*

been receiving such notices since 1998.[32] An interim rule has been issued that outlines the content for a notice submission.

If a manufacturer decides to make a submission, the components include information on the notifying party, the common or usual name of the substance that is subject to the GRAS exemption, the conditions of its use, the basis of the GRAS determination (i.e., through scientific procedures or based on common use experience), and a detailed description of the substance (e.g., chemical name, methods of manufacture, and characteristic properties).[33] If the exemption is based on scientific procedure, the notice must include a comprehensive summary of (and citations to) generally available and acceptable scientific data, information, methods, or principles that are being relied upon to establish the substance's safety, as well as a comprehensive summary of any data and information that appear to be inconsistent with a GRAS determination.[34] Similarly, if the exemption is based on common usage, a comprehensive summary of (and citations to) the history of consumption by a significant number of consumers that demonstrate that safety of the substance as well as any information that appears to be inconsistent with a GRAS determination must be included in the notice.[35] In either case, the notifying party must demonstrate that there is consensus among the experts that there is reasonable certainty the substance is not harmful under the intended conditions of use.

In an effort to reduce resources devoted to this process, in contrast to the "affirmation process," FDA does not review—and does not want to see—the raw data. FDA has indicated that it will complete its review within 90 days of receipt of notice. However, given that this is a voluntary submission, waiting for results from FDA is solely at the discretion of the manufacturer. Moreover, any communication to or from the agency will be publicly available—which is something the manufacturer should give significant consideration to before initiating communication with that agency.[36]

2. Food Additive Petition Requirements and Submissions If a developer or FDA determines that a food product is, in fact, a food additive, the manufacturer has a statutory duty to file a premarket petition.[37] Prior to the submission of a petition, FDA recommends a prepetition consultation and provides a guidance document outlining FDA's thinking on this topic. The petition[38] proposes the issuance of a regulation prescribing the conditions under which the additive may be safely used. The petition must contain data and information that supports the

---

32. 70 Fed. Reg. 75009 (Dec. 8, 2005).
33. 21 C.F.R. § 170.36(c) (proposed), *see supra* note 11, at 18961.
34. 21 C.F.R. § 170.36(c)(4)(i) (proposed), *see supra* note 11, at 18961.
35. 21 C.F.R. § 170.36(c)(4)(ii) (proposed), *see supra* note 11, at 18961.
36. 21 C.F.R. § 170.36(f) (proposed), *see supra* note 11, at 18961.
37. Section 409(a), 21 U.S.C. § 348(a).
38. Form FDA-3503 (9/07).

position that the food additive is "safe."[39] FDA's implementing regulations define *safe* as being a situation in which "there is reasonable certainty in the minds of competent scientists that the substances are not harmful under the intended conditions of use."[40] The statute provides that in order to evaluate safety, it is necessary for FDA to examine such relevant factors as the probable consumption of the substance, the cumulative effect of the substance in the diet (taking into account any chemically or pharmacologically related substance in such diet), and any safety factors that would generally be recognized by experts as being acceptable.[41]

The petition will include among other things: (a) the name of and all pertinent information about the food additive (e.g., chemical identity and composition of the food additive; its physical, chemical, and biological properties; the minimum content of the food additive; and identifying by-products); (b) a statement on the proposed use of the additive (e.g., directions, recommendations, and suggestions regarding proposed uses and proposed labels); (c) data establishing intended physical or technical effects that will be produced by the food additive as well as the quantity of the food additive; (d) a description of practical methods to determine the amount of the food additive in the raw, processed, and/or finished food; and (e) full reports of investigations made with respect to the safety of the food additive.[42]

The technical information required in a petition is the same quantity and quality of information as that necessary for a GRAS determination. Generally speaking, there are three types of technical information that are part of the safety evaluation: chemical, toxicological, and environmental. FDA has issued extensive guidance documents addressing each of these categories that applicants should consult prior to conducting studies or submitting a petition.[43] Applicants should also speak with the Office of Food Additive Safety before initiating any

---

39. 21 U.S.C. § 348(b)(1).

40. 21 C.F.R. § 170.3(i).

41. 21 U.S.C. § 348 (c)(5).

42. 21 U.S.C. § 348(b)(2); *see also,* 21 C.F.R. § 171.1(c)(A–H). In addition, the petitioner must address the issues of whether the food additive requires a tolerance for the food additive and whether it involves an environmental assessment or a categorical exclusion under NEPA.

43. For a list of the various guidance documents, *see* http://www.cfsan.fda.gov/~dms/opa-guid.html#cg. For example, with respect to toxicological information, the guidance information includes *Toxicological Principles for the Safety Assessment of Direct Food Additives and Color Additives Used in Food* (also known as Redbook I); *Toxicological Principles for the Safety Assessment of Direct Food Additives and Color Additives: 1993 Draft Redbook II*; *Toxicological Principles for the Safety Assessment of Food Ingredients*: Redbook 2000 (July 7, 2000; updated Oct. 2001, Nov. 2003, Apr. 2004, Feb. 2006, and July 2007) as well as Templates for Reporting Toxicological Data.

studies in order to understand any particular information that will be needed to support the application.

After the petition is submitted, FDA will notify the petitioner within 15 days of its acceptance or rejection, including the reasons for any rejection. The agency will publish a notice in the Federal Register within 30 days of the filing indicating, among other things, the name of the petitioner and a brief description of the proposal in general terms.[44] FDA may request additional information in the course of its evaluation of the petition such as "a full description of the methods used in, and the facilities and controls used for, the production of the food additive, or a sample of the food additive . . . or of the food in which the additive is proposed to be used."[45] The statute provides for a 90-day review period for a response to the petition. However, the time frame can be extended to a maximum of 180 days if additional information is requested.[46]

If FDA views the petition favorably, it will issue a regulation outlining the conditions under which the food additive may be used.[47] Any person adversely affected by such a regulation may file objections with FDA Secretary (specifying the provisions of the order deemed objectionable and the reasonable grounds for the objection) and request a public hearing upon such objections.[48] After the hearing, FDA will issue an order on the objection.[49] Ultimately, within 60 days after the entry of an order on the objection the objecting party may seek judicial review in the U.S. Court of Appeals for the circuit in which the person resides or has his principal place of business, or in the U.S. Court of Appeals for the District of Columbia Circuit.[50]

D. Documents on Genetically Modified Foods

This section will discuss the specific documents that refer to the way GM foods will be evaluated by FDA. These documents have to be understood within the context of the general requirements outlined above. These documents pertain only to foods derived from transgenic plants. However, FDA has indicated that its approach toward foods derived from transgenic animals is likely to be similar to its approach to transgenic plants. As discussed in Chapter 5, FDA still has not

---

44. The following information or data are typically available for public disclosure after the Federal Register notice: all safety and functionality data and information; protocols for tests that are not covered by trade secrets; adverse reaction reports; product experience reports; consumer complaints and other similar data; a list of all ingredients contained in the food additive; and certain assay methods or other analytical methods. 21 C.F.R. § 171.1(h).

45. 21 C.F.R. § 171.1(j).

46. *See id.*

47. M.R. Taylor, *Regulating the Products of Nanotechnology: Does FDA Have the Tools It Needs? Project for Emerging Technologies,* WOODROW WILSON SCHOOL (October 2006) at 36.

48. 21 U.S.C. § 341(f).

49. *See id.*

50. 21 U.S.C. § 341(g).

approved a transgenic animal whose meat would be consumed, and thus, any distinctions that FDA may ultimately decide upon remain to be seen.

I. **FDA's 1992 Policy Statement on Foods Derived from New Plant Varieties** In 1992,[51] FDA published its "Statement of Policy: Foods Derived from New Plant Varieties," which provided the agency's view on how it planned to regulate foods derived from all methods of plant breeding, including varieties developed using genetic modification such as recombinant DNA techniques (rDNA).[52] The impetus for the statement was not only the inquiries from regulated entities, academia, and the public on how rDNA plants would be regulated, but also industry's request that FDA provide appropriate oversight to ensure public confidence in foods produced by these new techniques.[53] In a nutshell, the articulated policy was that most GM foods are presumed to be GRAS; that FDA reserves the right to regulate any rDNA-developed food on case-by-case basis; that GM foods, as a class, are not required to be labeled; and that manufacturers should (but are not required to) to have a premarketing meeting with FDA.[54]

The 1992 statement was in line with the agency's position as outlined in the 1986 Coordinated Policy—namely, that regulation of genetically engineered foods would be based on the characteristics of the food products themselves as opposed to the process by which they were produced. Thus, FDA asserted that genetic modification constitutes the next step along the continuum of traditional hybridization of plants, and that both methods seek to accomplish the same purpose: enhancing agronomic characteristics (e.g., yield, resistance to disease, insects, herbicides, etc.) and quality characteristics (e.g., preservation, nutrition, and flavor).[55] From FDA's perspective, the distinction between traditional breeding techniques and rDNA techniques is that the latter are "more precise, and increase the potential for safe, better-characterized, and more predictable foods."[56]

The primary legal tool for regulating the safety of genetically modified foods is the postmarket authority under the adulteration standard of section 402(a)(1) of FFDCA. As noted above, the burden here falls on the manufacturer to ensure that the foods presented to the customer are not injurious to health. Thus, a manufacturer of a genetically modified food must ensure that no new toxicants arise from and that no existing toxicants increase due to the modification. To assist

---

51. See supra note 4, at 22989 (citing to 51 Fed. Reg. 23302 (June 26, 1986)). The approach was also consistent with the *White House interest in assuring the safe, speedy development of the U.S. biotechnology industry*. D. Kessler (Commissioner of Food and Drug Administration) Letter to Secretary of Health and Human Services re: *FDA Proposed Statement of Policy Clarifying the Regulation of Food Derived From Genetically Modified Plants—Decision* (Mar. 20, 1992).

52. See supra note 4, at 22984.

53. See id.

54. *Alliance for Bio-Integrity, supra* note 15, at 170.

55. See supra note 4, at 22986.

56. See id.

a manufacturer in determining whether the adulteration standard may be triggered, FDA recommended consulting the guidance section in the policy statement (which is summarized below).

A second mechanism of regulatory control is through the application of section 409 of the Act (i.e., submitting a food additive petition or making a GRAS determination). The 1992 policy statement noted that, "it is the *transferred genetic material* and the *intended expression product* or products that could be subject to food additive regulation, if such material or expression products are not GRAS."[57] However, the clear presumption is that most substances that are introduced into food by genetic modification have been safely consumed as food or are substantially similar to such substances.[58] Thus, *transferred genetic material* (nucleic acids) is presumed to be GRAS because nucleic acids are present in all animals and plants used for food by humans.[59] On the other hand, some *intended expression product* (i.e., proteins or substances that are produced by the action of carbohydrates, fats, and oils) might not be GRAS because they are not present in foods in substantially equivalent quantities and type as those present in currently consumed foods.[60]

To assist a manufacturer in making these determinations about adulteration, GRAS, and food additive petitions, FDA outlined the questions and issues that need to be addressed. While the guidance does not identify all safety and nutritional questions that could arise in any given situation, among the issues that need to be considered are: (a) the toxicants known to be characteristic of the host and donor species; (b) the potential that food allergens will be transferred from one food source to another; (c) the concentration and bioavailability of important nutrients for which a food crop is ordinarily consumed; (d) the safety and nutritional value of newly introduced proteins; and (e) the identity, composition, and nutritional value of modified carbohydrates, fats, and oils.[61] The guidance section also provides a series of flowcharts that address particular scenarios that need to be reviewed by the manufacturer prior to making a determination to proceed. These flowcharts provide three endpoints for the manufacturer indicating whether (a) there are no concerns, (b) the new variety is not acceptable, or (c) there is a need to consult with FDA. The necessity to consult with FDA is underscored by the fact that it may be necessary to use nontraditional approaches rather than simply the traditional evaluation of food safety with toxicological tests.[62] For instance, a multidisciplinary approach may be warranted, including

---

57. *See supra* note 4, at 22990.

58. FDA Center for Food Safety and Applied Nutrition, *FDA's Policy for Foods Developed by Biotechnology, available at* http://vm.cfsan.fda.gov/-lrd/biopolcy.html.

59. *See id.*

60. *See id.*

61. *See supra* note 4, at 22991–22992.

62. *See supra* note 4, at 22993–23005. The toxicological tests are conducted in accordance with the principles outlined in the *Toxicological Principles for the Safety Assessment of*

examination of the agronomic and quality attributes of the plant, genetic analysis of the modification and stability of the expected genomic traits, evaluation of toxicity and allergenicity of newly introduced proteins, and chemical analyses of important toxicants and nutrients.[63] The discussion below on the FLAVR SAVR™ tomato illustrates how these various components were examined.

The policy statement also addressed—and rejected—the notion that because the process used to create genetically modified plants was a new method, foods derived from such plants should be specifically labeled as such. The agency's rationale was that it was not aware of any information showing that foods derived from these new methods differed in any meaningful or uniform way from foods developed using traditional plant breeding. However, labeling would be appropriate if the food differed from its traditional counterpart such that the "common or usual name" no longer applied to the new food or if the new food had safety or usage issues.

The positions outlined in this policy statement came under criticism and legal challenges from environmental and consumer groups. The statement, however, was upheld in the decision *Alliance for Bio-Integrity v. Shalala*, 116 F.Supp.2d 166 (D.D.C. 2000). The petitioners challenged the policy statement both on its procedural development and its substantive positions. Specifically, they argued that the presumption that most foods are GRAS was arbitrary and capricious as was the decision not to impose mandatory labeling on all genetically modified foods.

As to the GRAS presumption, the court rejected the petitioners' argument, concluding that the administrative record (at the time the policy statement was made) contained sufficient information for the court to find the agency's presumption decision reasonable and consistent with the statutory scheme, and thus, the court would defer to the agency's determination.[64] As to the labeling requirement, the court rejected the idea that consumer interest alone was sufficient reason to require mandatory labeling. Rather, the court found that FDA has authority to require a label only if a factual determination is made that the product is materially different from the type of product it claims to be.[65]

The 1992 policy statement and the 1997 guidance document (discussed immediately below) remain the effective policies on foods that contain genetically modified plants. However, FDA did propose a new rule in 2001 that indicates the degree to which its thinking has been influenced by the events and technological advances that occurred in the intervening years.[66] While this proposed

---

Direct Food Additives and Color Additives Used in Food (referred to as the "Redbook"). *See infra* note 66, at 4708.

63. *See supra* note 58.

64. 116 F.Supp.2d 166, 176 (court applied the ruling in *Chevron U.S.A. v. Natural Resources Defense Council*, 467 U.S. 837 (1984)).

65. *See id.* at 179.

66. Fed. Reg. 4706, 4707 (Jan. 18, 2001).

rule has not been finalized, some companies have sought to comply with some aspects of them.

2. 1997 Guidance on Consultation Procedures under FDA's 1992 Statement of Policy In June 1996, the again in October 1997, FDA issued guidance documents regarding the consultation process discussed in the 1992 policy statement. As outlined in the 1992 policy statement, consultation with FDA is not required, but it is recommended for a developer wishing to avoid being tagged with violating the adulteration or misbranding standards.

When having discussions with FDA, a developer must identify and discuss relevant safety, nutritional, and other regulatory issues prior to marketing. "During the consultation process, [the] FDA does not conduct a comprehensive scientific review of the data generated by the developer."[67] Instead, FDA considers, based on internal scientific evaluation of available information, whether there are any unresolved issues that would necessitate legal action by the agency if the product were introduced into the marketplace.[68]

The consultation process FDA describes involves two parts: initial and final. The initial consultation refers to those discussions that occur early in the development process.[69] There can be multiple such discussions with the aim of identifying the relevant scientific, regulatory, and policy issues—thereby enabling the developer to focus on the type of data it must generate to satisfy the requirements of GRAS or to submit a food additive petition. The final consultation process begins when the developer accumulates sufficient data with the belief that it can demonstrate the safety of the product. When this occurs, the developer should submit to FDA a summary of the safety and nutritional assessment[70] that has been developed. Additionally, if the developer believes it is necessary to explain the submission or that FDA may not be familiar with the product, the developer should meet with FDA's scientists to discuss in detail the scientific data and information that supports the summary assessment.

---

67. *See* FDA, *Guidance on Consultation Procedures: Foods Derived from New Plant Varieties* (October 1997) at Section II.

68. *See id.*

69. The Office of Premarket Approval of the Center for Food Safety and Applied Nutrition and the Office of Surveillance and Compliance of the Center for Veterinary Medicine have established a Biotechnology Evaluation Team.

70. *See* FDA, *supra note* 67, at Section II. The safety and nutritional assessment summary should contain sufficient information on: the name of the bioengineered food and the crop from which it is derived; a description of the various uses for the bioengineered food; the sources, identities and functions of the introduced genetic material; the concentration of the bioengineered material in food; the purpose or the intended technical effect of the modification; the potential to induce an allergic reaction; the known or suspected allergenicity or toxicity of the product; the distinctions between the genetically modified foods and similar nongenetically modified foods; and any other information relevant to the safety and nutritional assessment of the bioengineered food.

The Biotechnology Evaluation Team (BET) (e.g., consumer safety officer, molecular biologist, chemist, environmental scientist, and toxicologist from the Center of Food Safety and Applied Nutrition and Office of Surveillance and Compliance) oversees the consultation process, identifies scientific and regulatory issues that must be addressed, reviews the submission, and provides closure to the process by notifying the developer of the agency's position. The agency can then decide whether (a) there are any further questions, (b) the genetically modified food is subject to the food additive petition provisions (which were described above), or (c) there are other regulatory issues such as labeling that need to be addressed.

3. **2001 Proposed Rule and Draft Guidance** On January 18, 2001, FDA a proposed rule and a draft guidance document that respectively provide for: (a) a stringent review process for genetically modified foods and a mandatory premarket notice, and (b) a voluntary labeling scheme. To date, neither the proposed rule nor the draft guidance document have not been finalized. However, both of these documents testify to the evolution in thinking that has occurred at the agency since 1992. The underlying position remains wedded to the 1986 Coordinated Framework perspective—namely, that the individual product should be regulated, not the process by which it was created. Nonetheless, as demonstrated below, the 2001 proposals (especially the one addressing premarket notice) indicate generic concerns about genetic engineering and view those concerns differently from those arising under conventional manipulation. In order to understand the context for the new proposed rules, it is necessary to consider some of the events that occurred in the intervening years.

First, the mid-1990s was a period of global concern about the safety of the food supply. Specifically, there was extensive coverage of "mad cow" disease, including human deaths and the slaughter of livestock to prevent its spread. These events led to restrictions in many countries on the importation of meat, and more generally, to a larger concern about the safety of imported foods. As FDA noted in its premarket notification proposed rule, approximately 45 percent of the U.S. plant-derived food is imported.[71] Although FDA asserted its belief that all companies that had grown GM food in the United States had consulted with them about their foods, it expressed concern that, due to the lack of labeling requirements, an importer could place GM food in the marketplace without the knowledge of FDA.[72]

Second, the 1992 policy came under intense fire from consumer and environmental groups for failing to impose stringent premarketing and labeling requirements on GM foods, as shown by the lawsuit filed by the Alliance for Bio-Integrity (noted above) as well as the one filed in relation to FDA's approval

---

71. *See supra* note 66, at 4712.
72. *See id.*

of recombinant growth hormone that remained in milk (noted below). In 1999, FDA commenced public meetings to discuss its experience under the 1992 policy statement and the approximately 40 consultations it had completed.[73] As a result of these public meetings and approximately 35,000 comments, FDA acknowledged that there was a general consensus that more information should be made available to the consumer.[74] However, this was only the minimum that consumer and environmental groups were seeking from FDA.

Third, FDA recognized that technological advances had occurred in genetically engineered foods. Not surprisingly, the 2001 proposed rule referred to the 1992 policy statement as a historical document by noting that the policy adequately addressed both the scientific and regulatory issues involving the products that were being made at that time.[75] The new proposed rule further built on the 1992 policy statement in addressing the potential allergenic characteristics of bioengineered foods as well as the increased development of new traits among GM foods (e.g., altered protein content, increased carotinoid content, increased fruit solids, altered fiber quality, and increased fruit sweetness).[76]

Fourth, in 2000 FDA conducted a series of consumer focus groups to provide insight into the general public's perceptions of biotechnology with regard to food.[77] As FDA recognized, the participants had well-developed and nuanced opinions. On the one hand, the participants noted the potential benefits of GM foods, including the improvement of agricultural production by increasing yield and reducing the costs associated with growing crops as well as the development of foods with desirable characteristics such as improved taste, appearance, or nutritional characteristics.[78] On the other hand, the participants expressed concern about the unknown long-term health consequences and skepticism as to whether regulatory controls were designed to protect the public or to shield industry.[79]

(a) Proposed Premarket Notice Concerning Bioengineered Foods The recommendation made in the proposed rule—requiring the data be submitted in a premarket biotechnology notice (PBN)—has not yet either been adopted or withdrawn by the agency.[80] Nonetheless, the proposed rule does highlight some changes in the

---

73. *See id.* at 4708.

74. *See id.*

75. *See id.* at 4709.

76. *See i.e.* 4720.

77. U. S. Food and Drug Administration, Center for Food Safety and Applied Nutrition, Office of Scientific Analysis and Support *Report on Consumer Focus Groups on Biotechnology,* (Oct. 20, 2000) *available at* http://www.cfsan.fda.gov/~comm/biorpt.html.

78. *See id.*

79. *See id.*

80. Based on a conversation with Dr. Linda S. Kohl of the FDA, it appears that after the FDA issued its proposed rule, it received numerous comments challenging its authority to issue such a rule. As the FDA commenced its review of these comments, the

agency's view toward plant-derived bioengineered foods. Moreover, some growers have even voluntarily complied with the data requirements noted below, thus submitting information to FDA in accordance with this proposed rule.

FDA recommended that a developer consult with the agency prior to submitting the pre-manufacturing biotechnology notice.[81] The agency outlined a more formalized process in the 2001 proposed rule by requiring that the prospective notifier ask in writing for a meeting and include a synopsis giving sufficient detail about the bioengineered food to allow FDA to engage in meaningful dialogue.[82] Presently, a telephone call to the agency initiating the dialogue is sufficient. The data submitted in this pre-submission consultative process (as well as all correspondence between the agency and the notifier, and the agency's notes of the meetings) would be placed in an administrative file that is subject to disclosure under the Freedom of Information Act. However, the prospective notifier could attempt to shield certain data by affirmatively asserting confidentiality protection for business and trade secrets.

The PBN submission would contain much of the same information that is currently needed to ensure that the food does not fall within the scope of the adulteration or misbranding provisions.[83] Thus, for example, the information would cover the process involved in developing the bioengineered plant, how it would be incorporated into food, its resistance to antibiotics, its creation of antibiotic-resistant strains of bacteria in the human digestive system, and its potential to be an allergen. The notifier would also be required to provide data comparing the composition and characteristics of the bioengineered foods to comparable foods (e.g., levels of significant nutrients and naturally occurring toxicants and antinutrients) in order to address whether the name of the food adequately describes the food or whether the food is adulterated.

(b) 2001 Draft Guidance for Industry: Voluntary Labeling Indicating Whether Foods Have or Have Not Been Developed Using Bioengineering In 1992, FDA asserted that there were no data to indicate that by itself, the fact a product was genetically modified crossed the "material" threshold as set forth in sections 403 and 201(n) of the Act so as to justify special labels for such foods. FDA reaffirmed this decision in its 2001 draft guidance. However, in its proposal to create

---

September 11th attacks required it to direct its focus to setting procedures for food safety through bioterrorism detection. *See* Interview with Dr. Linda S. Kahl (FDA), June 27, 2008.

81. The consultation does not have to take place face to face, and thus, the notifier can save the costs of the travel. *See supra* note 66, at 4716.

82. *See supra* note 66, at 4714–4715. The request would be made to the CFSAN, which coordinates with the CVM as necessary.

83. The approach proposed by the FDA is a case-by-case evaluation of the adequacy of the data. The FDA would use its experience with the GRAS notification program (which, as discussed above has not been finalized, but is still used by parties) as the basis for administrating this program.

a voluntary labeling mechanism, FDA acknowledged that the process of bio-engineering or genetic engineering was relevant to consumers and to certain manufacturers who wanted to respond to consumer desires.[84]

As noted above, with certain exceptions, if foods are fabricated from two or more ingredients, a label is necessary. Moreover, the labels placed on food products cannot be misleading or false, as otherwise they would run afoul of section 403. To avoid being misleading, a label must not omit material information. As the statute does not define the term *materiality*, FDA relies upon historical precedent to argue that material information is that which if absent would (a) pose special health or environmental risks, (b) mislead the consumers in light of other statements made on the label, or (c) lead a consumer to assume that a food has nutritional or functional characteristics of the food that it resembles when in fact it does not.[85] On the other hand, historically consumer demand for a label, in the absence of any of the factors cited above, is not sufficient to justify its imposition.

Despite FDA's view that, as a class, bioengineered foods are not materially different from conventional foods, its drafting of the proposed guidance document was an acknowledgment of the importance of public pressure. But the draft guidance document was as much focused on giving assistance to companies who wanted to advertise the use of biotechnology as admonishing those who sought to distinguish their products by asserting that they were "Genetically Modified Organisms-free" or "GMO-free." Thus, for companies who wanted to insert information about bioengineering, FDA stated:

1. The simple statement that the "food was produced using genetically engineering" would not be misleading, but also would not be very informative.

2. Regardless of this draft guidance document, developers have an obligation to reveal how the common or usual name may not be adequate to describe the product, or if the nutritional content is different, or if new allergens were inserted. In the course of such descriptions, a developer has the option to insert language that the food was created through the use of biotechnology (e.g., "product contains high oleic acid soybean oil from soybeans developed using biotechnology to decrease the amount of saturated fat.")[86]

---

84. United States Food and Drug Administration, Center for Food Safety and Applied Nutrition, *Guidance for Industry: Voluntary Labeling Indicating Whether Foods Have or Have Not Been Developed Using Bioengineering* (Draft Guidance, January 2001), *available at* http://www.cfsan.fda.gov/~dms/biolabgu.html.

85. *See id.*

86. *See id.*

On the other hand, FDA was especially critical of those who want to claim that their food is "GMO-free" by adopting a literal definition of the terms *genetic modification*, *organism*, and *free*.

- The term *genetic modification* refers to alteration of the genotype of a plant using any technique, new or conventional. Thus, from FDA's perspective, because most crops, if not all, have been genetically modified (in the broad sense), it would be inaccurate to state that a food that had not been developed using biotechnology was not genetically modified without providing context.[87]
- The term *organism* is considered misleading because most foods do not contain "organisms."[88]
- The term *free* is problematic because it implies a zero level when the technology is not yet available to establish thresholds.[89]

FDA also is critical of any attempt to indicate that a food is not bioengineered if the implication is that such food is superior to foods that are not so labeled.[90] Also, a statement that a particular ingredient is not bioengineered if there is another ingredient that is, is viewed as misleading because consumers may incorrectly assume that the entire product is free of bioengineered ingredients.[91] Notwithstanding FDA's referral to this draft guidance document in the context of transgenic animals, it should, however, be recalled that this is a draft guidance document. Thus, it is subject to significant alterations if and when it is finalized. Moreover, even if it is finalized this is not binding on the agency.

III. CASE STUDIES

A. FLAVR SAVR™ Tomatoes

The development of the FLAVR SAVR™ tomato serves as an early example of regulatory oversight under the 1992 policy.[92] Calgene, Inc., a California-based company, designed FLAVR SAVR™ as a tomato that was genetically modified so that it could ripen slowly on the vine, develop a full flavor, and yet remain firm when it was put on supermarket shelves.[93] In contrast, conventional tomatoes

---

87. *See id.*
88. *See id.*
89. *See id.*
90. *See id.*
91. *See id.*
92. *See supra* note 66, at 4708.
93. *See id.*

have to be picked while still green and firm and then treated with ethylene gas to induce ripening rather than ripening on the vine.[94]

The genetic modification involved the suppression of the polygalacturonase gene (PG) that occurs naturally in tomatoes. Typically, PG is involved in the breaking down of pectin.[95] Pectin is found in the cellular walls, and its breakdown causes ripe tomatoes to soften. The PG is suppressed by introducing a reverse copy, an antisense polygalacturonase gene (antisense PG).[96] However, as noted in Chapter 3, the transformation of plant cells by introducing exogenous DNA is an inefficient process as only a small percentage of cells will successfully take up, integrate, and express the new genetic information.[97] Thus, in order to distinguish which plant cells have taken up the antisense-PG, a selectable marker—kan[r] gene—is linked to the antisense PG. Kan[r] gene encodes a protein enzyme called aminoglycoside 3'-phosphotransferase II (APH(3')II). As APH(3')II is resistant to the antibiotic kanamycin, it is possible to detect which plants have taken up the antisense PG and the kan[r] gene by growing the plants in a kanamycin-containing medium.[98]

In 1990, after initial consultation with FDA, Calgene submitted a request for an advisory opinion as to whether the kan[r] gene could be used in the production of a genetically engineered tomato.[99] As this request was being considered, FDA announced its 1992 policy, which noted that selectable markers that provide antibiotic resistance (such as the kan[r] gene), unless removed, are expected to be present in foods made from such plants. As such, these foods would be considered as adulterated, and the gene producing the antibiotic resistance would be considered a food additive.

To address these issues, in 1993 Calgene requested FDA convert the advisory opinion into a food additive petition for the safe use of APH(3')II. FDA approved

---

94. *See, e.g.*, A.L.S. Chaves & P.C. de Mello-Ferias, *Ethylene and Fruit Ripening: From Illumination Gas to the Control of Gene Expression, More than a Century of Discoveries*, 29(3) GENETICS & MOLECULAR BIOLOGY 508–515 (2006); L. Alexander & D. Grierson, *Ethylene Biosynthesis and Action in Tomato: A Model for Climacteric Fruit Ripening*, 53(377) J. OF EXPERIMENTAL BOTANY 2039–2055 (October 2002).

95. *See supra* note 66, at 4708.

96. *Id.*

97. 59 Fed. Reg. 26700, 26702 (May 23, 1994).

98. *See id.* Plants that are not resistant to kanamycin will die when exposed to the antibiotic. In 1992, the FDA noted that the kanamycin-resistant gene has been used as a selectable marker in more than 30 crops to develop varieties that exhibit improved nutritional and processing properties, tolerance to chemical pesticides, resistance to pests and diseases, and other agronomic properties. *See supra* note 4, at 22988.

99. The Calgene request applied to the use of the kan[r] gene for genetically engineered tomatoes, cotton, and oilseed rape plants. However, this discussion specifically focuses on tomatoes. *See supra* note 66, at 4708.

the use of APH(3')II the following year.[100] In its analysis, FDA examined the food safety and environmental considerations associated with the use of APH(3')II and the kan^r gene (i.e., potential allergenicity, inactivation of antibiotics, and impacts on microorganisms in the gastrointestinal tract or the environment).

A brief examination of how Calgene handled the major health and environmental issues provides insights. After evaluating all data submitted by Calgene for the use of APH(3')II and kan^r gene, FDA concluded that the FLAVR SAVR™ tomato was as safe as nongenetically modified tomatoes. The following are some of the issues covered by Calgene:[101]

- Calgene addressed the issue of allergenicity of APH(3')II by demonstrating that (a) the protein degraded under simulated gastric conditions, (b) the protein did not share characteristics of allergenic proteins (e.g., proteolytic stability, glyosylation, or heat stability), and (c) the protein DNA sequence does not have significant homology to any proteins listed in various national and international DNA databases.[102]
- Calgene provided information from in vitro degradation studies that APH(3')II would not interfere with the effectiveness of the orally administered antibiotics because APH(3')II requires the presence of the substance adenosine-5' -triphosphate (ATP) in order to inactivate the antibiotics kanamycin and neomycin. But uncooked fruits and vegetables (practically the only source for ATP) do not contain enough ATP to allow the inactivation reaction to proceed. Moreover, the antibiotics are usually given to preoperative patients who are unlikely to be eating a tomato before going for a surgery.[103]
- Calgene provided theoretical and experimental data on the potential for the kan^r gene to impact microorganisms in the human digestive system and in the soil. The company was able to demonstrate that most, if not all, of the kan^r gene would be degraded in the stomach and upper small intestine, and thus would not have an impact on gut microorganisms.[104]

---

100. See supra note 66, at 4708.

101. See 59 Fed. Reg. 26700 (May 23, 1994). There were other issues examined, such as potential impacts to wild relatives; however, the issues of allergenicity, resistance to orally administered antibiotics, and impacts to microorganisms in the gut and soil were the main concerns. See also, J. H. Maryanski, Center for Food Safety and Applied Nutrition, U.S. Food and Drug Administration, Washington, DC 20204 U.S.A., *FDA'S Policy for Foods Developed by Biotechnology, in* GENETICALLY MODIFIED FOODS: SAFETY ISSUES (Engel, Takeoka, and Teranishi eds 1995) American Chemical Society, Symposium Series No. 605, Chapter 2, pp. 12–22 (1995), *available at* http://vm.cfsan.fda.gov/~lrd/biopolcy.html#eval.

102. See id.

103. See supra note 101, at 26706.

104. See supra note 101, at 26704–26705. The soil microorganism studies also demonstrated the lack of impacts.

After addressing the safety of genetic components, FDA turned to the issue of whether the presence of APH(3')II needs to be noted on a label. FDA rejected this idea by noting that: first, APH(3')II is exempted from disclosure because it is not an ingredient, but rather an inherent part of the plant as well as of all foods derived from the plant, or alternatively, it is a processing aid. Second, FDA said that APH(3")II is not material because the agency determined that ingestion of food containing this protein would not compromise the clinical efficacy of orally administered antibiotics.[105] The epilogue of the story, however, was that despite obtaining FDA approval, Calgene's tomato failed as a commodity.

B. rbST Milk

FDA's decision to permit dairy farmers to give their cows a milk production-enhancing, synthetic bovine growth hormone drug generated—and continues to generate—enormous controversy. It led not only to protests, but also to legislation at the state level, imposing restrictions that were not imposed by FDA. This is not surprising given that the states traditionally are responsible for milk production with FDA relying on them to ensure that milk label claims are truthful and not misleading.[106] Thus, an examination of the issues involving synthetic bovine growth hormone could easily include an extensive discussion of various state-level efforts. Instead, this section will examine only the narrow issue of FDA's decision that milk produced from cows given this drug does not have to be labeled.

In the 1930s,[107] scientists discovered that cows injected with the naturally occurring hormone produced by other cattle would increase milk production. Because this process was not cost-effective, it was not pursued. However, once scientists could isolate the gene responsible for the hormone, they were able to produce a synthetic version referred to as recombinant bovine somatotropin (rbST) or by Monsanto's product name (Posilac®).[108]

FDA's involvement with Posilac began in the early 1980s, when Monsanto sought an investigative drug application, and concluded in 1993 when the agency approved Posilac as the first milk production enhancement drug for sale. During that period of time, FDA received thousands of letters and comments from scientists, consumers, environmental and animal rights organizations, farmers, and economists requesting that it either deny approval or require labeling.[109] In fact, after FDA approved the drug, Congress stepped in and imposed a moratorium until an interagency task force reviewed FDA's decision. It was only after the task force concluded FDA's position was justified that Posilac could be marketed.

---

105. *See supra* note 101, at 26709.

106. 59 Fed. Reg. 6279 (Feb. 10, 1994).

107. The historical summary in this paragraph is based on the information provided in Stauber v. Shalala, 895 F.Supp. 1178, 1183 (W.D. Wisc. 1995).

108. *See id.*

109. *See id.*

FDA's decision on labeling had two components: whether labels were necessary on the drug itself, and whether labels were needed on milk produced from cows given the drug. FDA decided that it was appropriate that the drug itself be labeled so as to inform farmers that its use might result in such adverse effects on cows as "reduced pregnancy rates, cystic ovaries, disorders of the uterus, decreases in length of gestation, increased twinning rates, decreased calf birth weight, an increased risk of clinical and sub-clinical mastitis, digestive disorders and infection site reactions."[110] However, FDA decided that the scientific testimony and data did not justify the conclusion that there was any material difference between the milk from cows treated with Posilac versus those that were not so treated.[111] As a result, FDA issued a guidance document entitled, "Interim Guidance on the Voluntary Labeling of Milk and Milk Products from Cows that Have Not Been Treated with Recombinant Bovine Somatotropin," which held that farmers could not label their milk "BST-free" because no milk is BST-free.[112] Rather, states could allow farmers who wanted to make the distinction to say, "from cows not treated with rbST," provided that they included some context, such as the statement: "No significant difference has been shown between milk derived from rbST-treated and non-rbST-treated cows."[113]

FDA's decision was challenged in court by a group of consumers of commercially sold dairy products, who alleged, among other things, that the underlying data did not justify FDA's conclusion that there was no material difference between milk from rbST-treated cows and from non-rbST-treated cows. The plaintiffs claimed that there were organoleptic differences (i.e., differences that could be detected by a human sense organ) between these two types of milk and that there was widespread consumer demand for mandatory labeling of rbST-milk.[114] The court rejected the plaintiffs' arguments, finding that the administrative record did not support the contention that there were any physical properties, flavor characteristics, nutritional quality, functional properties, or length of shelf life that made rbST-milk materially different.[115] The court also noted that, absent actual material differences, a product could not be labeled as different simply because consumers perceived the product as different.[116] In fact, the court found that to do so would in itself be misbranding.

---

110. *Stauber*, 895 F.Supp. at 1185.

111. *See id.*

112. *See supra* note 106.

113. *See id.*

114. *Stauber*, 895 F.Supp. at 1193.

115. *See id.* One specific argument centered on the insulin-like growth factor (IGF-1), which is a protein hormone whose production is regulated in part by somatotrophin. rbST increases the amount of IGF-1 in milk. The FDA evaluated a Monsanto two-week study on rats and concluded that even if IGF-1 from milk was in fact absorbed, its addition would be physiologically insignificant. The court rejected the argument that increased IGF-1 constituted an organoleptic difference.

116. *See id.*

As an epilogue, the issue of rbST milk continues to generate controversy. States are still wrestling with how to regulate such products. For example, in 2007 Pennsylvania reversed its previous stance that milk producers could not insert a voluntary label on r-BST.[117] Moreover, also in 2007, a group of consumer and environmental groups petitioned FDA to reconsider its rbST decision.

IV. EUROPEAN REGULATION OF GENETICALLY MODIFIED FOODS

A. The Context for European Regulations

The European Union's regulations governing genetically modified foods are based on an amalgam of influences. These include "green parties" that are part of coalition governments in certain member countries, powerful agricultural unions, well-organized environmental groups, and media coverage of the risks and benefits of such foods. Each of these influences (along with others)[118] is relevant in different degrees to the shaping of public perceptions and, ultimately, to the creation of the regulatory framework. Nonetheless, there is general aversion toward GM foods in Europe. For example, in a 2002 poll, 89 percent in France, 81 percent in Germany, and 74 percent in Italy said it was "bad" to scientifically alter fruits and vegetables.[119]

One of the reasons for this public hostility may lie in the timing of when GM foods were introduced to European consumers. In 1996, the first GM crops were planted, and the first GM foods were placed on store shelves. However, in the preceding years, European agriculture had been rocked by media and public outrage over "mad cow" disease.

"Mad cow" disease or bovine spongiform encephalopathy (BSE) was first discovered in the United Kingdom in 1986.[120] The disease was the result of feeding

---

117. A. McNally, *Growth Hormone Free Milk Labels Adopted by Pennsylvania*, Food USA Navigator.com (Jan. 18, 2008), *available at* http://www.foodnavigator-usa.com/Legislation/Growth-hormone-free-milk-labels-adopted-by-Pennsylvania.

118. Differences in cultural attitudes towards food and food shopping between the United States and Europe (e.g., Europeans are used to purchasing locally grown foods from small retailers while Americans typically purchase in supermarkets foods that have been brought in from different regions) have also been offered as an explanation for differing public attitudes towards GM crops and foods. *See U.S. v. EU: An Examination of the Trade Issues Surrounding Genetically Modified Foods*, Pew Initiative on Food and Biotechnology (Dec. 2005) at 8.

119. *Broad Opposition to Genetically Modified Foods*, Pew Research Center for the People & the Press (Commentary) (June 20, 2003), *available at* http://people-press.org/commentary/?analysisid=66.

120. *BSE (Bovine Spongiform Encephalopathy, or Mad Cow Disease), About BSE*, Department of Health and Human Services, Centers for Disease Control, *available at* http://www.cdc.gov/ncidod/dvrd/bse/.

meat and bonemeal that contained BSE prion protein to young cattle.[121] As these cattle became diseased, they were sent to slaughterhouses and then served as food. Initially, when reports were received of diseased cattle, the British regulatory authorities asserted that disease in cattle was simply a form of scrapie, which (as with sheep scrapie) could not be transmitted to humans even if they ate meat from infected animals.[122] Nonetheless, the first human death occurred in 1994.[123]

Then, in 1996 a link was discovered between BSE and a variant of Creutzfedlt-Jacob disease ("CJD")—a fatal degenerative brain disease in humans.[124] There was an intense media frenzy over each death as it predominantly struck otherwise healthy young adults and teenagers. There was also the incineration of thousands of sick cattle and the treatment of farms as toxic hot spots.[125] The European Union placed a ban on the importation of British beef in 1996 that was only lifted in 2006.[126] European countries such as France, Germany, Denmark, Belgium, and the Netherlands also discovered BSE in their respective cattle, and in some cases, individuals contracted CJD.[127]

Before the first death and before the link was established, the public had been given assurances by governmental regulatory officials regarding the safety of beef production and the consumption of beef.[128] Thus, the public upon learning that their fears had been justified were understandably distrustful of and angry at their respective governments. Many believed their governments did not

---

121. See id.

122. See id.

123. See id.

124. See id.

125. J. Darnton, *British Beef Banned in France and Belgium*, N.Y. TIMES, Mar. 22, 1996. It was feared that the consumption of over 1.8 million infected cattle might result in hundreds of thousands of human deaths. As of July 1, 2007, there were 161 deaths as a result of the CJD in the UK. The largest number was in 2000, with there being a rapid drop-off in cases starting in 2004. *See* http://www.cjd.ed.ac.uk.

126. Associated Press, *After BSE Ban, British Beef Producers Will Seek to Win Back EU*, HIGH PLAINS MIDWEST AG J. (May 25, 2006).

127. As the purpose of this section is to discuss how European attitudes toward GM foods have been shaped, it will not discuss the bans imposed in the United States or Japan or the impact of BSE on public perceptions in countries other than Europe. For example, since 1989 the USDA has prohibited the importation of live animals and animal products from BSE-positive countries. "Subsequently, USDA expanded the ban to include both countries with BSE and countries at risk for BSE. Since 1997, FDA has prohibited the use of most mammalian protein in the manufacture of ruminant feed (e.g., cattle, sheep, goats). In 2004, FDA issued a rule prohibiting the use of certain cattle materials in human food and cosmetics, and USDA issued a rule prohibiting certain cattle materials from use as human food." *Available at* http://www.cfsan.fda.gov/~comm/bsefaq.html.

128. *See generally* D. T. MAX, THE FAMILY THAT COULDN'T SLEEP: A MEDICAL MYSTERY (2006).

sufficiently regulate the beef industry and that industry had failed to take adequate safety precautions in favor of increasing profits.[129]

Given that GM foods and crops were being introduced in the midst of this environment, and given that the public was being assured about the safety of crops and foods by the same regulatory agencies that had assured them about BSE-tainted meat, it is not too surprising that the public was skeptical. This public skepticism and opposition was covered by the media and echoed in the policy positions and rhetoric of certain political parties. It is in this context that the following regulatory policies must be understood.[130]

B. Early Years of Regulation

The European Union began to regulate the deliberate release of genetically modified organisms in 1990 with the issuance of Council Directive 90/220/EEC,[131] which applied to all EU countries. The Directive required notification prior to (a) deliberate releases of GM organisms for research and development purposes, and (b) deliberate releases for placing in the marketplace products that contained GM organisms.[132] This Directive was subsequently repealed by the issuance of 2001/18/EC. However, much of the same formulation was adopted in the subsequent directive, with a number of strengthened provisions such as mandatory information to the public and mandatory labeling and traceability.[133] One key provision in the Directive that was subsequently incorporated was the "safeguard provision," which states that if a Member State has justifiable reasons to consider that a genetically modified product constitutes a risk to human health or

---

129. *Id.*

130. This book does not purport to examine European laws or regulatory structure in detail, and there are countless books that would be more appropriate for such a task. However, in discussing the particular EU-wide laws that address genetically modified foods (and it should be noted that most of these laws also address genetically modified animal feed), it is necessary to make clear the distinction between the terms "Directive" and "Regulation." EU-wide Directives must be incorporated into the Member State's national law in order to take effects. Thus, different Member States may effectuate their national laws at different dates. Moreover, a Member State may add provisions to the Directive when it actually incorporates it into its national laws. "Regulations," on the other hand, come into effect on the effective date in each EU country without modification and at the same time. Because this section is not intended to be a country-by-country examination, it will assume that the relevant Directive was incorporated in the Member State's national laws as it was drafted and issued by the European Commission.

131. Council Directive 90/219/EEC was also issued in 1990, which addressed only contained research regarding genetically modified microorganisms (e.g., GM viruses and bacteria). EC, *Questions and Answers on the Regulation of GMOs in the European Union*, Memo 07/117 (Mar. 26, 2007) at 3.

132. *See* Article 5 and Article 11 of EC, Council Directive 90/220/EEC (Apr. 23, 1990) *available at* http://www.biosafety.be/GB/Dir.Eur.GB/Del.Rel./90.220/TC.html.

133. EC, *Questions and Answers on the Regulation of GMOs in the European Union*, Memo 00/277 (July 24, 2001) at 2–3.

the environment, it can provisionally restrict or prohibit the use and/or sale of even an approved GM product from its territory.[134] This provision was invoked by a number of Member States and resulted in there being a de facto moratorium on the introduction of GM crops or foods from the late 1990s to 2004.[135]

In 1997, the European Union adopted the Novel Food and Novel Food Ingredients Regulation. Under this Regulation, "novel foods" and "novel food ingredients"—that is, foods and food ingredients that had not been used for human consumption to a significant degree within the European Commission before May 15, 1997 (including, but not limited to, foods or ingredients containing or consisting of GM ingredients, or which had been produced from but did not contain GMOs)—could not be marketed unless approved.[136]

To obtain approval, an applicant must demonstrate through the use of studies that the food or food ingredient meets the following safety criteria: the novel food or novel food ingredient must not present a danger for the consumer, mislead the consumer, or differ from foods or food ingredients they are intended to replace to such a degree that eating them would be nutritionally disadvantageous.[137] To meet these safety criteria, the applicant can use "substantial equivalence" to compare the potential new food with its conventional counterpart.[138] The components of a substantially equivalent evaluation include an examination of composition, nutritional value, metabolism, intended use, and level of undesirable substances contained therein.[139] Between May 1997 and May 2004, 53 applications were made with 14 novel foods approved for placement in the marketplace, 2 products were refused, and the remaining 37 were not acted upon until after the change in regulatory controls in 2004.

C. The Current Regulatory Structure
A new EU legal framework went into effect in April 2004.[140] The main documents addressed: (a) the deliberate release of GMOs into the environment for experimental purposes or for placing the GMO in the marketplace (Directive 2001/18/EC), (b) the placement in the marketplace of GMO food or products

---

134. *See* Article 16 of EC, Council Directive 90220/EEC (Apr. 23, 1990); *see also, supra* note 115.

135. *See supra* note 131, at 8.

136. *Regulation (EC) No. 258/97 of the European Parliament and of the Council of 27 January 1997 Concerning Novel Foods and Novel Food Ingredients*, OFFICIAL J. L 043 (Feb. 14, 1997).

137. *See id.* at Article 3.

138. *See Commission Recommendation of 29 July 1997 Concerning the Scientific Aspects and the Presentation of Information Necessary to Support Applications of the Placing on the Market of Novel Foods and Novel Food Ingredients and the Preparation of Initial Assessment Reports under Regulation (EC) No. 258/97 of the European Parliament and of the Council,* 97/618/EC OFFICIAL J. L 253 (Sept. 16, 1997).

139. *See id., supra* note 136, at Section 4 of Article 3.

140. *See supra* note 131, at 3.

containing or consisting of GMOs (Regulation 1829/2003), and (c) the labeling and traceability of GMOs and of food and feed produced from GMOs (Regulation 1830/2003).[141]

1. **Directive 2001/18/EC** Directive 2001/18/EC addresses the "deliberate release"[142] of GM organisms for two activities: (a) experimental testing, and (b) placement in the marketplace (i.e., making the genetically modified organisms available to third parties either in return for payment or free of charge) for cultivation, importation, or transformation into different products. With respect to both of these activities, prior to undertaking a deliberate release, a person must submit an application (referred to as a "notification"), must receive written consent, and must operate in conformity with any conditions set forth by the consent.

(a) Risk Assessment Guidelines Before submitting a notification for either type of authorization, a manufacturer must perform a risk assessment. This information must be included in the notification package that is submitted to the competent national authority from which the manufacturer is seeking approval. The risk assessment is based on a detailed examination of the technical information on the GMO that the manufacturer must collect before conducting the assessment. Annex III to Directive 2001/18/EC notes the type of technical information that is necessary to conduct such an assessment.[143] Generally speaking, the developer will submit information about the genetically modified organism (e.g., characteristics of the recipient organism and the donor organism, the inserted genetic material, and the final organism that is produced), the conditions of the release; the recipient environment, and the interaction between the GMO and the environment.[144] This information is then evaluated in order to:

1. Identify any of the characteristics of the GMO(s) that may cause adverse effects;
2. Evaluate the potential consequences of each adverse effect;

---

141. *See id.* at 4.

142. The term *deliberate release* is defined as "any intentional introduction into the environment of a GMO or a combination of GMOs for which no specific containment measures are used to limit their contact with and to provide a high level of safety for the general population and the environment." *Directive 2001/18/EC of the European Parliament and of the Council of 12 March 2001 on the Deliberate Release into the Environment of Genetically Modified Organisms and Repealing Council Directive 90/220/EEC* ("Directive 2001/18/EC"), 106(1) OFFICIAL J. L 5–6, Article 4.

143. *See id.* at 23, Annex III.

144. *See id.* Annex III provides detailed guidelines on the specific information that must be collected before an environmental risk assessment is conducted. Specifically, Annex III draws a distinction between the technical information necessary for "genetically modified higher order plants" and those necessary for all other genetic modification. The term *higher order plants* is defined as "plants which belong to the taxonomic group Spermatophytae (Gymnospermae and Angiospermae)." For all non-higher order plants,

3. Evaluate the likelihood of the occurrence of each identified potential adverse effect;
4. Estimate the risk posed by each identified characteristic of the GMO(s);
5. Apply management strategies for risks resulting from the deliberate release or placing on the market of GMO(s); and
6. Determine the overall risk of the GMO(s).[145]

In the process of risk assessment, the developer has to take into account direct or indirect effects, immediate or delayed effects, and any cumulative and long-term effects on human health and the environment that may result from the deliberate release or placing on the market of the GMO(s).[146] Of specific interest, as in the United States, is whether the GMO has potential to cause an allergenic reaction or generate a resistance to antibiotics.

The actual documentation that is provided in support of a notification depends on whether it is being submitted to conduct experimental testing or to place the food in the marketplace. Specifically, for an experimental testing notification, the applicant must comply with the provisions of Article 6 of Directive 2001/18/EC,[147] while a notification for placing the GMO into the marketplace must comply with Article 13 of Directive 2001/18/EC.[148]

---

the requirements include (but are not limited to) information on and the methods used for the modification; methods used to construct and introduce the insert(s) into the recipient or to delete a sequence(s); description of the insert and/or vector construction; description of genetic trait(s) or phenotypic characteristics of the donor and recipient organisms as well as the final GMO (and in particular any new traits and characteristics that may be expressed or are no longer expressed in the final GMO); stability of the final GMO in terms of genetic traits; and the history of previous releases or uses of the GMO (e.g., considerations for human, animal, and plant health; toxic or allergenic effects of the GMOs and/or their metabolic products; comparison of the modified organism to the donor, recipient or (where appropriate) parental organism regarding pathogenicity; capacity for colonization; and the impacts on humans whose immunology is compromised. *See supra* note 141, at 25, Annex III(A)(II)(C).

145. *See supra* note 142, at 19, Annex II; *see also, supra* note 130, at 7–8.

146. *See id.* at 5, Article 4.

147. A notification for an experimental release must contain the technical information as outlined in Annex III(A) and the environmental risk assessment conducted based on the technical information. *See id.* at 24.

148. A notification to obtain authorization to market requires not only technical information under Annex III and the environmental risk assessment, but also information outlined in Annex IV. This Annex requires the notifying party to provide the following information: proposed commercial names of the products and names of GMOs contained therein, and any specific identification, name, or code used by the notifying party to identify the GMO; name and full address of the person established in the Community who is responsible for the produce being placed on the market (*e.g.*, the manufacturer, the importer, or the distributor); name and full address of the supplier(s) of control samples; description of how the product and the GMO in the product are intended to be used;

(b) Article 6: Deliberate Releases for Experimental Testing Because the experiments can be limited to geographic areas, the applicant makes the notification to the competent national authority in the Member State in which the experiment is to occur. This agency then has exclusive authority to authorize or reject the notification. Other Member States and the European Commission may provide comments, but if the national authority believes that the notification satisfies the requirements of Directive 2001/18/EC, it can authorize the release for experimental testing. Under the Directive, the national authority will issue its determination within 90 days of receipt of the notification, indicating either that the notification is in compliance with this Directive and that the release may proceed or that the release does not fulfill the conditions of this Directive and that the notification is therefore rejected.[149]

(c) Article 13: Deliberate Releases for Placing the GMOs in the Marketplace The procedure for obtaining authorization to market GM foods requires not only the approval of the competent national authority where the notification is first submitted, but that of all Member States. The reason for this distinction is that once a product is authorized for marketing in one Member State, it may cross the jurisdictional boundaries of all Member States. The process can be rather complex if there are objections to the notification along the various stages of review and approval. However, in its simplest form, once a notification is submitted, the national authority forwards a summary of the dossier file to the competent authorities of the other Member States and the European Commission.[150] After receiving all the information it has requested from the notifying party, the original authority then completes an assessment report. This report, which is to be produced within 90 days of a complete notification, will state whether the

---

description of the geographical area(s) and types of environment where the product is intended to be used within the Community, including, where possible, estimated scale of use in each area; intended categories of users of the product (*e.g.*, industry, agriculture, and consumers); information on the mechanisms for detecting and identifying the particular GMO products necessary to facilitate postmarketing control and inspection; and the proposed wording on a label or in an accompanying document (*e.g.*, a statement that "this product contains genetically modified organisms"). *See supra* note 141, at 32, Annex IV(A). Additionally, the notifying party may be asked to provide other information such as measures that would be taken to address an unintended release, appropriate handling instructions, and proposed packing mechanisms. *See id.* at 32, Annex IV(B). The notifying party also has to provide information on the monitoring plan as outlined in Annex VII. *See id.* at 9, Article 13.

149. *See id.* at 6, Article 6(5). However, for the purpose of calculating the 90-day period, the time will not be counted during which the competent authority: (a) is awaiting further information it may have requested from the notifier, or (b) is carrying out a public inquiry (but this public inquiry or consultation shall not prolong the 90-day period by more than 30 days). *See id.* at 7, Article 6(7).

150. *See id.* at 9, Article 13(1).

national authority has made a favorable determination toward the notification and the conditions for the product's placement in the marketplace.[151]

If it has made such a favorable determination, it then submits the assessment report along with its ruling to the European Commission, which according to the Directive, must provide it to the Member State's national authorities within 30 days of receipt.[152] The other Member States review the notification and conduct their own analysis of the assessment to make a determination on whether to object. If there are no objections, the national authority that carried out the original assessment can then authorize the marketing of the product.

Once the product has been given approval, it can then be placed on the market throughout the European Union in conformity with any conditions set out in the authorization.[153] Significant conditions that are applicable to all GMOs in the marketplace include: approval for a 10-year period with an option to renew, a label on the product clearly indicating that the product contains GMOs, and provision for a monitoring plan that tracks the releases.[154]

(d) Safeguard Provision As noted above, when Council Directive 90/220/EC was repealed and superseded by Directive 2001/18/EC, the new directive incorporated within its terms a "safeguard provision" that a Member State may prohibit or restrict the use and/or sale of an *approved* genetically modified product from its territory if the State has justifiable reasons—based on new or additional information or scientific knowledge—to consider that a GMO constitutes a risk to human health or the environment.[155] In the event that the Member State considers the presence of the GMO to be a "severe risk," emergency measures such as suspending or terminating the placing of the GMO on the market shall be used along with warning the general public.[156]

Procedurally, the Directive outlines the following steps a Member State would need to take: A Member State that has decided against a particular GMO shall immediately inform the Commission and the other Member States of the actions it has taken and give its reasons, including supplying its review of the environmental risk assessment, indicating whether and how the conditions of the consent should be amended or the consent should be terminated, and, where appropriate, identifying the new or additional information on which its decision is based.[157] The European Commission is supposed to review the Member State's decision within a 60-day time frame,[158] but this period can be considerably longer

---

151. *See id.* at 9, Article 14.

152. *See id.* at 9, Article 13(1).

153. *See id.* at 10, Article 15.

154. *See id.* at 9, Article 13(2).

155. *See id.* at 13, Article 23(1).

156. *See id.*

157. *See id.* at 13, Article 23(2).

158. *See id.*

given that a number of activities are not counted toward this limit. For instance, the 60-day clock does not run when the Commission awaiting on information from the notifying party, the Scientific Committee, and/or the European Food Safety Authority to which it has communicated in order to obtain further information.[159] The Commission's Regulatory Authority then must review all the available information. There must be a qualified majority in favor of the Member State's decision to invoke the safeguard clause.[160] If this does not occur, the decision making is left to the European Council. If the Council votes against the Commission's decision, the Commission has various options, including submitting an amended proposal to the Council on the GMO.

Various Member States that have incorporated the safeguard provision into their national laws have exercised the rights against different GMOs.[161] The Member States provided data in support of their respective decisions in each case, and the Scientific Committee, which evaluated this information, concluded that the justification proposed by the Member State was insufficient to overturn the decision that the GMO should be approved. When the European Union adopted Regulation 1830/2003, these safeguard provisions were incorporated.

2. **Regulation (EC) 1829/2003** Recognizing that the Novel Foods and Food Ingredients regulation (Regulation (EC) 258/97) requirement of a "substantial equivalence" test was not sufficient to demonstrate the safety of GM foods (and also recognizing that Regulation (EC) 258/97 did not make use of the environmental risk assessment framework as outlined in Directive 2001/18/EC), the European Union decided to issue a new regulation governing the marketing of GM foods and feeds. Regulation 1829/2003 was adopted in September 2003 and became effective in April 2004. It addresses GM foods as well as feeds in two separate sections. However, because the feed and food requirements are similar, this section will focus on the food requirements. Moreover, in those cases in which the feed can also be used as food (e.g., maize), both uses must be approved for it to be marketed even solely as a feed.

This Regulation applies to GMO foods; foods containing or consisting of GMOs; and foods produced from, or containing, ingredients produced from GMOs.[162] In order to market such food products that fall within this scope,

---

159. *See id.*

160. *See id.* at 13, Article 23.

161. For a list of GMOs that have been subject to safeguard clauses, *see* http://ec.europa.eu/environment/biotechnology/safeguard_clauses.htm. There was one additional GMO that had originally been subject to the safeguard clause, but the safeguard was subsequently removed.

162. *Regulation (EC) No. 1829/2003 of the European Parliament and of the Council of 22 September 2003 on Generically Modified Food and Feed.* 268 OFFICIAL J. 6, Article 3(1). Additionally, the meat, eggs, and milk obtained from animals that have been given GM animal feed do not need to be labeled. Regulations are also not applicable to products created with the assistance of GM microorganisms such as beer and cheese.

a manufacturer must submit an application for authorization as outlined in Article 5 of the Regulation. Aside from the technical data on the product and the production methods, the developer has to submit studies (preferably independent peer-reviewed ones) demonstrating that the food does not have an adverse effect on human health, animal health, or the environment; that it does not mislead the consumer; and that it does not differ from the food which it is intended to replace to such an extent that its normal consumption would be nutritionally disadvantageous to the consumer.[163] This standard is similar to the criteria for approval outlined in the 1997 Novel Foods and Food Ingredients Regulation, but with two significant distinctions: the inclusion of adverse impacts on the environment, and the position that substantial equivalence is not sufficient to achieve this standard.

There is a distinct set of requirements applicable to foods that contain or consist of GMOs. Specifically, the developers of such foods have a choice of either (a) obtaining approval under Directive 2001/18/EC and then supplying that authorization when making a submission under this Regulation, or (b) conducting the environmental risk assessment under Regulation 1829/2003 at the same time as the safety assessment. The second option is referred to as the "one-door, one-key" policy, meaning that a developer can file a single application for all intended uses of a food that contains or consists of GMOs and thereby obtain a single review and a single authorization rather than expend the time and expense to complete two separate review processes. However, to avail itself of this option, the developer needs to submit the same information as required under Directive 2001/18/EC (i.e., a complete technical dossier supplying the information required by Annexes III and IV of Directive 2001/18/EC and conclusions about the risk assessment carried out in conformity with guidelines set forth in Annex II to Directive 2001/18/EC).

The process for approval commences when the application is sent to the appropriate national authority of a Member State where the product is first to be marketed.[164] The national authority will acknowledge the receipt of the application and immediately notify the European Food Safety Authority.[165] Without delay, the Authority will inform the other Member States and the European Commission of the application and will make the submitted documents available to these entities. The Authority is tasked with preparing an opinion on the application. During this process, among other things the Authority has the responsibility for ensuring that a food safety assessment and an environmental risk assessment are completed.[166] The opinion, along with these assessments, is then made available for public comment. The Authority is supposed to have the

---

163. See id. at 7, Article 5.
164. See id.
165. See id.
166. See id.

opinion completed within six months unless it asks the applicant for additional information.[167]

Once the Authority completes its opinion, it will submit it to the European Commission, the Member States, and the applicant. Within three months of receiving the opinion, the Commission will submit a draft decision on the application.[168] This decision will take into account, but not defer to, the recommendation reflected in the Authority's opinion. The Commission's draft proposal will then be submitted to the Standing Committee on the Food Chain and Animal Health. A qualified majority of Member States in the Committee must approve the application in order for the Commission to notify the applicant of the decision.[169] If the Committee rejects the application, the Council of Ministers has to make the decision by a qualified majority. If the Council does not act or does not meet the qualified majority requirement, the Commission's recommendation is adopted as the default option.[170]

As noted above, the labeling of GM foods has been a part of the regulations that preceded the enactment of this Regulation (EC) 1829/2003. Labeling requirements were part of Directive 2001/18/EC, which required that the label must not mislead a purchaser about the characteristics of the food—in particular, as to its nature, identity, properties, composition, method of production, and manufacturing.[171] This Regulation reiterates that position by noting that foods delivered to the final consumer or mass caterers (e.g., restaurants, hospitals, etc.) that contain or consist of GMOs or are produced from or contain ingredients produced from GMOs must be labeled.[172] The purpose of the labeling is to provide clear, objective information on the GMO that will not mislead the consumer and will facilitate informed decision making. The label should give information about any characteristic or property that renders a food different from its conventional counterpart with respect to composition, nutritional value, or nutritional effects; intended use of the food; and implications for the health of certain sections of the population.[173] In addition, if the food may give rise to ethical and religious issues, this should be noted in the label.

Recognizing that there may be minute traces of GMOs even when operators are striving to prevent the accidental presence of such substances, the Regulation exempts foods that contain, consist of, or are produced in a proportion of

---

167. *See id.* at 8, Article 6.

168. *See id.* at 9, Article 7.

169. *See id.*

170. Questions and Answers on the Regulation of GMOs in the European Union, MEMO/05/104 (Mar. 22, 2005), *available at* http://europa.eu/rapid/pressReleasesAction.do?reference=MEMO/05/104.

171. *See supra* note 142, at 13, Article 21.

172. *See supra* note 162, at 11, Article 12.

173. *See id.* at 11, Article 13.

0.9 percent or less of GMO.[174] However, this presence must be adventitious or technically unavoidable; and thus, operators must be in a position to demonstrate that they have taken all actions to avoid the accidental presence.

3. Regulation (EC) 1830/2003 On the same day Regulation (EC) 1829/2003 was issued, the European Union also issued Regulation (EC) 1830/2003. This Regulation addresses mainly the issue of traceability, and to a lesser degree, labeling.

Traceability refers to the ability to trace a GMO or a product produced from a GMO through its production and distribution chains. For example, it places an obligation on a seed manufacturer to inform a farmer, and the farmer to inform the crop buyer, who then has an obligation to inform the food manufacturer, who ultimately has an obligation to inform the general public.[175] Thus, each operator has to know who sold it the GMO and to whom it sold the GMO. This concept was noted in Regulation (EC) 1829/2003 (as well as in Directive 2001/18/EC), but was not discussed in any detail. The concept is not dissimilar from the management of hazardous wastes in the United States in which each shipment of hazardous waste can be traced from generator to transporter to disposal site through the use of manifests. The ability to trace a product can be of use in removing GMOs that are determined after approval to have adverse impacts on human health or the environment from all the various sources in which they are found.

The traceability requirements vary depending on whether (a) the product consists of or contains GMOs, or (b) the product is produced from GMOs.

- In the case of the former, the party transmitting the GMO must inform the party receiving the product in writing: (a) that the product contains or consists of GMOs, and (b) of the unique identifiers assigned to those GMOs.
- In the case of the latter, the party transmitting the GMO must inform the party receiving the product in writing: (a) of each of the food ingredients that is produced from GMOs, (b) of each of the feed materials or additives that is produced from GMOs, and (c) in case of products for which no list of ingredients exists, that the product is produced from GMOs.[176]

---

174. *See id.* at 11, Article 12.

175. *See supra* note 162, at 6, Article 2(2), *citing, Regulation (EC) No. 1830/2003 of the European Parliament and of the Council of 22 September 2003 Concerning the Traceability and Labeling of Genetically Modified Organisms and the Traceability of Food and Feed Products Produced from Genetically Modified Organisms and Amending Directive 2001/18/EC*, 268 Official J. L. (Oct. 28, 2003). An *operator* is defined as any person who places a product on the market or who receives a product that has been placed on the market in the Community, either from a Member State or from another country, at any stage of the production or distribution chain, but does not include the final consumer. *See supra* note 161, at 6, Article 2(3).

176. *See supra* note 175, *Regulation (EC) No. 1830/2003* at 26, Article 4.

In both instances, for each transaction, a record must be maintained and kept available for a period of five years. Additionally, as with the labeling requirements set forth in Regulation (EC) 1829/2003, traceability does not apply to products that contain trace amounts of GMOs (i.e., no higher than 0.9 percent) that were present because it was technically unavoidable or adventitious.[177]

An applicant must comply with *labeling* requirements to obtain approval for marketing a GMO. Labeling requirements are set forth in Directive 2001/18/EC and in Regulation 1829/2003.[178] In addition, this Regulation also contains labeling requirements. Specifically, these labeling requirements draw a distinction between whether the product is prepackaged or not. For prepackaged products, the label must contain the words: "This product contains genetically modified organisms" or "This product contains genetically modified [name of organism(s)]."[179] If a non-prepackaged product if it is offered to final consumers or mass caterers (e.g., restaurants or hospitals), the same words must appear on, or in connection with, the display of the product.

D. The Significant Distinction between U.S. and European Regulatory Structure

As described above, there are significant differences between the American and European systems for regulating GM foods. These distinctions have, in fact, sparked a trade dispute. From 1997 to 2004, a number of Member States employed the safeguard provisions outlined above and the precautionary principle adopted by the European Commission as a general framework for risk analysis to prohibit the sale of GM seeds or foods. They argued that because labeling and safety requirements needed to be strengthened, they could permit the use or sale of these products.[180] The European Commission, however, asserted that the precautionary principle should not be used as a justification for disguised protectionism. Initially, the United States complained that the de facto moratorium and the Commission's failure to enforce its own rules against its Member States constituted illegal trade barriers.[181] Ultimately, the United States, Canada, Argentina, and nine other nations filed a complaint with the WTO alleging that the de facto moratorium constituted a violation of international trade agreements. The complainants and the EU argued their respective cases to the WTO Dispute Settlement Body. In 2006, that body ruled that banning GM crops was tantamount to an illegal trade barrier.[182] The WTO decision was issued, however,

---

177. *See id.* at 27, Article 5.

178. *See supra* note 142, at 13, Article 21; *see also, supra* note 162, at 11, Section 2.

179. *See supra* note 162, at 11, Article 13(b).

180. See Pew Initiative on Food and Biotechnology, *US v. EU: An Examination of the Trade Issues Surrounding Genetically Modified Food* (Dec. 2005) at 10–11.

181. *See supra* note 142, at 9–10.

182. E. Rosenthal, *A Genetically Modified Potato, Not for Eating, Is Stirring Some Opposition Europe*, N.Y. TIMES, July 24, 2007.

after Member States already begun approving GM foods under the laws adopted in 2003.[183] Underlying this dispute, however, were two different ways of looking at GM foods.

FDA's 1992 policy statement sets forth the view that GM foods may be placed in the marketplace without the need for any premarket approval from FDA provided that such foods are generally recognized as being safe. To meet this standard, the manufacturer of such food must have technical information that the GM food is substantially equivalent to a food that is already in use. If the manufacturer is able to cross this threshold, it does not even need to consult with FDA prior to making the food available for consumer use (although as noted above, meeting with FDA may be advisable to ensure that one's determination is accurate). The underlying view that supports this mechanism for introducing new foods is FDA's perspective that as a technique for creating new food products, genetic modification remains wholly within the continuum of manipulating natural products that has been practiced by man for millennia.

In contrast, while the concept of "substantial equivalence" is a part of the European regulatory structure, nonetheless, as is apparent from the preceding discussion, that Europeans have a more skeptical and cautious analysis toward GM foods. The European model, as a practical matter, perceives GM foods as being created from process that is inherently different from conventional hybridization and thus, the products are subject to a different level of scrutiny. FDA 2001 proposed rule on notification would have moved the American regulatory system closer to the European model. However, as discussed above, the 2001 proposed rule has not been finalized. Thus, the question remains, to what degree will the American regulatory system come to resemble the existing European system?

V. PUSHING NEW FRONTIERS: FOOD FROM CLONED ANIMALS

The concept of cloning animals is highly controversial for both ethical and social reasons.[184] There are also health implications for animals that are bred using cloning as well as for humans who may ultimately eat food made from

---

183. Agricultural interests in the United States and Canada wanted to continue the case because they desired a decision not only demonstrating such barriers were illegal, but that the scientific evidence was sufficient to indicate that GM crops and foods were as safe as conventional foods.

184. *Cloning* is the term commonly used to describe the process of somatic cell nuclear transfer. Center for Veterinary Medicine and Food & Drug Administration, *Animal Cloning: A Draft Risk Assessment* (Dec. 28, 2006), *available at* http://www.fda.gov/cvm/Documents/Cloning_Risk_Assessment.pdf at 3. *See also* J. Zhang et al., *FDA is Posed to Clear Cloned Food*, WALL ST. J., Jan. 4, 2008 at B2.

cloned animals. However, in January 2008, FDA's Center for Veterinary Medicine (CVM) issued a final rule that found that meat and milk products from cloned cattle, pigs, and goats or their offspring are as safe as products from animals naturally bred or bred using other means of artificial reproduction. Despite the CVM's decision, and despite the views of some companies that consumer reluctance can be overcome when consumers obtain superior products (e.g., having leaner and larger cuts of meat), it appears that meat and milk products from cloned animals or their progeny will not be widely available anytime soon. For example, surveys have consistently found that a majority of consumers are wary of food from clones, with many saying they would avoid them.[185] Moreover, food manufacturers have shied away from food produced from cloned animals. For example, the coalition of milk producers has opposed the sale of milk from any cloned animals out of concern that consumers will not purchase such milk because of the "yuck factor" surrounding cloned animals.[186] This section will examine the steps FDA has taken to address the issue of cloning.

From a regulatory perspective, the policy began to evolve in 2001. In the wake of the cloning of "Dolly the Sheep," and when it became evident that commercial ventures were developing clones for use in breeding food-producing animals, the CVM issued an update indicating to all stakeholders its intention to work with them to assess potential risks presented by cloning food-producing animals.[187] The CVM also requested that companies voluntarily refrain from introducing animal clones, their progeny, or their food products (such as milk or meat) into the human or animal food supply pending completion of the risk assessment process.[188]

At the end of 2006, the CVM issued a draft risk assessment, a proposed risk management plan, and draft guidance for industry.[189] In these documents, the CVM concluded that milk and meat from healthy cloned cattle, swine, and goats do not pose consumption risks greater than those of foods derived from their conventional counterparts. The CVM's analysis of whether cloned foods were safe was based on a comparison with products that are currently in the marketplace using other assisted reproductive technologies (such as artificial insemination) that have been used extensively for over a century.[190] However, the CVM

---

185. *See id, FDA is Posed to Clear Cloned Food.*

186. R. Weiss, *FDA Says Clones are Safe to Eat: Voluntary Ban on Food Sale Still in Effect,* WASH. POST, Dec. 29, 2006 at A01.

187. *July 13, 2001 UPDATE ON LIVESTOCK CLONING,* FDA Center for Veterinary Medicine, *available at* http://www.fda.gov/cvm/CVM_Updates/clones.htm.

188. 72 Fed. Reg. 136–137 (Jan. 3, 2007).

189. The documents the FDA produced were a draft risk assessment, proposed risk management plan, and draft guidance for industry.

190. *See supra* note 184, at 4.

requested that manufacturers and breeders continue to refrain from introducing meat or milk from cloned animals until the final rule was issued.

In 2008, after more than six years of consideration of whether meat and milk from cloned animals and their offspring are safe for consumption, the CVM concluded in a final set of documents that such foods are, in fact, safe. In January 2008, the CVM issued a 978-page analysis. In its final risk assessment, the CVM used a two-prong approach. Specifically, the agency developed the Critical Biological Systems Approach, which evaluated all the available data (e.g., physiological, health, and when available, behavioral) on animals involved in cloning at five functional developmental stages. The conclusion reached was that these animals met all the developmental milestones appropriate for their species and become otherwise indistinguishable from sexually reproduced animals.[191] The second prong of the analysis was the Compositional Analysis Method, which examined the gross composition of the animal (e.g., percent of fat and protein), as well as analyses of the vitamins and minerals, fatty acids profiles, and protein characteristics of meat and milk produced by clones. The risk assessment noted that while there were some inherent risks associated with cloning, those risks were not unique to cloning as they exist with other artificial reproduction technologies.[192]

In the risk management program and guidance document for industry, the CVM noted that the current rules and regulations applicable to noncloned meat or milk from cattle, swine, and goats are equally applicable to cloned animals, and that there is no need for additional or special regulation. Additionally, the CVM rejected calls to have cloned animals regulated under the new animal drug requirements despite the Center for Food Safety's filing of a petition with the CVM seeking such relief.[193] However, as it recognized that the state of the science is in flux, the CVM indicated it would: (a) monitor and review additional animal health and food composition data on animal clones and their progeny as they become available, (b) monitor and review changes in animal cloning techniques and technologies, (c) continue to consult with clone producers to review changes in the technology, and (d) monitor and maintain a knowledge base on the evolving scientific literature regarding the biology of animal clones.[194]

For critics of cloning, aside from the ethical concerns there are health concerns, including that an animal clone may develop with apparently normal functions, but with subclinical physiological anomalies, which in turn could alter the

---

191. U.S. Food and Drug Administration, Center for Veterinary Medicine, *Animal Cloning: A Risk Assessment* (Jan. 8, 2008) pp. 50–54, *available at* http://www.fda.gov/cvm/CloneRiskAssessment_Final.htm.

192. *See id.*

193. *See supra* note 184, WALL ST. J., at B2.

194. *See supra* note 191, at 41–55.

expression of key proteins that affect the nutritional content of food and possibly lead to dietary imbalances.[195] As a result of all these concerns, critics of making consumer products from cloned animals have demanded that there be a tracking system to distinguish cloned animals from noncloned animals.[196] Currently, some of the major livestock cloning companies have a voluntary tracking system to help food makers, slaughterhouses, and marketers prove they are not selling food from cloned animals. However, this voluntary system does not apply to the offspring of clones. This is problematic to critics who note that, given the high costs of producing a cloned animal (i.e., $15,000 to $20,000),[197] most cloned animals would be used for breeding, and therefore, the milk and meat would actually be from their offspring.[198] Thus, from the critics' perspective, without coverage of offspring, the effectiveness and rationale for having a tracking system is severely compromised.

VI. NANOMATERIALS AND FOODS

The controversy over—and public reaction to—GM foods should be seen as lessons for those involved in the manufacture of nanomaterials. The respective regulatory structures that have developed both in the United States and Europe to address GMOs are also instructive as to the degree to which nanomaterials, especially those involved in food production, will be regulated.

There are foods already produced that claim to contain nanomaterials. For example, there is a canola oil that allegedly contains "nanodrops" that inhibit the transportation of cholesterol from the digestive system to the bloodstream; a tea that allegedly boosts the absorption of viruses, free radicals, cholesterol, and fat; and a chocolate milk that allegedly enhances flavors without the need for excess sugar.[199] These products do not appear to have entered the American marketplace. But, if and when these products or similar products do, the issues

---

195. *See supra* note 184, at 3.

196. It also remains to be seen what requirements will be imposed due to considerations of foreign trade. It is likely that the European Union will ban the importation of any food derived from cloned animals or their offspring, which is likely to increase pressure in the United States to have a robust tracking system. Otherwise the United States will risk there being a reduction in sales of American foods even if those foods are from non-cloned sources if there is consumer fear that cloned and noncloned animals cannot be distinguished.

197. *See supra* note 184, WALL ST. J., at B1.

198. *Id.*

199. The Project on Emerging Nanotechnologies, *available at* http://www.nanotechproject. org/inventories/consumer/browse/products.

of whether the nanomaterials in such products constitute a food additive[200] or whether they are GRAS will need to be explored by FDA in detail.

As noted above, it is the company's decision as to whether to file a food additive petition prior to marketing the product or to conclude instead that the product is GRAS and market it without FDA approval. In determining whether an ingredient is GRAS, a developer can consult FDA's list of ingredients it considers to be GRAS in 21 C.F.R. Part 182, as well as those ingredients that are "affirmed as GRAS" that are listed in 21 C.F.R. Parts 184 and 186. The ingredients listed in the latter regulations provide chemical composition and specifications, but these parameters do not address size. As discussed in Chapter 3, nanoscale versions of substances may not display the same properties as the conventionally sized products, and thus a manufacturer should consult these chemical specifications and determine if the nanoscale version operates within these parameters. In any event, it would be advisable for a company to consult with FDA prior to marketing; otherwise, the company could potentially be confronted with an enforcement action if FDA can demonstrate that the nanomaterial is not generally recognized as safe.

The other mechanism by which nanoscale products may influence foods is through food packaging, with such substances being referred to as *food contact substances*. A number of products such as storage bags and food containers are infused with nanomaterials (e.g., nanosilver) that allegedly lead to better quality foods (e.g., through reduction of the growth of bacteria and mold).[201] Food contact substances are regulated through section 409 of the FFDCA.

Prior to 1997, food contact substances[202] (or "indirect food additives" as they were then referred to) were regulated under the food additive petition program (which is set forth in detail in Section II(C)(2) above). However, in 1997, the enactment of the Food and Drug Administration Modernization Act (FDAMA) amended the FFDCA to create a new procedural mechanism for approving food contact substances through a notification process rather than through the food additive petition process. As with the petition process, the applicant would need to submit chemical, toxicological, and environmental information about the

---

200. A *food additive* is defined as any substance the intended use of which results or may reasonably be expected to result, directly or indirectly, in its becoming a component or otherwise affecting the characteristics of any food (including any substance intended for use in producing, manufacturing, packing, processing, preparing, treating, packaging, transporting, or holding food, and including any source of radiation intended for any such use), if such substance was not generally recognized as safe or sanctioned prior to 1958.

201. *See supra* note 47, at 37.

202. The term *food contact substance* is defined as any substance intended for use as a component of materials used in manufacturing, packing, packaging, transporting, or holding food if such use is not intended to have any technical effect in such food. 21 U.S.C. § 341(h)(6).

product to demonstrate its safety.[203] Moreover, as with direct food additives, there are specific guidance documents that should be consulted as well as the need for a meeting with FDA's Office of Food Additive Safety prior to testing or submission of a notification. However, unlike the petition process—in which a party may not market the product until a regulation is issued—with the notification process, the applicant may market the product after 120 days if FDA has not objected. The product, however, may be subject to the petition process rather than the notification process under two scenarios: (a) if FDA Secretary makes a discretionary decision that submission and review of a petition is necessary to provide adequate assurance of safety,[204] or (b) if FDA and any manufacturer or supplier agree that such manufacturer or supplier may submit a petition

Another difference from the petition process is that, unlike food additive regulations that are issued through the petition process, approvals under the notification process are proprietary in nature. As the statute and implementing regulations note, the notification is effective for the food contact substance manufacturer or supplier identified in the notification submission.[205] If another manufacturer or supplier wishes to market the same product for the same use, that manufacturer or supplier must also submit a notification.[206] It remains to be seen how these regulations will be enforced with respect to food contact substances that contain nanomaterials.

---

203. The contents of a notification are set forth in 21 C.F.R. § 170.101. *See also supra* note 47 at 38.

204. The regulations note that a petition is likely to be requested in either of the following circumstances: (a) the food contact substance increases the cumulative dietary concentration of a substance that is not a biocide to 3 mg per person per day, or for a biocide, to 0.5 mg per person per day; or (b) there exists a bioassay on the food contact substance, the FDA has not reviewed the bioassay, and the bioassay is not clearly negative for carcinogenic effects. 21 C.F.R. § 170.100(c).

205. 21 U.S.C. § 341(h)(1)(C).

206. 21 C.F.R. § 170.100(a).

7. DRUGS, BIOLOGICS, AND MEDICAL DEVICES

I. INTRODUCTION

The first biotech drug to be approved and marketed was "Humulin"—an insulin.[1] Humulin is manufactured by genetically modifying *Escherichria coli* bacteria (a bacterium that resides in the human digestive tract) to create a product identical to human insulin.[2] Prior to the development of this product, insulin was obtained from the pancreas of domesticated animals.[3] However, not only does Humulin's molecular structure more closely resemble insulin produced by a normal human pancreas, but an unlimited supply can be produced.

Today there are over a hundred biotech drugs in the marketplace, with hundreds more in the research pipeline or being subjected to clinical testing. Further, as nanotechnology develops, research and development is focusing on whether it can help produce more effective drug delivery systems as well as more accurate detection and diagnostic medical devices.

This chapter will explore the complex regulatory structure that is applicable to biotech and nanotech drugs, biologics, and medical devices. Specifically, in 1986 FDA stated, "The agency need not establish new administrative procedures to deal with generic concerns about biotechnology."[4] Therefore, biotech drugs, biologics, and devices are subject to the same regulatory structure that governs their conventional counterparts. Moreover, there is no indication that FDA is thinking there is a need to change the requirements as applied to nanotech drugs or devices.[5] Even those who have questioned the adequacy of FDA's controls over foods and cosmetics have generally found that the agency's regulatory controls over drugs and medical devices to be sufficiently strong.[6] It is only after highly publicized events (e.g., the deaths resulting from the use of Elixir Sulfanilamide

---

1. *See* L. Altman, *A New Insulin Given Approval for Use in the U.S.*, N.Y. TIMES, Oct. 30, 1982 *available at* http://www.nytimes.com.

2. *See id.*

3. *See id.*

4. *See* FDA, "Statement of Policy for Regulating Biotechnology Products," 51 Fed. Reg. 23309 (June 26, 1986).

5. *See* FDA, "FDA Regulation of Nanotechnology Products," *available at* http://www.fda.gov/nanotechnology/regulation.html.

6. *See, e.g.*, Michael Taylor, "Regulating the Products of Nanotechnology: Does FDA Have the Tools It Needs?" Project for Emerging Technologies, Woodrow Wilson International Center for Scholars (October 2006).

that pushed Congress to adopt the 1938 law, the death and deformities resulting from the use of thalidomide that caused the adoption of the 1962 amendments, and the suffering and deformities resulting from the use of the Dalkon Shield that resulted in the medical devices amendments to the statute) that there have been significant changes to the regulatory structure. Thus, unless there is a major incident with respect to biotech or nanotech drugs or devices, the regulations described below are likely to remain the mechanism for controlling these emerging technologies.

II. UNDERSTANDING THE DISTINCTIONS BETWEEN DRUGS AND BIOLOGICS

As the manufacture of medicine using recombinant DNA technology involves a biologically derived product, a question arises as to whether the product is a *drug* or a *biologic* (*biological product*). Under the current regulatory system, some biologically derived products are classified as drugs while others are classified as biologics.

The term *drugs* is primarily defined as "articles recognized in the official United States Pharmacopoeia, official Homoeopathic Pharmacopoeia, or official National Formulary" as supplemented; "articles intended for use in the diagnosis, cure, mitigation, treatment, or prevention of disease in man;" or "articles (other than food) intended to affect the structure or any function of the body of man."[7] The term *biologic product* is defined as "virus, therapeutic, serum, toxin, antitoxin, vaccine, blood, blood component or derivative, allergenic product, or analogous product . . . applicable to the prevention, treatment, or cure of a disease or condition of human beings."[8] Typically, a drug is chemically synthesized, has a small number of molecules, is well defined, and can be thoroughly characterized.[9] Biologics, on the other hand, are generally derived from a living organism or cell, are heterogeneous in nature, elicit an immunogenic response from the human body, are complex in molecular structure, and are not usually fully characterized by current analytical techniques.[10] Notwithstanding these rather clear distinctions, certain biologically derived products are regulated as drugs rather than

---

7. 21 U.S.C. § 321(g)(1).

8. *See* 42 U.S.C. § 262.

9. *See, e.g.,* Celia Henry, *FDA Reform and the Well-Characterized Biologic,* 68 ANALYTICAL CHEMISTRY 674A–677A, *available at* http://www.pubs.acs.org/hotartcl/ac/96/nov/fda.html.

10. *See, e.g.,* Tam Q. Dinh, *Potential Pathways for Abbreviated Approval of Generic Biologics under Existing Law and Proposed Reforms to the Law,* 62 FOOD & DRUG L.J. 77, 82; *see also* Henry, *supra* note 9.

biological products due to tradition (i.e., FDA began regulating them as drugs) and bureaucratic authority rather than any specific product distinctions.[11]

III. THE EVOLUTION OF FDA'S AUTHORITY OVER DRUGS AND BIOLOGICS

A. Drugs

Modern federal efforts to regulate the sale of drugs commenced with the passage of the Progressive era legislation known as the Pure Food and Drug Act of 1906 ("the 1906 Act").[12] This act, which was in reaction to prominent articles on the harms caused by fraudulent medicines, did not have a premarket approval process, but rather focused on postmarket control through the ability to seize and/ or penalize those manufacturers of adulterated or misbranded drugs.[13] In 1938, in the wake of the death of more than one hundred people (including many children) due to the ingestion of a drug known as the "Elixir of Sulfanilamide" that had not been clinically tested prior to sale, public outrage forced Congress not only to repeal the 1906 Act but to enact the Federal Food, Drug, and Cosmetic Act ("the 1938 Act") that had been stalled in the Congress since 1933.[14] The 1938 Act, which serves as the general framework for the current regulatory structure, significantly expanded the powers of FDA to regulate drugs prior to their entry into the marketplace by requiring, among other things, the submission of an investigational new drug application (IND) or a new drug application (NDA).[15] However, the 1938 Act permitted automatic NDA approval unless FDA objected within a specified period of time.[16]

In the wake of another health scare—the use of thalidomide by pregnant women in Europe which resulted in pronounced birth defects in the children that were born along with the therapeutic trials being conducted in the

---

11. In criticizing the logic of FDA's taxonomic system, it is noted that the classification of some types of compounds is determined by how they are made rather than by what they are or what they do, while other classifications are based on source of the compound (*e.g.*, tissue-derived products are regulated as drugs as opposed to blood-derived products are regulated as biological products). *See* Dinh, *supra* note 10, at 83.

12. The first federal drug law was the Import Drug Act, which was enacted in response to the discovery of the gross adulteration and inadequate potency of the antimalarial medication used by U.S. troops in Mexico. *See* Sheryl Lawrence, *What Would You Do with a Fluorescent Green Pig? How Novel Transgenic Products Reveal Flaws in the Foundational Assumptions for the Regulation of Biotechnology*, 34 ECOLOGY L. Q. 201, 213 (2007).

13. *See id.* at 213–215.

14. *See* Lawrence, *supra* note 12, at 215.

15. *See* D.L. Stepp, *The History of FDA Regulation of Biotechnology in the Twentieth Century*, FOOD & DRUG L. 9–10 (Winter 1999).

16. *See id.*

United States—Congress in 1962 again expanded FDA's powers.[17] Significantly, the revised statute prohibited the sale of new drugs that had not been approved by FDA, and thus, affirmative approval by FDA was now necessary to the marketing of any drug. In evaluating whether a drug should be approved, drug manufacturers were required to demonstrate its safety and effectiveness through "substantial evidence" consisting of adequate and well-controlled investigations, including clinical investigations. Thus, the 1938 Act as amended by the 1962 amendments is the basis for the current system for regulating drugs.

B. Biologics

In contrast, biologics or biological products are regulated under the Public Health Services Act, 42 U.S.C. § 262 (PHSA) as well as the FFDCA. The PHSA enacted in 1944 defines a biological product as a "virus, therapeutic, serum, toxin, antitoxin, vaccine, blood, blood component or derivative, allergenic product, or analogous product . . . applicable to the prevention, treatment, or cure of a disease or condition of human beings."[18] There are many categories of biological products such as vaccines that can include live attenuated viruses and inactivated viruses, products made from bacteria and other microorganisms, products made from cells (human or other), and protein products made using biotechnology.[19] With respect to this last category, certain products are expressly classified as biological products including therapeutic DNA plasmid products and monoclonal antibody products for use in humans, as well as many cellular products.[20]

As discussed in further detail in this chapter, in order for a manufacturer to market a biological product, it is necessary to submit a biologics license application (BLA). A BLA will be approved if the applicant demonstrates the biologic is safe, pure, and potent via data derived from nonclinical laboratory and clinical studies and that the manufacturing facility is operated in such a manner as to consistently produce such a biologic. This dual focus on product and establishment is a function of the very nature of biologics. Unlike chemically synthesized drugs, which can be produced in uniform quantities, the biologic production process is complex, and there can be discrepancies between batches of the product due to physical (e.g., temperature), enzymatic, or operating conditions. Thus, FDA seeks to ensure that once approval has been granted, the result will be a consistent product.

Due to historical reasons as well as administrative convenience, a few biological products have been regulated as a drug under FDA's NDA process

---

17. *See id.* at 12–13.

18. 42 U.S.C. 262(i).

19. *See* Statement of Janet Woodcock, M.D., Deputy Commissioner, Chief Medical Officer, FDA, before the Committee on Oversight and Government Reform (Mar. 26, 2007) at 3.

20. *See* Dinh, *supra* note 11, at 83–84.

(e.g., insulin, hyaluronidase, menotropins, and human growth hormones).[21] However, most recombinant-DNA therapeutic proteins have been approved under the PHSA. In 2002, there was shift in which division within FDA had jurisdictional authority for regulating most therapeutic biologics. The responsibility was transferred to the Center for Drug Evaluation and Research—FDA division that handles NDA submissions. However, despite the change in organizational authority, the CDER still requires that the production of a biologic be in compliance with PHSA requirements rather than under the FFDCA.[22]

The CDER regulates monoclonal antibodies for in vivo research; most proteins intended for therapeutic use whether derived from plants, animals, humans, or microorganisms; and recombinant versions of these products (except for products such as blood or vaccines); immunomodulators (i.e., nonvaccine and nonallergenic products intended to treat disease by inhibiting or down-regulating a preexisting, pathological immune response); and growth factors, cytokines, and monoclonal antibodies intended to mobilize, stimulate, decrease, or otherwise alter the production of hematopoietic cells in vivo.[23] On the other hand, the Center for Biologics Evaluation and Research regulates gene and cellular therapies, blood products, vaccines, antitoxins, and allergenics and those antibodies, cytokines, and proteins used solely in manufacturing processes or as reagents.[24]

IV. INVESTIGATIVE PROCESS FOR BIOLOGICS AND DRUGS

This section will examine the process for investigating a new biologic or drug. The manufacturer of any recombinant DNA drug or biologic must comply with the provisions set forth in this section. Moreover, any drug or biologic that contains nanoscale materials must also comply with these provisions. Naturally, the specific application of the steps outlined will depend on the particular drug or biologic in question. Additionally, for the sake of convenience, this section will refer to the process in terms of drugs rather than make reference in each instance to both biologics and drugs.

A. Preclinical Investigation

A manufacturer of a new drug must first conduct a preclinical investigation to determine whether the drug is reasonably safe to test on humans. Preclinical

---

21. *See id.* at 84. Woodcock Statement, *supra* note 19, at 3.

22. *See* Dinh, *supra* note 11, at 83.

23. U.S. Food & Drug Administration, *Frequently Asked Questions about Therapeutic Biological Products* (Nov. 26, 2007), *available at* http://www.fda.gov/CDER/biologics/qa. htm.

24. U.S. Food and Drug Administration, *About CBER, available at* http://www.fda.gov/ Cber/about.htm.

testing refers to tests conducted in laboratories and/or on animals. In the preclinical phase, the researcher will examine, among other things, the pharmacological effects and mechanisms of the drug in animals, and information on the absorption, distribution, metabolism, and excretion of the drug (if known). The researcher will also examine the toxicological effects of the drug in animals and in vitro (e.g., results of acute, subacute, and chronic toxicity tests; tests of the drug's effects on reproduction and the developing fetus; and toxicity tests related to the drug's particular mode of administration). While such tests must be conducted in accordance with Good Laboratory Practices, it is not necessary to obtain approval from FDA prior to conducting them. Good Laboratory Practices for nonclinical studies are set forth in 21CFR Part 58.

B. Investigational New Drug Applications

The next step is for a company to conduct clinical trials on humans.[25] Except under limited circumstances, a new drug may not be shipped lawfully for clinical trials, unless the sponsor of the research has filed for and received approval for an investigational new drug (IND).[26]

In the context of clinical investigations, it is necessary to understand who is being regulated and how they are being regulated. The regulations apply to sponsors, investigators, and sponsor-investigators. The term *sponsors* refers to any person or entity (e.g., pharmaceutical company, private organization, or academic institution) who takes responsibility for and initiates the clinical investigation, but the sponsor does not actually conduct the investigation.[27] An *investigator*

---

25. An investigational new drug application may also be submitted for reasons other than to get a product to market, including as an *emergency use investigational new drug application* or *treatment investigational new drug application*. An emergency use investigational new drug application is submitted when there is a need for an investigational drug, but there is not enough time for submission of an IND application. In such cases, FDA may authorize shipment of the drug for a specified use in advance of submission of an application. 21 C.F.R. § 312.36. (Except for extraordinary circumstances, such authorization will be conditioned on the sponsor making the appropriate IND submission as soon as practicable after receiving the authorization.) A treatment investigational new drug application is a drug that may be used in a clinical investigation for a serious or life-threatening disease in patients for whom no comparable or satisfactory alternative drug is available. During the course of the clinical investigation, it may be appropriate to use the drug in the treatment of patients not in the clinical trials for the purpose of making the drug available to those who are desperately ill, and to acquire additional safety and effectiveness information, provided the investigation is conducted in accordance with the treatment protocol or treatment IND. *See* 21 C.F.R. §§ 312.34, 312.35.

26. The term *investigational new drug* (IND) applies to any new drug that is used in an experiment in which the drug is dispensed to or used involving human subjects. *See* 21 C.F.R. §§ 312.2, 312.3. Additionally, in the case of biological products, it also refers to products used in vitro for diagnostic purposes.

27. *See* 21 C.F.R. §§ 312.3, 312.50.

is the individual or responsible leader under whose immediate direction the drug is administered or dispensed to the subject.[28] A *sponsor-investigator* combines both these functions (i.e., both initiates and conducts the investigation).[29]

FDA divides clinical investigations into three sequential phases, though these may overlap. An IND may be submitted for one or more phases of an investigation. The review FDA conducts on an IND application depends on which phase is being evaluated—for example, an IND application for a phase 1 clinical study is reviewed for safety and the rights of the participants, while an IND application for a Phase 2 or 3 will also focus on the scientific quality of the study and the likelihood the investigation will yield data capable of meeting the marketing approval standards.[30] The three phases are:

Phase 1: The purpose of Phase 1 is to: (a) determine the metabolism and pharmacologic actions of the drugs in humans, (b) identify the side effects associated with increasing doses, and, (c) if possible, gain early evidence on effectiveness. Typically, Phase 1 studies are closely monitored with the drug/product administered to 20 to 80 patients or volunteers. A Phase 1 study should lead to sufficient information about the drug/product's pharmacokinetics and pharmacological effects to design a well-controlled Phase 2 study.[31]

Phase 2: The purpose of Phase 2 is to: (a) evaluate the effectiveness of the drug based on a particular reaction of patients with the disease or condition under study, and (b) determine the common short-term side effects and risks associated with the drug. The study is usually conducted with a relatively small number of patients, usually involving no more than several hundred subjects.[32]

Phase 3: The purpose of Phase 3 is to: (a) gather additional information about the effectiveness and safety of the drug in order to evaluate the drug's overall risk-benefit, and (b) provide a basis for an appropriate label based on the effectiveness of the drug. The study usually involves several hundred to several thousand participants.[33]

FDA regulations note that although there is a specific application form that must be completed (Form 1571) to obtain an IND permit, a sponsor has "considerable discretion" regarding the content of information provided in each section

---

28. *See* 21 C.F.R. § 312.3. The regulations set forth the basis for disqualification of an investigator. *See* 21 C.F.R. § 312.70.

29. *See* 21 C.F.R. § 312.3.

30. *See* 21 C.F.R. § 312.22(a).

31. *See* 21 C.F.R. § 312.21(a). Phase 1 studies also include studies of drug metabolism, structure–activity relationships, and mechanism of action in humans.

32. *See* 21 C.F.R. § 312.21(b).

33. *See* 21 C.F.R. § 312.21(c).

of the form depending on the kind of drug being studied and the nature of the available information.[34] The content of the IND application will also be dependent on, among other things, what phase of clinical investigation is being conducted, the novelty of the drug, the extent to which it has been studied previously, and the known or suspected risks involved.[35]

While recognizing that there will be significant differences among the IND applications that have been submitted, FDA does provide a list of documents that are required from the sponsor, in the following order:

- A cover sheet (Form 1571) that contains certain basic identification information of the sponsor, investigator, or research organization as well as the phase of the investigation; and certain commitments to FDA (e.g., not to begin the investigation until the IND is in effect or there has been compliance with Institutional Review Board requirements).[36]
- A table of contents.[37]
- A brief introductory statement that contains the name of the drug, all active ingredients, the drug's pharmacological class, the structural formula for the drug (if known), the formulation of the dosage form(s) to be used, the route of administration, and the broad objectives and planned duration of the investigation. A brief summary is also included of prior human experiences with the drug, including any information on other FDA-approved investigations of the drug, as well as any marketing or investigations conducted in other countries (or the withdrawal from investigation or marketing in other countries) that would be informative about the safety of the drug.[38]
- A brief description of the overall plan for investigating the drug, including the rationale for the drug or the research, the drug responses to be studied, the general approach for evaluating the drug, the estimated number of participants in the clinical trial, the structure of the clinical trial, and any serious known risks.[39]
- A copy of the brochure that the sponsor has created and given to all investigators. The brochure should contain the following information: a brief description of the drug, including its formulation and structural formula (if known); a summary of the known pharmacological and

---

34. *See* 21 C.F.R. § 312.21(d).

35. *See* 21 C.F.R. § 312.22(b).

36. *See* 21 C.F.R. § 312.23(a)(1). The terms *institutional review board* or *institutional review committee* refers to "any board, committee, or other group formally designated by an institution to review biomedical research involving subjects and established, operated, and functioning in conformance with FDA requirements."

37. *See* 21 C.F.R. § 312.23(a)(2).

38. *See* 21 C.F.R. § 312.23(a)(3).

39. *Id.*

toxicological effects of the drug on animals and humans; a summary of the known pharmacokinetics and biological disposition in animals and humans; a summary of safety and effectiveness information available on the drug; and the description of possible risks and anticipated side effects.[40]

- A protocol must be created that reflects the type of investigation to be conducted. Regardless of how detailed the protocol may be, it must cover the following topics: a statement on the objectives and purpose of the study, information on and the qualifications of each investigator, the criteria for the inclusion or exclusion of patients; a description of the design of the study (e.g., methods used to reduce bias on the part of participants or investigators), the methods used to determine the appropriate dosage and duration, a description of the observations and measurements to be made to fulfill the objectives of the study, and a description of the clinical procedures and other methods to determine the effects of the drug.[41]

- A section to demonstrate compliance with NEPA either through an environmental assessment or through a categorical exclusion (Section III of Chapter 5 generally discusses the NEPA process).

- A section describing the composition, manufacture, and control of the drug. This mandatory section is dependent on the phase of investigation and the scope of investigation within that phase.[42] As the investigation proceeds, the sponsor should submit amendments to FDA informing the agency about the chemistry, manufacturing, and control processes that are commensurate with the level of the investigation. The categories of information that must be provided include:
  - A description of the drug substance (e.g., physical, chemical, or biological characteristics), how it was prepared, and methods used to determine its acceptable limits;
  - A description of all the components (including inactive compounds) used in the manufacture of investigational drug product, as well as information on the manufacture and packaging of the product;
  - Information about the placebo used in the trial; and
  - Copies of the labels that were provided to investigators.

- A section needs to describe information about the pharmacological and toxicological tests that were conducted during the preclinical phase that served as the basis for the applicant's conclusion about the reasonable safety of conducting the proposed clinical investigation.[43]

---

40. *See* 21 C.F.R. § 312.23(a)(5).
41. *See* 21 C.F.R. § 312.23(a)(6).
42. *See* 21 C.F.R. § 312.23(a)(7).
43. *See* 21 C.F.R. § 312.23(8).

- A section that summarizes previous human experiences with the drug that are known to the applicant, including whether the drug has been investigated or marketed in any other country, and details on the experience as to safety; details on any controlled trials; and any relevant published materials.[44]
- For certain applications, FDA will require information on such specialized topics as drug dependence, radiation-absorption dosages, and pediatric studies. With respect to all applications, FDA has the ability to ask for any other information.

I. **Procedures for Approval and Clinical Holds** Once the IND application is submitted and received by FDA, unless the agency issues a clinical hold, the IND goes into effect after 30 days (whether or not FDA contacts the sponsor).[45] A *clinical hold* is an order to delay a proposed clinical investigation or suspend an ongoing investigation[46] The numerous reasons FDA may issue a clinical hold are outlined at 21 C.F.R. § 312.42. There are more reasons for placing a hold on Phase 2 or 3 studies than a Phase 1 study; however, the reasons for a Phase 1 study clinical hold are equally applicable to Phase 2 or 3 studies. For example, a Phase 1 clinical hold may be issued if FDA finds that the clinical investigators are not qualified, that the investigator brochure is misleading or materially incomplete, or that there is not sufficient information to assess the risks to the participants.[47] Additionally, clinical holds may be issued regardless of phase if there is reasonable evidence that an investigation that is not well designed would interfere with a well-designed investigation, that the drug has been studied in a well-controlled study that strongly suggests the drug is ineffective, or that another drug that is being investigated or approved has a demonstrated better risk–benefit balance for the same patient population and problem.[48]

Prior to instituting a clinical hold, the agency will have a discussion with the sponsor to attempt to reach a satisfactory resolution, such as the sponsor providing

---

44. *See* 21 C.F.R. § 312.23(9). If the drug is a combination of drugs previously investigated or marketed, then information should be provided about each active ingredient. However, no information needs to be provided for any component that is lawfully marketed in the United States.

45. FDA may provide an earlier notification to the sponsor to commence the investigation. *See* 21 C.F.R. § 312.40.

46. If there is a suspension of an ongoing investigation, the subjects will not be given new doses of the drug and new subjects will not be recruited. If it is a clinical hold on an ongoing study, except under limited circumstances, the patients will no longer be given the investigational drugs.

47. *See* 21 C.F.R. § 312.42(b)(1)(i)–(iv).

48. *See* 21 C.F.R. § 312.42(b)(4).

additional information.[49] Once a clinical hold is issued, FDA must provide the sponsor with a written set of comments within 30 days.[50] Upon receiving the comments, the sponsor has three options: respond, appeal, or ignore. If the sponsor provides a response to the issues raised in the clinical hold order, FDA can respond either by removing or maintaining the hold and providing an explanation for its decision.[51] If the sponsor disagrees with the reasons cited for the clinical hold, the sponsor can appeal and proceed through FDA's dispute resolution mechanism.[52] The third alternative is for the sponsor to ignore the hold for a year, in which case the investigational new drug is placed on the inactive status list.[53]

2. Reporting Requirements The sponsor of an IND has two reporting obligations: an annual reporting obligation and an obligation to report adverse experiences. In the annual one, the sponsor should provide a brief report on the progress of the investigation during the previous year. The sponsor also has an obligation to promptly review all the data relevant to the safety of the drug regardless of the source of the information and to investigate all safety information in order to notify FDA about all adverse experiences associated with the use of the drug that are both serious and unexpected.[54] FDA may terminate the IND if a sponsor fails to submit an annual report or to promptly investigate and inform FDA and all investigators of serious and unexpected adverse experiences.[55]

3. Advice and Modification, Protocol Amendments, and Information Amendments FDA may at any time communicate with the sponsor during the course of the investigation about any deficiencies in the IND or about FDA's need for additional data. However, unless the communication is accompanied with a clinical hold order, these communications are solely advisory and do not require any modification in any planned or ongoing clinical investigation.[56] If, on the other hand, the discussions are held because FDA has concluded that a deficiency exists in the clinical investigation and is intending to issue a clinical hold order, the parties may be able to negotiate modifications.[57]

---

49. *See* 21 C.F.R. § 312.42(c). FDA will immediately issue a hold if the patients are exposed to immediate and serious risk.

50. *See* 21 C.F.R. § 312.42(d).

51. *See* 21 C.F.R. § 312.42(e).

52. *See* 21 C.F.R. § 312.42(f).

53. *See* 21 C.F.R. § 312.42(g).

54. *See* 21 C.F.R. § 312.32(c). For example, any lab animal test results that suggest a significant risk for human subjects (such as mutagenicity, teratogenicity, or carcinogenicity) and any unexpected fatal or life-threatening experience associated with the use of the drug. *See id.*

55. *See* 21 C.F.R. § 312.44(b). There are other reasons for terminating a IND, including an unreasonable and significant risk of illness or injury.

56. *See* 21 C.F.R. § 312.41.

57. *See* 21 C.F.R. § 312.42.

Once an IND is in effect, the sponsor has an ongoing obligation to amend the submitted protocols.[58] Whenever a sponsor intends to conduct a study that is not covered by a protocol already contained in the IND, the sponsor must submit a protocol amendment containing a protocol for the particular study.[59] Also, a sponsor shall submit a protocol amendment describing any change in a Phase 1 protocol that significantly affects the safety of the participants. With respect to Phase 2 or 3 protocols, not only do changes that significantly affect the safety of the participants need protocol amendments, but so do changes in the scope of the investigation or the scientific quality of the study (e.g., increase in drug dosage or duration of exposure).[60] If these are *essential* (e.g., new toxicology, chemistry, or other technical information; or a report regarding the discontinuance of a clinical investigation), then an *information amendment* must be submitted.[61]

V. NEW DRUG APPLICATION

A. Application Process

Since the enactment of the 1938 Act, prior to the commercialization of any new drug, it must be approved under the "New Drug Application" or NDA process. Under section 505(b)(1) of the Act, an application must be submitted that contains: (a) full reports of investigations which have been made to show whether the drug is safe and effective for the uses proposed; (b) a full list of the components in the drug; (c) a full statement of its composition; (d) a full description of the methods used in, and the facilities and controls used for, the manufacture, processing, and packaging of the drug; (e) samples of the drug and/or the components of the drug (if requested by FDA); (f) drafts of the labeling proposed to be used; and (g) any assessments concerning pediatric use.[62]

FDA regulations, 21 C.F.R. Part 314, provide further detail as to the content of the NDA.[63] Even though the exact requirements will be determined by the specific drug, an application for a new drug will generally contain an application form,

---

58. *See* 21 C.F.R. § 312.30(a).

59. *See* 21 C.F.R. § 312.30(b).

60. *Id.*

61. *See* 21 C.F.R. § 312.31(a).

62. *See* 21 U.S.C. § 355(b)(1). As discussed below, in 1984 Congress enacted the Drug, Price Competition and Patent Restoration Act (commonly referred to as the "Hatch–Waxman" Act), which provided two alternative methods—505(b)(2) and 505(j)—for the introduction of drugs into the marketplace.

63. *See* 21 C.F.R. § 314.1(a). These requirements are also applicable to abbreviated applications as well as any amendments and supplements to the NDA and abbreviated applications.

an index, a summary,[64] technical sections,[65] case report tabulations of patient data, case report forms, drug samples, and labeling. The submission must contain reports of all investigations of the drug product sponsored by the applicant and any other pertinent information about the drug received by the applicant from any other source. Additionally, the technical sections must contain data and information that is sufficient in detail to allow the agency to make a knowledgeable decision on whether to approve the application.

The agency may refuse to approve an application if any of the following seven conditions are applicable:

1. The investigations (reports of which are required to be submitted) do not include adequate tests by all methods reasonably applicable to show whether such drug is safe for use under the conditions prescribed, recommended, or suggested ("appropriate conditions") in the proposed label;
2. The results of tests show that such drug is unsafe for use under the appropriate conditions or do not affirmatively show that the drug is, in fact, safe for use under appropriate conditions;
3. The methods used in, and the facilities and controls used for, the manufacture, processing, and packaging of the drug are inadequate to preserve its identity, strength, quality, and purity;
4. FDA has insufficient information to determine whether such drug is safe for use under appropriate conditions;
5. There is a lack of substantial evidence that the drug will have the effect it purports or is represented to have under the conditions of use prescribed, recommended, or suggested in the proposed label;

---

64. The summary must have enough detail that the reader may gain a good general understanding of the data and information in the application. The summary must contain the proposed text of the labeling, as appropriate; a statement identifying the pharmacologic class of the drug and a discussion of the scientific rationale for the drug, its intended use, and the potential clinical benefits; a brief description of the marketing history of the drug (if any) outside of the United States; a brief summary of each of the technical sections of the application; and a concluding discussion of the risks and benefits. *See* 21 C.F.R. § 314.50(c).

65. The technical sections include information on the composition, manufacture, and specification of the drug substance and the drug product; a section describing use of in vivo and in vitro studies of the pharmacological actions and toxicological effects of the drug; a section describing the human pharmacokinetic data and human bioavailability data; a section describing the microbiology data (only applicable to anti-infective drugs); a section describing the clinical investigations (*e.g.*, a description and analysis of each controlled clinical study pertinent to a proposed use of the drug, an integrated summary of the data demonstrating substantial evidence of the effectiveness for the claimed indications, etc.); a section describing the statistical evaluation of clinical data; and a description of the investigations of the drug for use in pediatric populations. *See* 21 C.F.R. § 314.50(d).

6. The application fails to contain the required patent information; or

7. Based on a fair evaluation of all material facts, the label is false or misleading.[66]

Additionally, the applicant must submit information on the drug substance (i.e., ingredient) patents or drug product (i.e., formulation and composition) patents, or alternatively, indicate the belief that there are no applicable patents. The patent requirements for a section 505(b)(1) application are set forth in 21 U.S.C. 314.53, while the patent requirements for a section 505(b)(2) application (which is described below) are set forth in 21 U.S.C. 314.50(h)(i).

Once FDA receives an NDA, it will make a threshold determination as to whether the application is sufficient to be filed. Typically, FDA will refuse to file an application if one of three conditions exist: (a) omission of a required section to the application or presentation of the material in such a haphazard fashion as to render it incomplete on its face; (b) clear failure to include evidence about the effectiveness of the drug (e.g., lack of any adequate, well-controlled studies); and (c) omission of critical data, information, and analyses needed to evaluate the effectiveness and safety or provide adequate directions of use (e.g., omission without explanation of animal reproduction studies for drugs that will be administered to people of reproductive age).[67] According to FDA's own guidance document, the "refuse to file" process is not to be used by FDA as a means of determining close judgment calls about whether a drug should be approved[68] as that determination is made after FDA conducts a substantive review of the application. Given that an applicant should have had pre-NDA submission conferences with FDA, the application should not suffer from the defects that would cause FDA to refuse to file it.[69] However, if a refuse-to-file letter is transmitted outlining the deficiencies, the applicant can either make the appropriate revisions and resubmit the NDA or seek substantive review (a review referred to as a review under protest).

Prior to August 11, 2008, once FDA had completed its review of a filed application, it would transmit an approvable letter, non-approvable letter, or approval letter.[70] However, as of that date, in order to perform its obligations under the Prescription Drug User Fee Amendments of 2002, FDA replaced approvable

---

66. *See* 21 U.S.C. § 355(d); 21 C.F.R. § 314.50. The applicant must also submit information to comply with NEPA—that is, submit a categorical exclusion or an environmental assessment. Additionally, if FDA requests, an applicant must submit samples of the drug as required or any other information.

67. *See* FDA, "New Drug Evaluation Guidance Document: Refusal to File" (July 12, 1993) *available at* www.fda.gov/CDER/guidance/rtf.pdf, at 4–5.

68. *See id.* at 3.

69. *See id.* at 2.

70. *See* 73 Fed. Reg. 39588 (July 10, 2008).

and non-approvable letter components with a "complete response letter."[71] Thus, an applicant could receive either an approval letter or a complete response letter.

In a complete response letter, FDA states that it will not approve the application in its present form. The letter will usually describe all the specific deficiencies the agency has identified in the application and will recommend—when appropriate—the actions the applicant may take to address the deficiencies.[72] FDA is supposed to send an approval or complete response letter 180 days after the receipt of the application. However, the agency may seek an extension if it needs more time to conduct its review. Different discipline teams review different sections of the application, such as the clinical information, the chemistry, the nonclinical pharmacology and toxicology, the human pharmacokinetics, and bioavailability. Until these discipline teams complete their respective analyses, FDA is not in the position to send any response to the applicant.

Once an applicant receives a complete response letter, it can respond by: (a) resubmitting the application after addressing the deficiencies identified in the complete response letter,[73] (b) withdrawing the application,[74] or (c) requesting a hearing on whether there are grounds for denying approval of the application.[75] A resubmission of the application starts a new time period for review.

B. Post-Approval Reporting Obligation

Applicants, manufacturers, packers, and distributors have a mandatory requirement to report all serious adverse drug experiences or unexpected adverse drug experiences to FDA within 15 working days (referred to as the 15-day alert reports) while health-care professionals and consumers may do so on a voluntary basis.[76]

---

71. A complete response letter will also be issued in the context of an abbreviated new drug application. This is also similar to the process used for approval of biological products. *See id.*

72. *See* 73 Fed. Reg. 39589–90. During its course of its review, FDA may contact the applicant for additional information or to clarify the application. Additionally, FDA may convey early thoughts about deficiencies found at the end of its discipline review. These discipline review letters, however, have not been reviewed by supervisors, and therefore may not ultimately represent the position of the agency. Different discipline teams review different sections of the application such as clinical information, chemistry, nonclinical pharmacology and toxicology, human pharmacokinetics, and bioavailability.

73. Depending on the type of information that will need to be resubmitted, different additional reviews will be necessary.

74. The failure to take any action for a year will be construed as a withdrawal unless there has been a request for an extension of time to resubmit.

75. The type of hearings that are conducted for denial of an application are outlined in 21 C.F.R. Parts 10 through 16.

76. To avoid unnecessary duplication in submitting information to FDA that an applicant may have already provided to the agency, manufacturers, packers, and distributors may meet their obligation if they provide the information to the applicant. 21 C.F.R. § 314.80(c)(iii). Additionally, an applicant is not required to submit a 15-day alert report

However, an applicant only makes the 15-day report if the applicant concludes there is a reasonable probability that the drug has caused an adverse reaction.[77] A *serious adverse drug experience* is defined as one that causes a fatal or life-threatening reaction, causes a persistent or significant disability/incapacity, requires inpatient hospitalization, or causes congenital anomaly/birth defects.[78] An *unexpected adverse drug experience* is any adverse drug that is not listed in the current labeling for the drug, including any events that are listed on the label but differ in severity or specificity.[79] It should be noted that there is no indication that there can be a delay in submitting the 15-day alert report while the applicant makes an evaluation of casualty.[80]

Adverse drug experiences that do not meet the criteria of a 15-day report are still reportable to FDA through the submission of periodic reports each quarter for three years from the date of approval of the application and annually thereafter.[81] The periodic report must contain a narrative summary and analysis of the information in the report, as well as a history of actions taken since the last report because of adverse drug experiences (e.g., labeling changes or studies initiated).[82] Included in any periodic report is an analysis of any 15-day alert report that has been submitted since the last periodic request. The applicant must maintain records for all known adverse drug experiences, including raw data or correspondence, for a period of 10 years.[83]

Failing to submit an alert report, submitting inaccurate or incomplete information, repeatedly or deliberately failing to maintain or submit periodic reports, or failing to conduct a prompt and adequate follow-up investigation will result in the applicant receiving a warning letter and potentially a citation or

---

for an adverse drug experience obtained from a postmarketing study (whether or not conducted under an investigational new drug application) unless the applicant concludes there is a reasonable possibility that the drug caused the adverse experience. 21 C.F.R. § 314.80 (e). For reporting the adverse drug experience, FDA has established certain forms: Form 3500A to be used by applicants, manufacturers, packers, or distributors for mandatory reporting of domestic adverse drug experience; FDA Form 3500 that may be used by health-care professionals or consumers for voluntary reporting; and preapproved, computer-generated FDA Form 3500As to be used by drug firms.

77. *See* 21 C.F.R. § 314.80(e).

78. *See* 21 C.F.R. § 314.80(a). Furthermore, the applicant must promptly investigate all adverse drug experiences that are the subject of the 15-day alert report and submit follow-up reports within 15 calendar days of receipt of new information or as requested by FDA. If additional information is not available, the applicant has an obligation to maintain records noting the unsuccessful steps that were undertaken to obtain additional information.

79. *See* 21 C.F.R. § 314.80.

80. *See* 21 C.F.R. § 314.80(b).

81. *See* 21 C.F.R. § 314(c)(iv)(2).

82. *See id.*

83. *See* 21 C.F.R. § 314.80(i).

a fine. If there are repeated failures to submit reports or provide complete or accurate information, injunctive action may be pursued. On the other hand, if the agency determines that the *drug is no longer safe* as labeled, more stringent action is required such as a change in the product's label or even removal from the marketplace.

VI. REGULATION OF BIOLOGICS

Biological products (also known as biologics) are products that are derived from natural sources (e.g., human, animal, and microorganism) and are composed of sugars, proteins, nucleic acids, or a complex mixture of substances that are not easily classified or identified.[84] Biologics are regulated pursuant to the PSHA and the FFDCA. The PSHA requires the biological product be licensed and that the manufacturing facility be operated in such a manner as to produce a safe, pure, and potent biologic. The FFDCA is also potentially applicable to most biologics because the definition of drug broadly encompasses any product intended for use in the diagnosis, cure, mitigation, treatment, or prevention of disease. As a drug, a biologic would be subject to the same requirements as any other drug, such as obtaining an NDA prior to being introduced into the marketplace.

A. Permitting Process

Under the current regulatory structure, only a single biologics license application (BLA) must be submitted and approved before a biologic may be placed in the marketplace. Historically however, FDA had required the submission of a separate establishment license application (ELA) as well as a product license application (PLA).[85] In 1999, as part of FDA's "Reinventing Government" and implementation of certain sections of FDA Modernization Act of 1997, the agency eliminated these separate submissions and combined their requirements within the context of the BLA—thus, the BLA contains some ELA-type requirements and some PLA-type requirements.

Some types of biological products derived through the use of biotechnology are exempt from ELA-type requirements in the single BLA form. The biological products that fall into this category are therapeutic DNA plasmid products, therapeutic synthetic peptide products of 40 or fewer amino acids, monoclonal antibody products for in vivo use, and therapeutic recombinant DNA-derived products.[86]

---

84. *See, e.g.,* Dinh, *supra* note 11.

85. 64 Fed. Reg. 56441 (Oct. 20, 1999).

86. 61 Fed. Reg. 24227, 24228 (May 14, 1996), codified at 21 C.F.R. § 601.2(a). Initially, FDA first proposed to exclude certain biotechnology-produced biologics (i.e., the "well-characterized biotechnology product"). As FDA explained in its proposal, "After over

In order to grant approval for a BLA submission, the agency focuses on (a) whether the proposed biological meets its standards for safety, purity, and potency; and (b) whether the product was manufactured in accordance with good manufacturing practices. A manufacturer must use clinical and nonclinical data[87] to establish the safety, purity, and potency of the product. Specifically, the *safety* of a product refers to its relative freedom from harmful effects (direct or indirect) when the product is prudently administered, taking into consideration the character of the product in relation to the condition of the recipient at the time.[88] *Potency* refers to the specific ability or capacity of the product to affect a given result. This is determined by appropriate laboratory tests or by adequately controlled clinical tests using the product.[89] *Purity* refers to the standard that products shall be "free of extraneous material except which is unavoidable in the manufacture of the product."[90] In order to obtain the clinical and nonclinical information, the investigative process described in the preceding of this chapter is equally applicable to biological products, and as result, a sponsor or investigator of a potentially new biological product must consult those requirements.

Beyond submission of clinical and nonclinical data, a manufacturer must submit, among other things, a full description of manufacturing methods; data establishing stability of the product; identification of the product by lot number; representative samples of the product; summaries of test results concerning the lot(s) representing the submitted sample(s); and specimens of the proposed labels, enclosures, and containers.[91] If any of these elements are missing, FDA has sufficient grounds to conclude that the application is administratively incomplete and thus cannot be reviewed.[92]

Once a BLA has been submitted, FDA will determine if the application is sufficiently complete to proceed to substantive review, or whether the agency should refuse to file the application. FDA will refuse to file a BLA when there are

---

a decade of experience with these products, the agency has found that it can review the safety, purity, potency, and effectiveness of most well-characterized biotechnology products without requiring the submission of a separate EIA." 61 Fed. Reg. 24227 (May 14, 1996). However, by 1997, when it issued its final rule, FDA recognized that term *well-characterized biotechnology product* was too amorphous, and instead established the specified categories of exempted products.

87. In the event that clinical and nonclinical studies are not conducted in compliance with the requirements as set forth in Parts 56 and 58, respectively, then a brief statement explaining the noncompliance needs to be submitted. 21 C.F.R. § 601.2(a).

88. *See* 21 C.F.R. § 600.3(p).

89. *See* 21 C.F.R. § 600.3(r).

90. *See* 21 C.F.R. § 600.3(s).

91. 21 C.F.R. § 601.2(a).

92. FDA, *Refusal to File Procedures for Biologics License Applications*, Manual of Standard Operating Procedures and Policies, SOPP 8404 (Aug. 27, 2007), *available at* http://www.fda.gov/cber/regsopp/8404.htm.

clear omissions of information or sections of required information (see above for the required information that needs to be submitted); omissions of critical data, information, or analyses needed to evaluate safety, purity, and potency; failure to provide adequate directions for use; or inadequate content, presentation, or organization of the information in the application.[93] If the applicant does receive a letter indicating a refusal to file, the applicant may still demand that FDA review the application as drafted—a procedure referred to "filing over protest"—or the applicant may revise or supplement the application and resubmit it.[94] A refusal-to-file letter is to be issued within 60 days of receipt of the application.[95]

Once the application moves to the substantive review phase, FDA will inspect the facility to examine the product at all phases of the manufacturing process.[96] FDA must conclude that the establishment complies with good manufacturing practices and that the manufacturing process does not impair the applicant's assurances about the product's safety, purity, and potency.[97]

At FDA, different discipline teams review different sections of the application. As with an NDA application, discipline teams will review, among other things, the clinical information, chemistry, nonclinical pharmacology and toxicology, human pharmacokinetics, and bioavailability. During the course of the discipline review phase, FDA may contact the applicant for additional information, seek clarification, or convey early thoughts about deficiencies found by the discipline review.

Upon completion of this review process, FDA will issue either a complete response letter or an approval letter.[98] Although FDA had been issuing complete response letters pursuant to guidance documents, in July 2008, it codified the process into a final rule.[99] The complete response process for BLAs parallels the process for NDAs. The complete response letter is sent to indicate that FDA will not approve the BLA in its present form.[100] It will also usually describe all the deficiencies the agency has been able to identify with the application and, when possible, recommend actions the applicant may take to address those deficiencies so as to obtain approval. After receiving a complete response letter, the applicant can (a) withdraw the application, or (b) resubmit the application after addressing the deficiencies. Additionally, the applicant retains the ability to seek a hearing.

---

93. *Id.*
94. *See id.*
95. *See id.*
96. 21 C.F.R. § 601.20(b),(d).
97. 21 C.F.R. § 601.20(c),(d).
98. *See* 73 Fed. Reg. 39588 (July 10, 2008).
99. *See id.* at 39598–39600, codified at 21 C.F.R. §§ 600.3, 601.3.
100. *See id.* at 39599.

As part of the approval letter, a biologics license will be issued. If, on the other hand, FDA determines that that the establishment or the product does not meet the criteria, the applicant is informed of the denial and given an opportunity for a hearing on the decision. Once the application is approved, the licensee has an obligation to comply with reporting requirements (e.g., submit its annual report).

B. Postmarketing Notification for Changes

FDA monitors the production process of biologics from the early stages to ensure that the scale-up from laboratory quantities to production quantities means that larger-scale batches maintain the same product purity and potency as the smaller-scale batches upon which approval was given.[101] Therefore, not surprisingly FDA requires notification if there is any change in the product, production process, quality controls, equipment, facilities, responsible personnel, or labeling.[102] From FDA's perspective, changes in the production process could lead to changes in the biological molecule. Although these changes may not be detected by standard characterization, they can significantly alter the safety or efficacy profile of the product. FDA requires that the manufacturer assess the effects of the change and demonstrate through appropriate studies or validation that it has not had an adverse effect on identity, strength, quality, purity, or potency of the product.[103]

For any "major" change in the production process, product, quality controls, facility, equipment, or personnel that has a substantial potential to have an adverse effect on the identity, strength, quality, purity, or potency, the manufacturer will need to make a supplemental submission and obtain approval from the agency prior to distribution of the changed product. Among the changes that are considered major are quantitative or qualitative changes in the formulation or in the specifications provided in the BLA application, change in the source material, changes in the virus, and changes in the inactivation method. Other changes require a 30-day review by FDA, including: (a) increases or decreases in the production scale using different equipment; and (b) replacement of equipment that is of a similar, though not identical design but without changing the operating parameters or the process methodology. Once FDA receives notification about such change, it will inform the applicant if the change is in actuality a major change, a change that can be approved, or if additional information is necessary. A manufacturer cannot go forward in the process unless approval is given.[104]

---

101. Food & Drug Administration, *Frequently Asked Questions about Therapeutic Biological Products* (Nov. 26, 2007), *available at* http://www.fda.gov/CDER/biologics/qa.htm.

102. *See* 21 C.F.R. § 601.12(a).

103. *See* 21 C.F.R. § 601.12(b).

104. *See* 21 C.F.R. § 601.12(c).

Finally, certain changes are considered to be only "minor." These include the deletion or a reduction of an ingredient that is only intended to affect the color of the product, a change in the containment storage for nonsterile products, or changes in the size or shape of a container.[105] A manufacturer may submit one or more protocols describing the specific tests or studies that justify why the changes should be considered as a lower level change. There are also specific requirements that are applicable to changes in the label or package information, with most significant changes requiring FDA's approval prior to the distribution of the product with the changed label.[106]

C. Enforcement

Once a license has been issued, FDA has the option to take two different types of administrative actions under PSHA: it may suspend or revoke the license. In order for FDA to suspend a license, it must have a "reasonable grounds to believe" that one of the elements necessary for revoking a license has been met (see below for the elements for revoking a license) *and* that there is a danger to public health.[107] If a manufacturer receives a suspension order, it must notify selling agents and distributors that the license for the product has been suspended.[108] Given that a biological product cannot be commercially distributed without an effective license, this effectively stops all distribution. Once a license is suspended, FDA has one of three options: it can proceed to revoke the license, it can seek to resolve the matter while holding the revocation in abeyance (provided the manufacturer agrees), or it can reinstate the license upon the manufacturer showing compliance and the agency performing an inspection.[109]

FDA can also seek to revoke the license. The reasons for revocation include FDA being unable after reasonable efforts to gain access to an establishment or a location for carrying out an inspection; the manufacturing of the product having been discontinued; the manufacturer having failed to report a change as required; the facility at which the product is manufactured not conforming to good management practices (therefore potentially compromising the safety, purity, and potency of the product); the establishment or the manufacturing methods having significantly changed; or the licensed product not being safe or effective for all of the intended uses (thereby being misbranded with respect to any such use).[110]

---

105. *See* 21 C.F.R. § 601.12(d),(e).
106. *See* 21 C.F.R. § 601.12(f).
107. *See* 21 C.F.R. § 601.6(a).
108. *See id.*
109. *See* 21 C.F.R. §§ 601.6(b), 601.9(a).
110. *See* 21 C.F.R. § 601.5(b). A license can also be revoked at the request of the manufacturer if the manufacturer gives notice of its intent to discontinue the manufacturing of all products covered under the license.

FDA will give the manufacturer an opportunity to correct these issues or have a hearing by first issuing an "intent to revoke" letter before instituting proceedings for revocation.[111] The only exception is if there is a danger to health or willful noncompliance, no opportunity to correct will be given.[112] Otherwise the manufacturer has an opportunity to have a hearing on the revocation (unless the manufacturer has previously waived such right).[113] Also, as with a suspended license, FDA may reinstate the license upon a showing of compliance by the manufacturer and inspection by the agency.[114]

VII. OTHER MECHANISMS FOR DRUG AND POTENTIALLY BIOLOGICAL PRODUCT APPROVAL

In 1984, Congress amended the Food, Drug & Cosmetic Act through the passage of the Drug Price Competition and Patent Term Restoration Act of 1984 (commonly referred to as the Hatch–Waxman amendments).[115] The Hatch–Waxman amendments created two mechanisms for the approval of drugs without it being necessary to submit the full complement of reports and studies on the safety and effectiveness of the drug as necessitated by the NDA—sections 505(j) and 505(b)(2).[116] Section 505(j) applies when the proposed drug is identical to an already existing drug. The process is also referred to as the abbreviated new drug application (ANDA), but is more commonly known as the generic drug approval process. The section 505(b)(2) process applies when the drug is a follow-on to an existing drug, with the process being derivative of the NDA one. But, neither of these processes is in actuality currently applicable to most drugs derived with the use of modern biotechnology. Thus, this brief section describes these processes in the event that, in the future, they, or similar processes, have greater relevance to a manufacturer of a biotech drug as well as to nanotech drugs—especially the section 505(b)(2) process.

---

111. *See* 21 C.F.R. § 601.5(b).

112. *See id.*

113. *See id.*

114. *See* 21 C.F.R. § 601.9(a). Additionally, if there are multiple facilities that manufacture the biological product, FDA may suspend or revoke the license as to only that facility that is not in compliance and allow the other facilities to continue to manufacture. Also, if there are multiple biological products under one license, FDA may allow those products that are in compliance to continue to be manufactured even though the license is suspended or revoked for the product in violation.

115. Pub. L. No. 98-417 (1984).

116. The enactment of these statutory provisions ended FDA's "paper NDA policy" that had permitted an applicant to rely on studies published in the scientific literature to demonstrate the safety and effectiveness of generic versions of certain post-1962 pioneer drug products. *See* Center for Drug Evaluation and Research, *Guidance for Industry: Applications Covered by Section 505(b)(2)* (Draft Guidance) (October 1999) at 1.

A. Generic Drugs

Pursuant to section 505(j), generic versions of already approved drugs (i.e., listed drugs) can be manufactured if patent protection of, and market exclusivity for, the listed drug has expired, *and* if the proposed drug contains the same active ingredients has the same dosage requirement, strength, route of administration, label, and conditions of use as; and is bioequivalent to the listed drug.[117] In this process, the generic drug manufacturer can rely upon the data originally submitted by the listed drug manufacturer to establish the safety and effectiveness of the drug. The generic drug manufacturer, however, must submit test data demonstrating the generic is pharmaceutically equivalent and bioequivalent to the listed drug (also referred to as *sameness*).

As to drugs manufactured using biotechnology, there is a limited possibility for generic versions. The section 505(j) process is predicated on a finding of sameness between the two products, and as result, the process is not ordinarily available to drugs produced through biotechnology. Biologically derived drugs are much larger and more structurally complicated and variable than conventionally manufactured, chemically synthesized, well-defined small-molecule drugs.[118]

B. "Follow-On" Drugs

The process under section 505(b)(2), on the other hand, applies to a drug that is different from but still sufficiently similar to an approved drug.[119] A section 505(b)(2) applicant may rely on published, peer-reviewed scientific literature as well as FDA's previous findings of the safety and/or effectiveness for the approved version of the drug. But the applicant will be expected to provide safety and effectiveness data in support of the proposed modification.[120] With respect to all section 505(b)(2) applications, FDA may, on a case-by-case basis, require additional information from preclinical and/or clinical studies or from scientific literature to support the proposed changes. Thus, FDA recommends that the manufacturer have a pre-submission meeting to discuss the type of information that will need to be submitted (e.g., patent certification with respect to any listed drug, a statement on whether the listed drug(s) are subject to a period of market exclusivity,[121] any bioavailability–bioequivalence study comparing the proposed drug with the listed drug, and the pharmaceutical equivalent).

---

117. If any of these criteria are not met, then in order to submit an ANDA, the person must first seek permission from FDA to file an ANDA. FDA regulations describe the process for submission of such a petition and the format for such a submission.

118. *See* Henry, *supra* note 9.

119. 21 U.S.C. § 355(b)(2).

120. *See* 21 C.F.R. § 314.54.

121. A section 505(b)(2) application may itself be granted three years of market exclusivity if one or more of the clinical investigations (other than bioavailability and bioequivalence studies) were essential to the approval of the application and was conducted or sponsored by the applicant. A section 505(b)(2) application may be granted five years of

In a 1999 draft guidance document, FDA noted that section 505(b)(2) could be used to obtain approval for biologically derived drugs that were approved under FDCA.[122] The document stated that an application could be made for "a drug product containing an active ingredient(s) derived from animal or botanical sources or recombinant technology where clinical investigations are necessary to show that the active ingredient is the same as the active ingredient in a listed product."[123] However, most recombinant-DNA therapeutic proteins, except for a few historical anomalies such as human insulin and human growth hormones[124] that have been approved through a NDA, are excluded from the ambit of section 505(b)(2) because most biologics have been, at least thus far, approved under section 351 of the PHSA—a statute for which there is no analogous abbreviated process for generics.

C. Biologics

Proponents of an abbreviated process for biologics have noted that PHSA contains sufficiently broad language to justify such a procedure. For instance, a biologics license may be approved "on a demonstration" that the biologic is safe, pure, and potent, but there is no requirement that the clinical or nonclinical data underlying such a demonstration be obtained by the applicant. Therefore, the proponents have argued that FDA may allow a follow-on biologic to a pioneer biologic without requiring full clinical studies of the proposed biologic based on the manufacturer establishing biochemical and functional equality.[125]

Nonetheless, the question of when FDA approves generic biologics on a wider scale may have less to do with statutory or regulatory authority than on technological advances. As FDA noted, as technology advances, it may be possible to

---

market exclusivity if the application is for a new chemical entity. *See* 21 C.F.R. § 314.50(j); 314.108.

122. *See* Center for Evaluation and Research, FDA, Draft Guidance for Industry: Applications Covered by Section 505(b)(2) (October 1999) at 7.

123. *See id.* at 5.

124. As an example, FDA approved "Omnitrope"—a recombinant human growth hormone—through the section 505(b)(2) pathway after receiving data specific as to Omnitrope (but less new data than would be needed to support an NDA), also basing its approval on its reliance on the clinical and preclinical data for a previously approved version of a rDNA-derived version of human growth hormone. *See* Woodcock Testimony, *supra* note 19, at 13. In FDA's response to manufacturers who filed a citizen petition to oppose the approval of Omnitrope under section 505(b)(2), the agency noted that Omnitrope was distinguishable from other (often more complex and less well-understood) protein product. *See* FDA Citizen Petition Response, Docket Nos. 2004P-0231/CP1 and SUP1, 2003P-0176/CP1 and EMC1, 2004P-0171/CP1, and 2004N-0355.

125. *See generally,* ABN-ABRO. Generic Biologics: The Next Frontier (June 2001), Amgen, Inc. v. Hoechst Marion Roussel, Inc., 457 F. 3d 1293, *cert. denied,* 127 S.Ct. 2270 (2007), holding that simple amino acid changes in a biologic does not result in a different product unless the changes result in functional differences.

understand the relationship between the structural characteristics of the protein and its function as well as demonstrate the structural similarity between the "follow-on" protein and the referenced product.[126] Although it may be currently possible for some relatively simple protein products to make such comparisons, the technology is not yet sufficiently advanced to allow the type of comparisons necessary for more complex protein products.[127]

VIII. GENERAL OVERVIEW OF FDA'S MEDICAL DEVICE APPROVAL PROCESS

The purpose of this section is to generally discuss the requirements applicable to medical devices.[128] In general, a *medical device* is a health-care product that does not achieve any of its principal intended purposes by chemical action in or on the body or by being metabolized (e.g., reagents, antibiotic sensitivity discs, and test kits for in vitro diagnosis of disease).[129] It must be recognized that there are a number of requirements that are specifically applicable to certain types of devices.

A. Classification System

The degree of regulatory control (e.g., differing levels of premarket review and postmarket regulation) is based on how the particular device is classified. The purpose of the classification system is to provide reasonable assurance of the safety and effectiveness of the device. There are three classes of devices that are intended for human use: class 1 (general controls), class 2 (special controls), and class 3 (premarket approval).[130] The following is a brief summary of each classification category:[131]

- A class 1 device is typically a *low-risk* device (e.g., dental floss without fluoride).[132] *General controls*, such as facility registration, device labeling,

---

126. *See* Woodcock Testimony, *supra* note 19, at 9.

127. *See id.*

128. In 1976, Congress enacted the Medical Device amendments as a response in part to the Dalkon Shield incident in which many women were seriously injured due to the use of an intrauterine device that had not been adequately tested and was marketed without FDA approval. *See* Michael R. Taylor, *Regulating the Products of Nanotechnology: Does FDA Have the Tools It Needs?* Project of Emerging Technologies (October 2006) at 19.

129. 21 U.S.C. § 321(h)(3).

130. 21 U.S.C. § 360c.

131. There is a process for reclassification that is discussed in Subpart C of 21 C.F.R. Part 860.

132. A class 1 device is also a device for which there is insufficient information that the general controls are adequate, but the device is not for use in life-supporting, life-sustaining,

adverse event reporting, and penalties for adulteration and misbranding, are sufficient to meet the safety and effectiveness threshold.[133]

- A class 2 device is a *moderate-risk* device (e.g., pregnancy test kits). As general controls are not sufficient, *special controls* are also required. These special controls include the promulgation of performance standards, postmarket surveillance, submission of premarket notifications, and development and dissemination of guidelines and recommendations as well as the taking of any other action deemed necessary to provide assurance about the safety and effectiveness of the device.[134]

- A class 3 device is a *high-risk* device used for supporting or sustaining human life or for a use that is of substantial importance in preventing impairment of human health, that presents a potential unreasonable risk of illness or injury, or that cannot be regulated through general controls or special controls.[135] It is necessary to obtain premarket approval from FDA for such a device.

B. Premarket Application

Before a manufacturer may introduce into commerce any class 3 medical device that has not been previously been marketed, the manufacturer must submit and receive approval for a premarket application (PMA). Unless deemed to be not relevant to the particular device, a PMA applicant is typically expected to provide, among other things, a description of the device, performance standards (both in effect or proposed), the results of nonclinical laboratory studies, the results of clinical investigations with the device involving human subjects, a bibliography of all published reports, and all proposed labeling for the device.

In order for a PMA to be approved, the device must be safe and effective. To satisfy these criteria, there must be "reasonable assurance" based on "valid scientific"[136] information that the probable benefits to health from the intended uses of the device and conditions of its use (when accompanied by adequate directions and warnings against unsafe use) outweigh any probable risks.[137] In making this analysis, FDA will take into account: (a) the persons for whose use

---

or other situations in which it would be of substantial importance in preventing impairment of human health; in addition, the device does not present a potential unreasonable risk of illness or injury. 21 U.S.C. § 360c(1)(A).

133. 21 C.F.R. § 860.3(c)(1).

134. 21 U.S.C. § 360c(1)(B).

135. 21 U.S.C. § 360c(1)(C).

136. The regulations define *valid scientific evidence* as evidence from "well-controlled investigations, partially controlled studies, studies and objective trials without matched controls, well documented case histories conducted by qualified experts, and reports of significant human experience with a marketed device," from which a qualified expert could make conclusions as to safety and effectiveness. 21 C.F.R. § 860.7(c)(2). The principles of well-controlled clinical investigation are outlined at 21 C.F.R. § 800.7(f).

137. 21 C.F.R. § 860.7(d)(1).

the device is intended, (b) the conditions under which the device would be used, and (c) the reliability of the device. The device may not have an unreasonable risk of illness or injury associated with its intended use.[138]

From a procedural standpoint, once a PMA is submitted, FDA must first make a threshold determination that the application can proceed to a substantive review. There are a number of reasons FDA may refuse to allow the application to proceed (in FDA parlance, this is referred to as a "refusal to file"), including that the application is incomplete or that there is a pending PMA with respect to a similar device. The regulations provide that FDA is to notify the applicant within 45 days of receipt of the application of its initial determination whether it will proceed.[139] If FDA does permit the filing of the PMA, the applicant has 70 days from FDA's receipt of the application to when FDA may schedule a meeting to discuss the status of the application.[140] This meeting will take place no later than 100 days after FDA receives the application (referred to as the 100-day meeting). Then, after receiving the report and recommendations of the appropriate FDA advisory committees (and potentially of an FDA review panel),[141] FDA will send the applicant one of four letters or orders: an approval order,[142] an approvable letter,[143] a not approvable letter,[144] or an order denying approval.[145]

---

138. 21 C.F.R. § 860.7(b).

139. 21 C.F.R. § 814.42(a). FDA may refuse to file a PMA if any of the following applies: (a) the application is incomplete because it is missing information required by statute, (b) the application does not contain each of the items required for a complete PMA and justification for omission of any item is inadequate, (c) the applicant has a pending pre-market notification under section 510(k) of the act with respect to the same device, (d) the PMA contains a false statement of material fact, or (e) the PMA is not accompanied by a statement of either certification or disclosure as required. 21 C.F.R. 814.42(e).

140. 21 U.S.C. § 360e(d)(3).

141. FDA may refer the PMA to a panel for its own initiative, or will do so upon request of an applicant, unless the agency determines that the application substantially duplicates information previously reviewed by the panel. *See* 21 C.F.R. 814.44(a).

142. FDA will approve the PMA if none of the reasons for denying the application applies. The agency will also conditionally approve an application on the basis of the draft final labeling if its only deficiencies concern editorial or similar minor matters. FDA gives public notice of its order granting approval. *See* 21 C.F.R. § 814.44(d).

143. An approvable letter will describe the information FDA requires in order to grant approval, such as final labeling, an FDA inspection, or post-approval requirements. *See* 21 C.F.R. § 814.44(e).

144. The "not approvable letter" will describe the deficiencies in the application, including each applicable ground for denial, and where practical, will identify measures required to place the PMA is approvable form. In response to a not approvable letter, the applicant may amend the PMA as requested or alternatively consider the not approvable letter to be a denial of approval of the PMA and request administrative review by filing a petition in the form of a petition for reconsideration. *See* 21 C.F.R. § 814.44(f).

145. FDA may issue an order denying approval of a PMA because, among other reasons, the PMA may contain a false statement of material fact, the device's proposed labeling may not comply with the requirements, or the applicant may not have permitted an authorized

An approval or a denial order is self-explanatory. An approvable letter and the not approvable letter provide the applicant with an opportunity to amend or withdraw the application. Alternatively, an applicant could consider such a letter as a denial and make a request for administrative review.

C. Confidential Information

Similar to the provisions related to new animal drug applications (see Chapter 5), FDA has rules on the public disclosure of the filing of a PMA. FDA may not disclose the existence of a PMA file (including any information or data contained in such file) before the application is approved or denied. However, there are exemptions from this general nondisclosure provision if the PMA file has been previously disclosed or acknowledged.[146] Specifically, if the existence of a PMA file has been publicly disclosed before an order granting or denying the application, the data and information contained in the file is still not available for public disclosure; but FDA may disclose a summary of portions of the safety and effectiveness data that would be relevant for public consideration.[147] After FDA issues an order either granting or denying the PMA, all safety and effectiveness data and information previously provided may be disclosed to the public. Additionally, unless the information is subject to trade secret or confidential commercial information protection, the protocol for a test or study, adverse reaction reports, product experience reports, consumer complaints and similar documents, and all correspondence will be disclosed to the public.

D. Postmarketing Surveillance

For both class 2 and class 3 devices, postmarketing surveillance occurs if any of the following criteria are met: (a) the failure of the device would be reasonably likely to have a serious adverse health consequence; (b) the device is intended for implantation in the human body for more than a one year; or (c) the device is intended to be used outside a hospital or other facilities to support or sustain a life.[148] The investigator and/or manufacturer will be notified by FDA as to whether it has a postmarketing surveillance obligation, the types of questions it needs to address through this surveillance, and the period of time for which the surveillance must be conducted.[149]

Once the notification has been received, the investigator and/or manufacturer has 30 days to submit a surveillance plan,[150] which should be designed to

---

FDA employee an opportunity at a reasonable time and in a reasonable manner to inspect the facilities. *See* 21 C.F.R. § 814.45.

146. 21 C.F.R. § 814.9(b),(c).
147. 21 C.F.R. § 814.9(d)(1).
148. *See* 21 C.F.R. § 822.1.
149. *See* 21 C.F.R. § 822.5.
150. *See* 21 C.F.R. § 822.8.

address the questions FDA has posed.[151] The agency will review the surveillance plan within 60 days.[152] The failure to have an approved postmarketing surveillance plan or the failure to conduct postmarketing surveillance in accordance with the plan can result in the device being labeled as adulterated and misbranded. FDA has the authority to seize adulterated or misbranded products and seek civil penalties from and criminal prosecution of manufacturers who distribute adulterated or misbranded products or undertake other actions prohibited by the FFDCA.[153]

If the receiver of a surveillance order objects to its issuance or disagrees with the disapproval of the surveillance plan, the person may (a) request a meeting with the Director of the Office of Surveillance and Biometrics, (b) request an informal hearing, (c) seek an internal review of the order, or (d) request a review by the Medical Device Dispute Resolution Panel for the Medical Devices Advisory Committee.[154]

E. Section 510(k) Exemption Notification Process

If a medical device is *substantially equivalent* to an existing, legally marketed device (referred to as the predicate device),[155] the manufacturer may market such a device without a premarket approval provided the manufacturer notifies FDA at least 90 days in advance of its intent.[156] A device is substantially equivalent to a predicate device if: (a) the device has the same intended use as the predicate device; (b) the device has the same technological characteristics as the predicate device; or (c) the device has different technological characteristics, but the data submitted does not raise different questions of safety and effectiveness than the predicate device and is as safe and effective as the marketed device.[157]

Additionally, manufacturers of most class 1 and class 2 products, even if they are not substantially equivalent to any predicate device, can file a section 510(k)

---

151. *See* 21 C.F.R. § 822.9 (provides a list of the items that will need to be included in a submission). Section 822.10 of the Code of Federal Regulations, Title 21 specifically outlines the elements of postmarket surveillance. In addition, guidance documents to assist with designing surveillance plans are at the Web page for the Center for Devices and Radiological Health.

152. 21 C.F.R. § 822.17.

153. 21 C.F.R. § 822.20.

154. *See* 21 C.F.R. § 822.22. Guidance documents that discuss these mechanisms are available from the Center for Device and Radiological Health's Web site.

155. A new device would be compared to: (a) a device in commercial distribution before May 28, 1976, or (b) a device introduced for commercial distribution after May 28, 1976 that has subsequently been reclassified into class I or II. See 21 U.S.C. § 360(k).

156. *See* 21 C.F.R. § 807.81.

157. *See* 21 C.F.R. § 807.100. Additionally, this process is only available if the predicate device has not been removed from the market at the initiative of FDA or has not been determined to be misbranded or adulterated by a judicial order. *See id.*

notification to introduce the product for commercial distribution. Furthermore, a premarket notification is also applicable to devices that are being reintroduced into the marketplace, but with modifications as to their design, components, or method of manufacture—each of which could have an effect as to the safety and effectiveness of the device or its intended use.[158] In such cases, the premarket notification submission must include appropriate supporting data to show that the manufacturer has considered what consequences and effects the modification or new use might have on the safety and effectiveness of the device.[159]

Among the materials contained in a premarket notification submission is a statement indicating the device is similar to and/or different from other products of comparable type in commercial distribution. This should be accompanied by data (e.g., design considerations, energy expected to be used or delivered by the device, and a description of the operational principles of the device) as well as either a section 510(k) summary[160] or a 510(k) statement[161] as described in the regulations.[162]

After reviewing a premarket notification, FDA may either issue an order declaring the device to be substantially equivalent to a predicate device and thereby allow the device to be marketed,[163] issue an order declaring the device to be not

---

158. 21 C.F.R. § 807.81. A premarket notification is not necessary for any device for which a premarket approval application has been submitted or for which a petition to reclassify is pending with FDA. 21 C.F.R. § 870.81(b).

159. 21 C.F.R. § 807.81.

160. A section 510(k) summary is a certified document that summarizes the safety and effectiveness information that is used to make a "determination of substantial equivalence." "The regulations provide a detailed description of the type of information that must be contained in the 510(k) summary, including information on the intended use of the device, device functions, the scientific concepts that form the basis for the device, and the significant physical and performance characteristics of the device. See 21 C.F.R. § 870.3(n).

There are specific requirement that are applicable for submissions that claim substantial equivalence to a device that is classified as class III. See 21 C.F.R. § 870.87(j).

161. A 510(k) statement means a statement asserting that all information in a premarket notification submission regarding safety and effectiveness will be made available within 30 days of request by any person if the device described in the premarket notification submission is determined to be substantially equivalent. The information to be made available will be a duplicate of the premarket notification submission, including any adverse safety and effectiveness information, but excluding all patient identifiers, and trade secret or confidential commercial information. See 21 C.F.R. § 870.3(o). The content of a 510(k) statement is set forth at 21 C.F.R. § 870.93.

162. 21 C.F.R. § 807.87.

163. As FDA notes, the submission of a premarket notification, and the decision by the agency to find the device is substantially equivalent, does not in any way denote official approval of the device. "Any representation that creates an impression of official approval of a device because of complying with the premarket notification regulations is misleading and constitutes misbranding." 21 C.F.R. § 807.97.

substantially equivalent to any predicate device, request additional information, withhold its decision until a certification or disclosure statement is submitted, or advise the applicant that a premarket notification is not required.[164]

For 90 days from the date of receipt of the submission, FDA will not disclose publicly the existence of a premarket notification submission for a device that is not on the market and for which the intent to market the device has not been disclosed. Confidentiality will be maintained only if the person making the notification submits a request and provides a certification indicating, among other things, that the person considers his intent to market the device to be confidential commercial information (and FDA concurs with this conclusion), that the person has not disclosed his intent to nonaffiliated parties, and that the person has taken precautions to protect the confidentiality of his intentions.[165] On the other hand, if the device is already in the marketplace or has been publicly disclosed by the person making the notification, FDA will make a public disclosure.[166]

F. Investigative Device Exemption

In order to conduct clinical investigations of a device, FDA allows a device that would otherwise be required to comply with a performance standard or to have premarket approval to be shipped lawfully if the manufacturer obtains for it an investigational device exemption (IDE).[167] FDA divides investigational devices into two major categories: *significant risk devices* (i.e., devices that present a potential for serious risk to human health, safety, or welfare of the subject") and *non-significant risk device*.

1. **Significant Risk Devices** For clinical investigations involving significant risk devices, the sponsor must submit an application with information that is substantially similar to the type of information required for new drugs. Among the information that an applicant must submit is a summary or complete copy of an investigational plan,[168] a certification from each institutional review board

---

164. 21 C.F.R. § 807.100.

165. 21 C.F.R. § 807.95(a),(b).

166. 21 C.F.R. § 807.95(a).

167. The issuance of an IDE means the device is exempt from certain provisions of the FFDCA, including, inter alia, the provisions relating to misbranding, registration, listing, premarket notification, performance standards, and premarket approval. *See* 21 C.F.R. Part 812.

168. The investigational plan summary shall include, among other things, information on: the name and intended use of the device and the objectives and duration of the investigation; a written protocol describing the methodology to be used and an analysis of the protocol demonstrating that the investigation is scientifically sound; a description and analysis of all increased risks and information on the demographics of the patient population; a description of each important component, ingredient, property, and principle of operation of the device and of each anticipated change in the device during the course of

(IRB) that has reviewed the proposed device, a report on all prior investigations,[169] and copies of all labels for the device.[170]

Procedurally, FDA will notify the sponsor in writing of the date it receives an application. FDA's options are that it may approve the investigation as proposed, approve it with modifications, or disapprove it.[171] A sponsor may proceed with the investigation 30 days after FDA receives the application unless FDA specifically prohibits the commencement of the investigation. On the other hand, a sponsor may decide to wait until FDA provides an order approving the IDE application.

---

the investigation; the sponsor's written procedures for monitoring the investigation and the name and address of any monitor; copies of all labeling for the device; and names of all institutions that will review the investigation and those institutions at which the investigation will be conducted. *See* 21 C.F.R. § 812.25.

169. The report of prior investigations shall provide information on all prior clinical, animal, and laboratory testing of the device and shall be comprehensive and adequate to justify the proposed investigation. The report also shall include: a bibliography of all publications (whether adverse or supportive) that are relevant to an evaluation of the safety or effectiveness of the device, and copies of all published and unpublished adverse information; a summary of all other unpublished information (whether adverse or supportive) in the possession of, or reasonably obtainable by, the sponsor that is relevant to an evaluation of the safety or effectiveness of the device; if information on nonclinical laboratory studies is provided, a statement that all such studies have been conducted in compliance with applicable requirements in the good laboratory practice regulations, or if any such study was not conducted in compliance with such regulations, a brief statement of the reason for the noncompliance. *See* 21 C.F.R. § 812.27.

170. *See* 21 C.F.R. § 812.20. As to the labeling for an investigational device, the regulations provide among other things that the investigational device or its immediate package shall bear a label with the following information: the name and place of business of the manufacturer, packer, or distributor, the quantity of contents, if appropriate, and the following statement: "CAUTION—Investigational device. Limited by Federal (or United States) law to investigational use." The label or other labeling must also describe all relevant contraindications, hazards, adverse effects, interfering substances or devices, warnings, and precautions. *See* 21 C.F.R. § 812.5.

171. *See* 21 C.F.R. § 812.30(a). As with FDA's regulations governing other types of investigations, the agency may disapprove or withdraw its approval if it discovers that: (a) there has been a failure to comply with any requirement of this part or the act, any other applicable regulation or statute, or any condition of approval imposed by an IRB or FDA; (b) the application or a report contains an untrue statement of a material fact or omits material information required by this part; (c) the sponsor fails to respond to a request for additional information within the time prescribed by FDA; (d) there is reason to believe that the risks to the subjects are not outweighed by the anticipated benefits to the subjects and the importance of the knowledge to be gained, that the informed consent is inadequate, that the investigation is scientifically unsound, or there is reason to believe that the device as used is ineffective; (d) there are inadequacies to the report of prior investigations or the investigational plan; the methods, facilities, and controls used for the manufacturing, processing, packaging, storage, and, where appropriate, installation of the device; or the monitoring and review of the investigation. *See* 21 C.F.R. § 812.30(b).

Once the investigation commences, the sponsor will continue to collect data until it believes that there is sufficient information to establish the safety and effectiveness of its device necessary for filing a PMA.[172]

If FDA disapproves an application or proposes to withdraw approval of an application, the agency will notify the sponsor by providing in writing a complete statement as to the reasons for the disapproval; the sponsor will then have an opportunity for a hearing.[173] For an already approved application that FDA is seeking to withdraw, the agency will typically have the hearing before the withdrawal unless it determines that there may be an unreasonable risk to public health.[174]

2. Non-Significant Risk Devices For non-significant risk devices, if the sponsor complies with the following requirements, then FDA considers it to be an "approved application for an IDE." The sponsor must: (a) demonstrate that labels on the device are in accordance with the appropriate rules; (b) obtain IRB approval of the investigation after presenting the reviewing IRB with a brief explanation of why the device is not a significant risk device (and maintain such approval); (c) ensure that each investigator participating in an investigation of the device obtains (as applicable) informed consent from the participants; (d) comply with FDA requirements regarding monitoring the investigations that are being conducted;[175] (e) maintain records as required, and ensure that the investigator is maintaining records as well;[176] and (f) comply with the prohibitions against promoting the device.[177] If FDA seeks to withdraw approval for a non-significant risk device, it will follow the steps outlined above regarding withdrawal of approval.

G. Monitoring Obligations

As with investigations of new drugs, FDA expects the sponsor of the investigation to monitor the activities of the investigators. Specifically, the sponsors are responsible for selecting qualified investigators and providing them with the information (e.g., providing the investigational plan and a report on prior investigations) they need to conduct the investigation properly.[178] The sponsor must obtain from

---

172. 21 C.F.R. Part 812, subpart C.

173. *See* 21 C.F.R. § 812.3(c). If FDA requests additional information concerning an investigation or a revision in the investigational plan, the sponsor may consider that request to be a disapproval of the application and seek an administrative hearing.

174. *See* 21 C.F.R. § 812.30(c)(2).

175. *See* 21 C.F.R. § 812.46.

176. The record-keeping requirements that are applicable to the sponsor are set forth in 21 C.F.R. 812.140(b) (4) and (5) along with the obligation to make the reports required under 812.150(b) (1) through (3) and (5) through (10). The record-keeping requirements for participating investigators are set forth in 21 C.F.R. 812.140(a)(3)(i) with the obligation to make reports as required appearing under 812.150(a) (1), (2), (5), and (7).

177. *See* 21 C.F.R. Part 812.

178. *See* 21 C.F.R. § 812.40.

the participating investigators an agreement that includes the investigator's commitment to conduct the investigation in accordance with the investigational plan and the conditions imposed by the IRB or FDA.[179] Additionally, the sponsor must obtain the investigator's agreement that it will ensure that the requirements of informed consent are met and that there will be supervision of all testing of the device involving human subjects.[180]

A sponsor who discovers that the investigator is not complying with the signed agreement, investigational plan, FDA regulations, or any of the conditions imposed by FDA or IRB must promptly either secure compliance or discontinue shipment of the device and terminate that investigator's participation in the study.[181] The sponsor must also require the investigator to return or dispose of the devices unless this type of action would jeopardize the health, safety, or welfare of the subjects.[182]

If the sponsor discovers that there is an "unanticipated adverse device effects,"[183] the sponsor will immediately conduct an evaluation of such effects. If it is determined that the effects present an "unreasonable risk" to the subjects, the sponsor will terminate the entire investigation or those portions of the investigation that present the risk no later than 5 days after the sponsor makes the determination and no later than 15 days after the sponsor receives information about the effect.[184] Both the investigator and the sponsor have reporting obligations.

H. Treatment Investigation Device Exemption

In limited circumstances, an IDE device may be used for treatment in the hope that the new device will be of assistance to treat or diagnose *serious* or *immediately life-threatening* disease in patients when there is no comparable or satisfactory alternative device or other therapy available to treat or diagnose that stage of the disease.[185] In the case of a serious disease, typically a device is available for treatment use after all clinical trials have been completed but before a classification

---

179. *See* 21 C.F.R. § 812.43.
180. *See* 21 C.F.R. §§ 812.45, 812.46.
181. *See* 21 C.F.R. § 812.46(a).
182. *See id.*
183. An *unanticipated adverse device effect* means any serious adverse effect on health or safety or any life-threatening problem or death caused by, or associated with, a device, if that effect, problem, or death was not previously identified in nature, severity, or degree of incidence in the investigational plan or application (including a supplementary plan or application), or any other unanticipated serious problem associated with a device that relates to the rights, safety, or welfare of subjects. 21 C.F.R. § 812.3(s).
184. *See* 21 C.F.R. § 812.46(b)(2).
185. 21 C.F.R. § 812.36(a). In order for a device to be used, it must be under investigation in a controlled clinical trial for the same use, or such clinical trials must be completed; and the sponsor must be actively pursuing marketing approval or clearance for the device. 21 C.F.R. § 812.36(b).

has been determined (and in the case of class 3 products, before a PMA is filed). In the case of an immediately life-threatening disease (i.e., a stage of the disease in which there is a reasonable likelihood that death will occur within a matter of months or in which premature death is likely without early treatment), the device is made available for treatment use prior to the completion of all clinical trials.

To use an IDE device for treatment, a sponsor must complete an application that includes, but is not limited to, a summary or complete copy of an investigational plan, certification from the IRB indicating a review has occurred, and a report on all prior investigations. Treatment use may begin 30 days after FDA receives the treatment IDE submission unless the agency notifies the sponsor in writing earlier than the 30 days that the use in treatment may or may not begin. FDA may approve the treatment use as proposed or approve it with modifications. The agency may also disapprove or propose to withdraw approval of a treatment IDE if, among other reasons, a comparable device or therapy becomes available, there is insufficient evidence to demonstrate the safety and effectiveness of the device in treating serious conditions, or it would expose those who are suffering from life-threatening diseases to significant additional risk of injury.[186]

With a treatment IDE, the sponsor must submit progress reports on a semi-annual basis to all reviewing IRBs and FDA until the filing of a marketing application.[187] Once a marketing application has been filed, the reports must be filed in accordance with those requirements.

IX. THE DEVELOPMENT OF NANOMEDICINE

A. Drugs and Drug Delivery Systems
The role of different nanomaterials in the health-care field is being explored every day because of the belief that nanomaterials can have a significant impact on the detection and treatment of various diseases. The preceding sections have discussed the regulatory framework that governs "nano"-scale or "nano"-infused drugs or devices; that is, a new "nano"-infused drug would need to comply with INDA and NDA requirements, and a new "nano"-infused, class 3 medical device would need to meet the PMA requirements. Thus, this section will mainly focus upon the potential role of nanomaterials rather than restating the application of regulatory requirements.

The research and product development has focused on imaging and diagnostic technology, drugs and drug delivery vehicles, and surveillance systems as well as

---

186. 21 C.F.R. § 812.36(d)(2).

187. 21 C.F.R. § 812.36. These reports include information on the number of patients treated with the device under the treatment IDE, the names of the investigators participating in the treatment IDE, and a brief description of the sponsor's efforts to pursue marketing approval/clearance of the device. *See id.*

hybrid products that can perform more than one of these tasks.[188] Nanoparticles are seen as being able to perform all these functions because their size allows them to enter cells and the organelles inside to interact with the DNA and protein.[189] Most animal cells are 10,000 to 20,000 nanometers in diameter while nanoparticles are less than 100 nanometers in diameter.[190]

With respect to drug delivery, nanoparticles are seen as providing a mechanism by which time-released, targeted drug delivery is achieved, thereby reducing the adverse impacts to healthy, neighboring cells that may result from conventional treatment.[191] Many types of nanomaterials are being considered for drug delivery, including single-walled carbon nanotubes, nanoshells, and dendrimers. For example, carbon nanotubes have already been used to deliver drugs in a variety of cell culture systems.[192] The nanotube is attached to a peptide or antibody on its outer surface, and when administered, that peptide or antibody binds to its target.[193] Once the binding has occurred, the drug (which can be either inside or outside the nanotube) is released due to changes in pH or enzymes produced by the tumor, thereby delivering the drug directly to diseased areas of the body.[194] Nanoshells, on the other hand, are beads coated with gold that are linked to an antibody,[195] which binds to the target. Once the binding occurs, near-infrared light is applied causing the light-absorbing nanoshells to generate heat, thereby destroying the targeted cells.[196] In either case, nanodevices are used to minimize the exposure of healthy tissues while more precisely targeting those areas that need to be treated.[197]

In 2004, the National Cancer Institute launched the "Cancer Nanotechnology Plan," which is an attempt by the government to increase the visibility of nanomaterials and nanoscale devices and to facilitate research into this subject matter. The Plan is based on a belief that a concerted, multidisciplinary research will

---

188. U.S. Dept. of Health and Human Services, National Institutes of Health and National Cancer Institute, "Cancer Nanotechnology Plan" (July 2004) at 3.

189. National Cancer Institute, *Understanding Cancer and Related Topics: Understanding Nanodevices, available at* http://www.cancer.gov/cancertopics/understandingcancer.

190. *See id.*

191. *See id. See also*, National Cancer Institute, *Carbon Nanotubes Target Tumor Cells, Deliver Anticancer Drugs, available at* http://www.nano.cancer.gov/news_center/2008/ aug/nanotech_news_2008—08-14b.asp; National Cancer Institute, *Nanoparticles Deliver DNA-Drug Combos to Tumors, available at* http://www.nano.cancer.gov/news_center/ nanotech_news_2006-10-16a.asp.

192. Y. Liu & H. Wang, *Nanotechnology Tackles Tumours,* 2(1) NATURE NANOTECHNOLOGY 20.

193. *See id.*

194. *See id.*

195. National Cancer Institute, *supra* note 188.

196. *See id.*

197. Liu & Wang, *supra* note 191.

result in the development of detection and anticancer therapeutic drugs under the ambitious goal of achieving elimination of death from cancer by 2015. The Plan sets forth milestones as a concrete measure of determining whether the goal can be achieved. For example, within a three-to-five-year period, the Plan envisions clinical trials and the filing of a drug application for the first nanoscale imaging agent. To achieve these goals, eight regional Centers of Cancer Nano-technology Excellence have been established. Each of these centers has both advanced biocomputing capabilities and partnerships with not-for-profit and private technology entities.

The first therapeutic nanotech drug to be approved by FDA is a chemotherapy drug for those suffer from metastatic breast cancer. The drug is known as Abraxane™, which consists only of albumin-bound paclitaxel nanoparticles.[198] According to the manufacturers, the compound paclitaxel is considered one of the most effective anti-cancer compounds. However, because of its being insoluble in water, it has been necessary to use a toxic solvent to put the compound into solution in order to inject into a patient.[199] Abraxane™ uses human albumin (a protein in the body that is a carrier for hydrophobic nutrients and other compounds) rather than a solvent as a means of creating an injectable suspension that contains paclitaxel with a mean particle size of approximately 130 nanometers (or a hundredth the size of single red blood cell).[200] By not using solvents, this drug avoids the side effects such as difficulty in breathing, hives, swollen eyes and lips, flushed face, and severe allergic reactions (hypersensitivity reactions) along with the need for pretreatment steroids and antihistamines to address these issues.[201] However, not only does Abraxane™ eliminate the need for pretreatment, but clinical studies demonstrate that Abraxane™ is more effective (e.g., study patients with Abraxane™ had a 21.5 percent versus an 11.5 percent response rate with a solvent-based paclitaxel) and that it can be administered in a shorter period of time (30 minutes versus 3 hours).[202]

To obtain approval, the manufacturers filed an NDA in March 2004 with supporting documentation indicating the results from Phase I and Phase II studies as well as a randomized controlled Phase III study that compared the efficacy of the solvent-based paclitaxel and Abraxane™. The manufacturers submitted the NDA under FDA's Fast Track designation (which is meant for drugs that address

---

198. FDA, Patient Information Enclosed: "Abraxane® for Injectable Suspension (paclitaxel protein-bound particles for injectable suspension) (albumin-bound)," (May 2007) at 1.

199. S. Johnson., *FDA Approves ABRAXANE for Metastatic Breast Cancer*, HEALTHSTAR PR (Jan. 7, 2005).

200. *See id.*; FDA, *supra* note 197, at 1.

201. Abraxane, *Benefits of ABRAXANE®*, *available at* http://www.abraxane.com/benefits-cancer-treatment.aspx.

202. *See id.*

unmet medical needs) and sought priority review designation for the marketing application (typically a six-month review).[203] The drug was actually approved in approximately 10 months.

B. Devices

Nanoparticles are also viewed as being a tremendous benefit in the early and accurate detection of diseases. The current method for detection of cancer, for instance, is through physical examination, symptoms, or imaging.[204] However, nanoparticles could be used to enhance imaging or detect precancerous changes in cells,[205] for example by the use of magnetic nanoparticles such as manganese-doped magnetism-engineered iron oxide as a powerful contrast agent in high performance magnetic resonance imaging.[206] Finally, researchers envision the possibility of linking both detection and drug delivery into a single nanoscale device.

However, in tandem with the research on the beneficial aspects of nanomaterials, questions have been raised about their unintended adverse consequences. The most relevant route into and through the body for nanomaterials is the circulatory system. As noted in Chapter 3, for researchers there are questions as to the toxicity of the nanomaterial as well as their degree of persistence in the body. However, as the findings of new research examining toxicity are that some nanomaterials (such as carbon nanotubes) are more toxic than others. This is because researchers were not previously appreciating that carbon nanotubes also contain metals and "amorphous impurities" that can have an impact on the nanomaterial's electronic character, transformation, and toxicology.[207] As the nanomaterials are subject to clinical testing, further research on their potential for toxicity and the reasons that some materials may be toxic will have to be explored.

---

203. American Pharmaceutical Partners, *American Pharmaceutical Partners and American BioScience Announce Filing of a New Drug Application for FDA Approval for ABRAXANE(TM) for the Treatment of Metastatic Breast Cancer* (Mar. 8, 2004), *available at* http://phx.corporate-ir.net/phoenix.zhtml?c=130431&p=irol-newsabbi&nyo=4.

204. National Cancer Institute, *supra* note 188.

205. *See id.*

206. Y. M. Huh et al., *Hybrid Nanoparticles for Magnetic Resonance Imaging of Target-Specific Viral Gene Delivery*, 19 ADVANCED MATERIALS 3109 (2007).

207. M. Berger, *Comparing Apples with Oranges—The Problem of Nanotube Risk Assessment*, *available at* http://www.nanowerk.com/spotlight/spotid=5248.php (Apr. 10, 2008).

8. COSMETICS

I. OVERVIEW OF REGULATION OF COSMETIC PRODUCTS

Cosmetic companies have begun to manufacture products that contain nanoscale delivery systems or nanoscale ingredients (e.g., nanoscale gold, silica, titanium dioxide, and zinc oxide).For example, in 1998, L'Oreal, a cosmetic manufacturer, was among the first companies to market a product that incorporated a nanoscale delivery system—namely, by using a polymer nanocapsule to deliver its ingredients (i.e., retinol)—in the anti-wrinkle cream called Plentitude Revitalift™.[1] Unlike other anti-wrinkle creams, it was reported that these capsules would act like sponges, soaking up and holding the cream inside until the outer shell dissolved under the skin.[2] Further, it was reported that these capsules could penetrate the skin's protective barrier and interact beneath the skin's surface to produce new cells that "firm up" the skin.[3]

The description of the L'Oreal product, however, highlights the very issue that is central to this chapter—namely, whether the presence of nanoscale delivery systems or nanoscale ingredients results in a product not only qualifying under the FFDCA's definition of a "cosmetic" but also qualifying under the FFDCA's definition of a "drug." From a regulatory perspective, for a cosmetic product to be considered a drug, it must be, as with any other drug, intended for use in the diagnosis, cure, mitigation, treatment, or prevention of disease; intended to affect the structure or any function of the human body; listed in one of the official compendium noted in Chapter 7, or intended as a component in any article that falls within the other three categories. As discussed in detail in Chapter 7, FDA has a stringent set of requirements pertaining to the regulation of drugs both prior to marketing and post-marketing (e.g., reporting adverse incidents). FDA also has regulatory authority over cosmetics. However, cosmetics are primarily

1. T. Little et al., *Beneath the Skin: Hidden Liabilities, Market Risk and Drivers of Change in the Cosmetic and Personal Care Industry* (2007), *available at* http://www.iehn.org at 12. *See also, New Report Slams Nanotechnology in Cosmetics* (Feb. 22, 2007), *available at* http://www.nanowerk.com/news/newsid=1505.php; C. H. Deutsh, *Cosmetics Break the Skin Barrier*, N.Y. TIMES (Jan. 8, 2005), *available at* http://www.nytimes.com.

2. R. Paull, The Forbes/Wolfe Nanotech Report, 12.29.03, 12:16 PM, *The Top Ten Nanotech Products of 2003*, *available at* Forbes.com. http://www.forbes.com/home/2003/12/29/cz_jw_1229soapbox.html.

3. L. Rogers, *Safety Fears Over 'Nano' Anti-Ageing Cosmetics*, THE SUNDAY TIMES (July 17, 2005), *available at* http://www.timesonline.co.uk/tol/news/uk/article544891.ece.

regulated through requirements that, if followed by the manufacturer, would result in the company not triggering the adulteration and misbranding provisions of the statute.[4]

To date, FDA has not issued any rule or guidance document that specifically addresses how cosmetics infused with nanoscale delivery systems or ingredients will be regulated. Instead, FDA has issued a report and held public meetings to examine whether the infusion of nanomaterials will actually result in physiological effects. Thus, while FDA considers these issues, cosmetics containing such materials will be treated in the same manner as cosmetics comprised solely of conventionally-sized particles. Thus, this chapter first discusses the current set of regulations that govern all cosmetics and their manufacturing. Then, given the possibility that FDA will regulate cosmetics containing nanomaterials differently, the chapter will examine broadly the "drug versus cosmetic" debate examining the warning letters that have already been sent to companies that manufacture cosmetics with nanomaterials as well as the specific issue of sunscreen.

II. FDA STATUTORY AUTHORITY TO GOVERN COSMETICS

FDA regulates the marketing of cosmetics under two federal statutes: the Federal Food, Drug, and Cosmetic Act, 21 U.S.C. § 301 et seq., (FFDCA), and the Fair Packaging and Labeling Act, 15 U.S.C. § 1451 et seq., (FPLA). As noted in prior chapters, the original FFDCA was enacted in 1938 and has been amended several times with respect to foods, drugs and medical devices. Regarding cosmetics, however, there have not been significant amendments (except with respect to color additives). The FPLA was enacted in 1966 in response to growing consumer concern with deceptive packaging practices.[5] FFDCA, however, is the primary statutory authority because the FPLA includes a "savings provision" that holds that it will not repeal, invalidate, or supersede the FFDCA.

A. Authority under the Federal Food, Drug, and Cosmetic Act

The FFDCA defines the term *cosmetic* as "(1) articles intended to be rubbed, poured, sprinkled, or sprayed on, introduced into, or otherwise applied to the human body or any part thereof for cleansing, beautifying, promoting attractiveness,

---

4. M. Taylor, *Regulating the Products of Nanotechnology: Does FDA Have the Tools It Needs?* Woodrow Wilson International Center for Scholars. Project on Emerging Nanotechnologies (October 2006), p. 28. *See also,* G. Yingling and S. Onel, Chapter 8: Cosmetic Regulation Revisited, *in* Food and Drug Law and Regulation 249–270 (2009).

5. "When he signed the bill into law, President Johnson said that: 'The housewife should not need a scale, a yardstick, or a slide rule when she shops. This law . . . will protect her from being shortchanged by 'slack filling'—where a box is made bigger than its contents.'" (internal citations omitted). P.B. Hutt, *Development of Federal Law Regulating Slack Fill and Deceptive Packaging of Food, Drugs, and Cosmetics,* 42 Food Drug Cosm. L.J. 1–37 (1987).

or altering the appearance, and (2) articles intended for use as a component of any such articles; . . . "[6]

The statute's primary purpose is to prohibit the introduction, delivery, manufacture, or sale of *adulterated* or *misbranded* cosmetics into interstate commerce.[7] A cosmetic is deemed to be "adulterated" if, among other things:

(a) it bears or contains any poisonous or deleterious substances which may render it injurious to users under the conditions of use prescribed in the labeling or under conditions of use as are customary and usual (except as applicable to coal-tar hair dye);

(b) it consists in whole or in part of any filthy or putrid or decomposed substance;

(c) it has been prepared, packed, or held under unsanitary conditions whereby it may have become contaminated with filth, or whereby it may have been rendered injurious to health; or

(d) its container is composed, in whole on in part, of any poisonous or deleterious substance which may render the contents injurious to health; or except for hair dyes, it is, or it bears or contains, a color additive which is unsafe as defined by the statute.[8]

A cosmetic shall be deemed misbranded under, among other things, the following conditions:

(a) "its labeling is false or misleading in any particular;"

(b) if in package form unless it bears a label containing (1) the name and place of business of the manufacturer, packer or distributor; and (2) an accurate statement of the quantity of the contents in terms of weight, measure, or numerical count (with an exception for small packages);

(c) "if any word, statement, or other information required by or under authority of this act to appear on the label or labeling is not prominently placed thereon with such conspicuousness (as compared with other words, statements, designs, or devices, in the labeling) and in such terms as to render it likely to be read and understood by the ordinary individual under customary conditions of purchase and use;" or

(d) the container is so made, formed, or filled as to be misleading;[9]

---

6. *See* 21 U.S.C. § 321(i).

7. If a cosmetic is repackaged, labeled, or processed at a facility that is different from where it was originally processed, labeled, or packaged, and that is the typical practice in the industry, then that facility is not subject to labeling requirements, provided that the cosmetic was not adulterated or misbranded as result of the actions at this facility. *See* 21 U.S.C. § 363.

8. *See* 21 U.S.C. § 361.

9. *See* 21 U.S.C. § 362. There are also specific misbranding provisions that relate to color additives. 21 U.S.C. § 362 (e).

As noted in FDA's implementing regulations, the labeling of a cosmetic is misleading if it fails to reveal facts that are: (a) material in light of other representations made or suggested by statement, word, design, device, or any combination thereof; or (b) material with respect to consequences which may result from use of the article under (i) the conditions prescribed in such labeling or (ii) such conditions of use as are customary or usual.[10]

If FDA determines that the product has been misbranded or adulterated, the agency can take an enforcement action, including, requesting that the manufacturer institute a recall; seeking to enjoin the sale of the product; seizing the product; or seeking criminal prosecution.[11] However, FDA enforcement action against manufacturers for violations of the adulteration and misbranding provisions has been limited.[12]

B. Authority under the Fair Packaging and Labeling Act

The purpose of FPLA is to assist consumers by requiring that a manufacturer's packages, and their labels provide accurate information as to the quantity of the contents of the product as well as facilitate value comparisons with other like products.[13] With certain specific exemptions, the statute is applicable to *consumer commodities* such as cosmetics that are customarily produced or distributed for sale through retail establishments and that are used by individuals for personal care. The statute prohibits the distribution of any consumer commodity unless the label identifies the commodity; the name and place of business of the manufacturer, packer, or distributor; and the net quantity of contents (in terms of weight or mass, measure, or numerical count).[14] Any consumer commodity that is introduced or delivered for introduction into commerce in violation of any of the FPLA provisions or the implementing regulations will be considered misbranded within the meaning of FFDCA. However, the penalties set forth under the FFDCA shall not be applicable to any deceptive or unfairly packaged or labeled product.[15] Rather, the manufacturer that has allegedly violated the FPLA can be subject to a Federal Trade Commission action in which the Commission has the authority to, among other things, issue a cease and desist order to the manufacturer.

---

10. *See* 21 C.F.R. § 1.21(a).

11. *See* 21 U.S.C.A. § 332, 21 U.S.C.A. § 334, 21 C.F.R. § 7.40, 21 C.F.R. § 7.45.

12. *See* Yingling, *supra* note 4, at 253, 258.

13. *See* 15 U.S.C. § 1451.

14. *See* 15 U.S.C. §§ 1452(a), 1453. With certain limitations, the prohibition does not apply to persons engaged in business as wholesale or retail distributors of consumer commodities, provided that these distributors are not involved in the packaging or labeling of these commodities or involved in prescribing how the packaging or labeling should be done. *See* 15 U.S.C. § 1452(b).

15. 15 U.S.C. § 1456.

III. COSMETIC REGULATION

A cosmetic ingredient, except a color additive, is not subject to FDA premarket review.[16] Color additives that are used in cosmetics are the sole exception to this rule, and such additives now must be tested for safety and approved by FDA prior to marketing. For non-color additive ingredients, a manufacturer needs to "adequately substantiate" that the ingredient and the finished product is safe, and the manufacturer has to comply with the labeling obligation in order to ensure that the cosmetic is not deemed to be misbranded.

A. Substantiation for Safety

FDA requires manufacturers to adequately substantiate the safety of each ingredient used in the manufacture of a cosmetic, as well as the finished cosmetic product itself, before marketing the product.[17] Manufacturers use different test methods to examine the toxicity, dosage, and exposure of the ingredients and finished products in order to establish the safety of their products.[18] If, however, an ingredient or finished product is not adequately substantiated for safety, then a manufacturer must place a warning that conspicuously[19] states on the principal display panel that the safety of this product has not been determined.[20] In addition to their own testing, manufacturers may decide to rely upon determinations made by the Cosmetic Ingredient Review (CIR) Expert Panel. The CIR Expert Panel is comprised of seven voting members publicly nominated by consumer, scientific, and medical groups who are physicians and scientists, and three nonvoting members from FDA, the Consumer Federation of America, and the Personal Care Products Council (formerly known as the Cosmetic, Toiletry, and Fragrance Association).[21] Historically, FDA has deferred to the decisions made by this quasi-independent body as being an appropriate determination

---

16. *Color Additives: FDA's Regulatory Process and Historical Perspectives.* CFSAN/Office of Cosmetics and Colors (October/November 2003), *available at* http://www.cfsan.fda.gov/~dms/col-regu.html.

17. *See* 21 C.F.R. § 740.10(a).

18. *See* Yingling, *supra* note 4, at 262.

19. *Conspicuously* means that it will be appear prominently in comparisons to other words, statements, designs, or devices, such as in bold type on contrasting background, to render it likely to be read and understood by the ordinary individual under customary conditions of purchase and use. FDA may establish by regulation an acceptable alternative method (e.g., change in the font size). *See* 21 C.F.R. § 740.2.

20. *See* 21 C.F.R. § 740.10(a). There are specific warning statements that are applicable to cosmetics in self-pressured containers (21 C.F.R. § 740.11), to feminine deodorant sprays (21 C.F.R. § 740.12), to foaming detergent bath products (21 C.F.R. § 740.17), to certain coal tar hair dyes (21 C.F.R. § 740.18), and to sun tanning preparations (21 C.F.R. § 740.19).

21. *See How Does CIR Work?*, *supra* note 23; *see also, supra* note 22.

of safety.[22] The CIR Expert Panel's review is limited to ingredients with the man-ufacturer still being responsible for substantiating the safety of the finished cosmetic product.

The CIR Expert Panel reviews the safety of the proposed ingredient based on an examination of: (a) the chemistry, including physical properties, of the ingre-dient and how it is manufactured; (b) the uses for the ingredient, including its cosmetic and non-cosmetic uses; (c) the general biological impacts of the ingre-dient, including its absorption, distribution, and metabolism in humans, as well as its toxicity based on acute, short-term, sub-chronic, and chronic studies; (d) dermal irritation and sensitization caused by the ingredient; and (e) epidemi-ology studies and other clinical data.[23] The Expert Panel solely examines pub-lished scientific research; it does not conduct its own studies. However, if the existing research is deemed to be insufficient, the Expert Panel may make a request to the industry or to interested groups that further research be conducted or that unpublished research be provided.[24] The CIR classifies ingredients into four categories: safe as used, safe with qualifications, insufficient data, and unsafe.[25] The CIR Reports are available for purchase and are eventually pub-lished in the *International Journal of Toxicology*.[26] To date, it does not appear that a CIR Expert Panel has substantiated the safety of a nanoscale ingredient. However, the panels have examined the safety of some of their conventionally sized counterparts, and thus, the critical issue, as discussed in chapter 3, is whether the data and safety conclusions associated with the conventional-sized counterpart can be applied to the nanoscale version.

B. Labeling

Manufacturers are also required to comply with labeling requirements, which are similar to those applicable to drugs. These requirements are set forth in 21 C.F.R. Part 701. In particular, in the case of cosmetics, manufacturer must ensure compliance with the "ingredient declaration" requirement. The term *declaration* means a listing of a cosmetic's ingredients in their order of predominance.[27] FDA provides a detailed set of regulations as to how an ingredient should be designated.[28] If an ingredient is omitted, misidentified, or inaccurately described, the product could be deemed to be misbranded and thus subject to an enforcement action. The key points relevant to every cosmetic manufacturer are highlighted below.

---

22. S. Washam, *Safe Cosmetics Act Aims to Lessen Cancer Risk*, 98 (20) J. Nat. Cancer Inst. (Oct. 18, 2006).

23. *See id.*

24. *See id.*

25. *See* Cosmetic Ingredient Review, *Cosmetic Ingredient Findings: 1976–Current*, *available at* http://www.cir-safety.org/findings.shtml.

26. *See id.*

27. *See* 21 C.F.R. § 701.3.

28. *See id.*

Ingredient declarations are not applicable to "incidental ingredients that are present in the cosmetic at insignificant levels, *and* that have no technical or functional effect in the cosmetic."[29] The ingredients in this category are (a) substances that are present in the cosmetic because they are an ingredient in another cosmetic product, but serve no technical or functional purpose; and (b) processing aids (e.g., substances that are added to the cosmetic during processing but are removed in accordance with good manufacturing practices before the cosmetic is packaged in its finished form).[30] Where one or more ingredients are accepted by FDA as exempt from public disclosure, the label declaration may simply identify the ingredients as "and other ingredients" at the end of the declaration.[31] For ingredients that must be declared, the manufacturer must place the information in a prominent and conspicuous location so that it is likely to be read and understood by ordinary individuals under normal conditions of purchase.[32]

The regulations also specify how the ingredient may be listed. Specifically, a manufacturer may:

(a) *Name* each ingredient in descending order of predominance.[33] The name of each ingredient should be listed as established by FDA Commissioner, or in the absence of a FDA established name, then as listed by the following sources: (i) the *CFTA Cosmetic Directory*, 2nd Edition (1977) except as to or with qualifications respecting certain ingredients as enumerated in the regulations; (ii) *United States Pharmacopeia* (USP), *19th Ed.* (1975), and Second Supplement to USP XIX and National Formulary XIV (1976); (iii) *National Formulary* (NF) *14th Ed.* (1975), and Second Supplement to UPS XIX and NF XIV (1976); (iv) *Food Chemicals Codex, Second Ed.* (1972), First Supplement (1974) and Second Supplement (1975); and the (v) USAN and the USP dictionary of drug names, USAN 1975, 1961–1975 cumulative list.[34] In the absence of a listing in any of these sources, the manufacturer should rely upon a name as generally recognized by consumers, or the chemical, other technical name, or description of the cosmetic.

(b) *Group* the ingredients. The groups must be listed in the following order: (i) ingredients other than color additives, present in concentrations greater than 1 percent, in descending order of predominance; (ii) ingredients other than color additives present at a concentration of not more than 1 percent, without respect to order of predominance; and then (iii) color additives without respect to order of predominance.

---

29. *See* 21 C.F.R. § 701.3(l).
30. *See id.*
31. *See id.*
32. *See id.*
33. *See* 21 C.F.R. § 701.3. This requirement does not apply to fragrance or flavor, which may simply be listed as "fragrance" or "flavor."
34. *See id.*

C. Color Additives

As noted above, there are specific requirements applicable to color additives. The regulatory structure governing the use of color additives is based on § 721 of the FFDCA. Specifically, color additives are defined as ". . . any material . . . that is a dye, pigment, or other substance made by a process of synthesis or similar artifice, or extracted, isolated, or otherwise derived, with or without intermediate or final change of identity, from a vegetable, animal, mineral, or other source and that, when added or applied to a food, drug, or cosmetic or to the human body or any part thereof, is capable (alone or through reaction with another substance) of imparting a color thereto" (e.g., synthetic organic dyes, lakes, or pigments).[35] To use a color additive in or on foods, drugs, or cosmetics, or in coloring the human body, FDA must have first listed the color additive in the Federal Register for such a purpose through the batch certification process.[36] Then the color additive must be used under the conditions prescribed by FDA (e.g., limiting the quantity that can be used or the manner in which the color additive is, in fact, added).[37]

Typically, for a color additive to be listed in the Federal Register, it must be approved via FDA's certification process,[38] which involves a manufacturer or importer submitting a petition to FDA proposing the listing of the color additive. The content of a petition is set forth in 21 C.F.R. § 71.1(c). Alternatively, the Secretary of Human Services may on his own initiative seek to list an additive.[39]

For a color additive to be approved, the data must demonstrate that the substance will be safe under its intended uses. The statutory criteria for determining safety includes factors that are particularly relevant if the color additive is used in food or drugs rather than in a cosmetic. Specifically, the statute states that a color additive's safety will be determined by examining the: (a) the probable consumption of, or other relevant exposure from, the additive and of any substance formed in or on food, drugs or devices, or cosmetics because of the use of the additive; (b) the cumulative effect, if any, of such additive in the diet of man or animal, taking into account the same or any chemically or pharmacologically related substance or substances in such diet; (c) safety factors which, in the opinion of experts qualified by scientific training and experience to evaluate the safety of color additives for the use or uses for which the additive is proposed to be

---

35. See 21 C.F.R. § 70.3. See also, Color Additives: FDA's Regulatory Process and Historical Perspectives. CFSAN/Office of Cosmetics and Colors. Oct./Nov. 2003, available at http://www.cfsan.fda.gov/~dms/col-regu.html.

36. See 21 U.S.C. § 379e(a)(1); see also, 21 C.F.R. § 71.20. A color additive may also be listed without compliance with the batch certification process if the Secretary of Human Services exempts it from the certification process. See 21 U.S.C. § 379e(a)(1).

37. See 21 C.F.R. § 71.20(a)(1).

38. See 21 C.F.R. § 71.20(a)(2).

39. See 21 C.F.R. § 71.1(c).

listed, are generally recognized as appropriate for the use of animal experimentation data; and (d) the availability of any needed practicable methods of analysis for determining the identity and quantity of (i) the pure dye and all intermediates and other impurities contained in such color additive; (ii) such additive in or on any article of food, drug or devices, or cosmetic; and (iii) any substance formed in or on such article because of the use of such additive.[40]

IV. VOLUNTARY REGISTRATION SYSTEM

In the 1970s, as the number of cosmetic ingredients increased and the technology for testing chemicals in compounds and products became more sophisticated and accurate, questions were raised about the potential for dermal absorption, allergic responses, and other possible impacts on the human body.[41] Moreover, congressional efforts were being contemplated and debated that would have placed cosmetics under a regulatory regime similar to drugs—namely, the imposition of premarket testing and registration and adverse incident reporting. These efforts were resisted both by the cosmetic industry and FDA. In an effort to placate some of those concerned, FDA established a voluntary registration system that applies to (a) the manufacturing facility, and (b) the types of cosmetic product ingredients that are manufactured at the facility. This system is referred to as the Voluntary Cosmetic Registration Program (VCRP)—which is set forth in Part 710 of the regulations.[42] The main purpose of the system is to provide an informational center for those in the cosmetic industry so that manufacturers, packers, and distributors can share information with each other about the ingredients.[43] Additionally, FDA can use the contact information provided through the VCRP to inform a manufacturer if a particular ingredient or product failed to meet any safety tests. However, the VCRP has not been used very effectively by FDA.

A. Voluntary Registration of Establishments
Owners or operators may voluntarily register each facility that engages in the manufacturing or packaging of a cosmetic product (regardless of whether the product that it manufactures actually enters into interstate commerce).[44] According to the

---

40. *See* 21 U.S.C. § 379e(c)(5).

41. *See Regulating the Products of Nanotechnology:Does FDA Have the Tools It Needs?* Woodrow Wilson International Center for Scholars. Project on Emerging Nanotechnologies (October 2006).

42. *See* 39 Fed. Reg. 10059–10060 (Mar. 15, 1974).

43. *See generally* FDA's Voluntary Cosmetic Registration Program, *available at* http://www.cfsan.fda.gov/~dms/cos-regn.html.

44. 21 C.F.R. § 710.1. Certain classes of companies/entities are not requested to submit a registration, including beauty shops, retailers, physicians, hospitals, and clinics. Persons who manufacture compounds, prepare compounds, or process cosmetic products solely

voluntary program that has been established, the registration should occur within 30 days *after* the commencement of operations at the facility via submission of a Form FD-2511 (Registration of Cosmetic Product Establishment) to FDA. The Form requests information on the name and address of the cosmetic product establishment, all business trading name(s) used by the establishment, and the type of business (manufacturer and/or packer). The information requested should be given separately for each manufacturing facility. If there is any change, the registrant has an obligation to amend the registration within 30 days. FDA will provide the registrant with a validated copy of Form FD-2511 as evidence of registration. However, as FDA has noted, registration of an establishment or assignment of a registration number does not denote approval of the firm.

B. Voluntary Registration of Ingredients

A manufacturer, packer, or distributor of a cosmetic product may voluntarily file a cosmetic product ingredient composition statement (Form FDA 2512), regardless of whether the cosmetic product enters interstate commerce. The Form FDA 2512 should be filed within 60 days *after* the commencement of commercial distribution of any product. Form FDA 2512 requests that the applicant submit the following information: the name and address of the person designated on the label,[45] the brand name or names of the cosmetic product, the ingredients in the product,[46] and the cosmetic product category or categories.[47]

If there is a change in the ingredient(s) or the brand name(s) of the product, the applicant is asked to submit an amended Form FDA 2512 within 60 days after the product enters into commercial distribution. Any other change should be noted in an amended Form 2512 within a year after the change. An amended Form FDA 2512 should also be sent within 180 days of the submitter knowing that commercial distribution of the product has been discontinued.

When a Form FDA 2512 is submitted, FDA will assign either a permanent cosmetic product ingredient statement number or an FDA reference number and inform the submitter of said number. But the receipt of such a number does not indicate FDA approval of the product. Furthermore, if a product's labeling or advertising creates the impression of official approval because of

---

for use in research, pilot plant production, teaching, or chemical analysis and who do not sell these products are also not requested to submit a registration. 21 C.F.R. § 710.9.

45. 21 C.F.R. § 720.4(a). If the manufacturer or packer is different from the person designated on the label, and the manufacturer or packer submits the Form, then the name and address of the manufacturer or packer must be provided as well.

46. FDA indicates in 21 C.F.R. § 720.4(d) the way information on ingredients should be provided. For example, with certain exceptions, a list of each ingredient in the cosmetic product should be provided in descending order of predominance by weight. Additionally, an ingredient should be listed by the name adopted by FDA, or in the absence of such name, its common or usual name, and if that is not available, then its chemical or technical name.

47. FDA has an extensive list of categories at 21 C.F.R. § 720.4(c). The applicant should cite the appropriate category to indicate the product's intended use.

the voluntary filing and/or receipt of a number, it will be considered to be misleading the public.[48]

V. ENFORCEMENT POWERS

If FDA determines that a product has been misbranded or adulterated, the agency has the authority to request that the manufacturer institute recall or alternatively, it may seek to enjoin the introduction of the product into interstate commerce, seize the product and/or pursue criminal prosecution.[49] With respect to a recall action, FDA may make such a request if: (a) a product presents a risk of illness or injury or there has been a gross consumer deception, (b) the company has not initiated a recall, and (c) a recall is necessary to protect the public health and welfare.[50] If FDA makes a recall request, a manufacturer should consult FDA's published guidance in 21 C.F.R. Part 7 on how to deal with the mechanics of the recall (e.g., communication to both the public and the agency).

FDA has the authority to inspect a facility without prior notification at reasonable times and in a reasonable manner to determine if the cosmetics are safe and properly labeled and to identify possible health risks and other violations of the law.[51] During inspection, FDA can examine manufacturing processes, research activities, marketing materials (e.g., Web sites, brochures, videos, etc.)[52] and samples. If, due to an inspection or otherwise, FDA believes that the manufacturer is in violation of safety standards or other regulatory requirements, the agency will issue a warning letter, although such a letter is not required by the statute. The manufacturer then will typically have an opportunity to voluntarily correct the alleged violations. The failure to comply with a warning letter can lead to other enforcement action.

VI. DRUG VERSUS COSMETIC DEBATE AND THE ROLE OF NANOMATERIALS

As discussed above, the fulcrum of the debate on nanomaterials is whether they result in a product which would otherwise be considered a "cosmetic" falling within the definition of a "drug." The key issues are whether there are physiological

---

48. 21 C.F.R. § 720.9.

49. See 21 U.S.C.A. § 332, 21 U.S.C.A. § 334, 21 C.F.R. § 7.40, 21 C.F.R. § 7.45.

50. See 21 C.F.R. § 7.45.

51. 21 U.S.C. § 374.

52. See, e.g., FDA Warning Letter to Ms. Randi Schnider, President, Fusion Brands International SRL (Apr. 24, 2007) File #07-NWJ-11. ". . . FDA has reviewed your Internet web site . . . and the labeling for these products, including the literature that accompanies these products when shipped to customers."

effects associated with these materials and whether the manufacturer intends the product to have a use in the diagnosis, cure, mitigation, treatment, or prevention of disease, or as something that will affect the structure or any function of the human body. As to the physiological effects, research on the toxicology, exposure and dose of different nanomaterials is ongoing (see discussion in Chapter 3), especially with respect to nanoscale titanium dioxide and zinc oxide.

With respect to "intended use," *intended use* refers to the objective intent of the person who is legally responsible for the labeling of drugs.[53] The objective intent may be demonstrated by claims stated on the product label, in advertising, on web sites, in other promotional materials (e.g., audiocassettes), or through oral or written statements by the responsible person or his/her representatives.[54] Alternatively, objective intent may also be shown by the circumstances under which the product is offered and used for a purpose for which it is neither labeled nor advertised, provided that persons knowledgeable of the product knew about the use.[55] Thus, if a manufacturer asserts in its own marketing and advertising material that the ointment has therapeutic powers, it is likely that the product will be classified by FDA as a drug. However, critics of FDA's regulatory program point out that a company may simply avoid the scrutiny of drug regulation by not making any claims that could trigger the drug standard even though the product may have actual effects on the body.[56] There is also the issue of FDA's resources in discovering such activity as well as the data necessary to establish that the product is actually having an physiological effect.

The issue of drug versus cosmetic can be understood by examining two items: (a) warning letters that have been issued to companies that use nanoscale ingredients in their cosmetic products and (b) the issue of titanium dioxide and zinc oxide in sunscreens.

An illustrative example of when FDA considers a cosmetic to be a drug is a warning letter sent to a Canadian company, Fusion Brands International SRL, that manufactures a "face- lift." FDA reviewed the company's Web site, the

---

53. 21 C.F.R. § 201.128; *see also* CFSAN (Office of Cosmetics and Colors) Intercenter Agreement Between the Center for Drug Evaluation and Research and the Center for Food Safety and Applied Nutrition to Assist FDA in Implementing Drug and Cosmetic Provisions of the Federal Food, Drug, and Cosmetic Act for Products that Purport to be Cosmetics But Meet the Statutory Definition of a Drug (June 1, 2006), *available at* http://vm.cfsan.fda.gov/~dms/cos-mou.html.

54. 21 U.S.C. § 321(g)(1)(C).

55. *See id.* FDA notes that "intended use" may be established by consumer perception, which may be established through the product's reputation, and through ingredients that may cause a product to be considered a drug because of their well-known (to the public and industry) therapeutic uses. *See* CFSAN, Office of Cosmetics and Colors, *Is it a Cosmetic, a Drug, or Both? (or Is it Soap?)* (July 8, 2002), *available at* http://www.cfsan.fda.gov/~dms/cos-218.html.

56. *See* Little, *supra* note 1, at 8.

product labels, and the marketing literature that accompanied the product and FDA concluded that the company's intent was to persuade customers that the product would have an effect on how the body functioned (e.g., relaxed muscles) or the body's structure (e.g., firmed skin). The following are examples of the company's claims that FDA pointed to:

The first Topical-Injectable™ alternative to doctor-administered anti-wrinkle injections: proven more effective than Botox® in a clinical study.

LiftFusion™, a Topical-Injectable™ with . . . active ingredients, helps reduce existing wrinkles AND boost collagen to promote skin's natural defenses against new ones.

M-Tox™ formula features patented nanosphere technology that delivers a powerful blend of anti-aging elements to actively counteract fine lines and deep wrinkles.

[B]locks muscle contractions within 10 minutes of application to the skin which helps prevent new lines and wrinkles from forming.

Nanospheres-spheres that deliver Mtox by relaxing muscles.

Mtox helps to prevent new wrinkle from forming by reducing repetitive facial muscle contractions.[57]

---

57. Douglas I. Ellsworth, District Director, Public Health Service, FDA, Warning Letter to Randi Schinder, President, Fusion Brands International SRL, April 24, 2007. Other claims made were:

LiftFusion's™ . . . wrin kle-repairing results are . . . measurably proven better than Botox® in a clinical comparative study.

Until now, eliminating lines and wrinkles effectively has required painful, costly and regular injections of muscle inhibitors like Botox® and fillers like Restylane® . . . LiftFusion™ formula . . . delivers immediate, visibly transformational results . . . without the discomfort, side effects and unnatural loss of facial expressiveness associated with many syringe-administered anti-wrinkle products.

Hyaluronic acid-filling spheres capture and instantly swell with the body's water, plumping to fill and smooth even deep wrinkles, smoothing and lifting to restore skin's youthful firmness.

[V]ertical and horizontal forehead furrows, frown lines, crow's feet + nasolabial lines are . . . repaired.

Another warning letter was sent to Freedom Plus Corporation for its "Nano Cover Facial Spray" concerning its Web site marketing that stated "it increases copper dependent enzyme activity that is essential in the production of elastin," and that "copper complexes cause some types of cancer cells to revert to non-cancerous growth patterns." FDA also noted that some of the dietary supplements sold by the company, such as "Nano Zinc Dietary Supplement" and "Nano Potassium Dietary Supplement," would be considered drugs due to claims of healing and protective powers. *See* Barbara Cassens, District

The issue of nanoscale titanium dioxide and zinc oxide in sunscreen has generated a significant amount of interest. Before examining the particulars of the issue, it is necessary to outline the complicated history involving of the regulation of sunscreen. Generally, a manufacturer cannot market a drug product in the United States unless it has an approved New Drug Application or FDA has issued an Over-the-Counter Monograph (OTC Monograph) that covers the drug. An OTC Monograph provides the conditions under which FDA believes that a product would be "generally recognized as safe and effective" (GRASE) under § 201(p) of the FFDCA and not misbranded under § 502 of the FFDCA.[58] Products containing the active ingredients that are already included in a monograph and that bear the labeling published in a monograph may be marketed without product-specific premarket authorization.

Because consumers expect any product that uses the term *sunscreen* or similar sun protection terminology to protect against the harmful effects of the sun, FDA determined that these products be regulated as drugs. The process for setting standards for sunscreen products began more than 30 years ago. However, with respect to nanomaterials, the Final Monograph for OTC Sunscreen Products issued in 1999 is the relevant document.

In this Final Monograph, FDA explained that it would evaluate engineered nanomaterials by considering them *micronized* forms of their conventionally sized counterparts rather than new ingredients.[59] Specifically, FDA noted that it does not consider micronized titanium dioxide to be a new ingredient but rather

---

Director, PHS, FDA, Warning Letter to Harold Zander, Freedom Plus Corporation (Mar. 29, 2007).

58. *See* 21 C.F.R. § 330.10.

59. *See* 64 Fed. Reg. 27666, 27671–27672 (May 1, 1999). As consumer and environmental groups have argued, regulatory agencies should consider nanoscale substances as "new" ingredients or products because industry has already shown that it views these ingredients as novel by filing patents for them. Moreover, for the Patent Office to grant a patent, it also has the consider that substance, ingredient or product is new or novel. rather than simply different variations of already existing substances. In the case of titanium oxide and zinc oxide, these associations in their petition noted the following U.S. Patents:

U.S. Pat. No. 5,223,250 (Mitchell, 1993) a patent for a "substantially transparent sunblock comprised of micronized particles of zinc oxide;"

U.S. Pat. No. 5,573,753 (Tapley, 1996) for a method of preparing sunscreens containing zinc oxide particles of 5 nm to 150 nm or milling nanoparticles to be substantially transparent to visible light while screening UV radiation;

U.S. Pat. No. 5,531,985 (Mitchell, 1996) for a "visibly transparent UV sunblock composition and cosmetic products containing the same."

U.S. Pat. No. 5,587,148 (Mitchell, 1996) for "visibly transparent UV sunblock agents" comprised of substantially dispersed zinc oxide particles of a specific average particle size range less than about 0.2. micros.

a specific grade of the titanium dioxide already reviewed by FDA.[60] Moreover, FDA, at that time, concluded that there is no evidence that "micronized" titanium oxide and zinc oxide were unsafe. In 2007, FDA reopened the administrative record, and asked for public comments on various testing and labeling issues, but it also asked for public comments on whether nanoscale materials should still be considered micronized versions of conventionally sized particles. FDA noted that as the number of sunscreen products with nanoscale titanium dioxide and zinc oxide has increased, there has also been an increase in the number of questions about the safety of these products. Among those asking questions were a group of environmental and consumer groups that had filed a petition with the lead petitioner being the International Center for Technology Assessment. The petitioners outlined general requests in connection with FDA's regulation of nanomaterials, but they also presented specific requests concerning nanomaterials in sunscreen. The petitioners asked for the following:

1. Reopen the administrative record of the Final OTC Monograph on Sunscreen for the purpose of considering and analyzing information on engineered nanoparticles of zinc oxide and titanium dioxide currently used in sunscreens.

U.S. Pat. No. 5,498,406 (Nearn, 1996) for "titanium dioxide-based sunscreen compositions" having substantially uniform microfine TiO_2 having a particle size of less than about 100 nm.

U.S. Pat. No. 6,187,824 (Swank, 1999) for a "zinc oxide sol and method of making," with a mean particle size of less than 50 nm, that is characterized as clear and transparent; and

U.S. Pat. No. 6,171,580 (Katsuyama, 2001) for an "ultraviolet-screening zinc oxide excellent in transparency and composition" in which zinc oxide particles with an average particle diameter of 50-100 nm "effectively experts the above-described excellent characteristics; i.e. UV-screening effect and transparency and can be applied to a composition for external use such as make-up cosmetics or sunscreen cosmetics."

See generally The International Center for Technology Assessment, et al., Petitioners, filed with: Andrew C. Von Eschenbach in his official capacity as, Acting Commissioner, Food and Drug Administration (May 16, 2006). Petition Requesting FDA Amend Its Regulations for Products Composed of Engineered Nanoparticles Generally and Sunscreen Drug Products Composed of Engineered Nanoparticles Specifically ("Petition") (May 16, 2006).

In fact, the U.S. Patent and Trademark office came out with a new classification for nanotechnology, Class 977, that includes: (a) nanostructure and chemical compositions of nanostructure; (b) a device that includes at least one nanostructure; (c) mathematical algorithms, e.g., computer software, etc., specifically adapted for modeling configurations or properties of nanostructure; (d) methods or apparatus for making, detecting, analyzing, or treating nanostructure; and (e) specified particular uses of nanostructure. Class 977. Nanotechnology Classification, Oct. 2007, available at http://www.uspto.gov/go/classification/uspc977/defs977.pdf.

60. See 64 Fed. Reg. 27671–27672.

2. Amend the Final OTC Monograph on Sunscreen such that sunscreen products containing engineered nanoparticles are not covered under the Monograph and instead are considered new drugs for which manufacturers must complete an NDA.

3. Declare all currently available sunscreen drug products containing engineered nanoparticles of zinc oxide and titanium dioxide as an imminent hazard to public health and order entities using the nanoparticles in sunscreens regulated by FDA to cease manufacture until FDA's Sunscreen Monograph is finalized and broader FDA nanotechnology regulations are developed and implemented.

4. Request a recall from manufacturers of all publicly available sunscreen drug products containing engineered nanoparticles of titanium dioxide and/or zinc oxide until the manufacturers of such products complete new drug applications, those applications are approved by the agency, and the manufacturers otherwise comply with FDA's relevant nanomaterial product testing regulations.[61]

The Personal Care Products Council, an industry group, challenged the assertions in the petition. They point out because there is greater acceptance of clear sunscreen, there is greater use which in turn means more people are protected from harmful ultraviolet rays. Moreover, with respect to safety of the nanoscale materials, they argued that research studies indicate that nanoscale titanium dioxide and zinc oxide are non-toxic when used in sunscreen and cosmetics and that they are not absorbed into the body. Since the date of the submissions, there has been new data on nanoscale titanium dioxide and zinc oxide (which is discussed in chapter 3). To date, FDA has not yet made any determination if it plans to treat products with nanosized titanium dioxide or zinc oxide any differently from products that do not contain such substances. Among the options FDA could adopt are: (a) request additional data, (b) determine that nanoscale materials should be covered under the OTC Monograph, or (c) determine that nanoscale substances are not generally recognized as safe, and thus require the submission of an NDA.

---

61. *See id.*, *supra* note 60, Petition at 3–4. Generally, with respect to all products regulated by FDA, the petitioners are requesting that (a) FDA amend its regulations to include nanotechnology definitions necessary to properly regulate nanomaterial products, including the terms *nanotechnology, nanomaterial,* and *engineered nanoparticle*; (b) FDA issue a formal advisory opinion explaining its position regarding engineered nanoparticles in products it regulates; (c) FDA draft regulations on how to oversee nanomaterial products, establishing and requiring, inter alia, that nanoparticles be treated as new substances; nanomaterials be subjected to nano-specific paradigms of health and safety testing, and that nanomaterial products be labeled to delineate all nanoparticle ingredients; and (d) FDA should require compliance with NEPA by requiring an environmental impact statement instead of a categorical exclusion or submission of an environmental assessment. *See id., supra* note 60, Petition at 3–4.

9. CONSUMER PRODUCTS

I. THE NANOMATERIALS IN CONSUMER PRODUCTS

As each year passes, more and more consumer products infused with nanomaterials are being placed on store shelves. For example, by early spring of 2008, there were reportedly over 600 consumer products or product lines that incorporated nanomaterials.[1] In addition, three to four new nanotechnology consumer products were being placed in the marketplace per week during this same time period.[2] The manufacturers of these products describe the "nano" component as enhancing the strength, durability, flexibility, and performance of the product.[3] For example, in the field of textiles, beyond using nanomaterials to create strain and wrinkle-resistant clothing (which are already on store shelves), nanostrucutured composite fibers could be used in clothing to provide wound healing, self-cleaning, and self-repairing properties.[4]

This chapter focuses on the regulatory entity that has authority over a significant portion of the products purchased every day: the Consumer Product Safety Commission (CPSC).[5] The CPSC is responsible for protecting the public from unreasonable risks of serious injury or death from over 15,000 consumer products, including clothing, household cleaners, electronic devices, appliances, furnishings, building materials, recreational products, products used in schools, toys, and other juvenile products.[6] This chapter primarily examines the way the CPSC may regulate products containing nanomaterials by focusing on its authority under the statutes: the Consumer Product Safety Act, 15 U.S.C. § 2051 et seq., (CPSA) and the Federal Hazardous Substances Act, 15 U.S.C. § 1261 et seq., (FHSA).[7]

---

1. Nanotechnology Project, Analysis, *available at* http://www.nanotechproject.org/inventories/consumer/analysis_draft.

2. Nanotechnology Project, News, *available at* http://www.nanotechproject.org/news/archive/6697/.

3. T. Treye et al., *Research Strategies for Safety Evaluation of Nanomaterials, Part VII: Evaluating Consumer Exposure to Nanoscale Materials*, 91(1) TOXICOLOGICAL SCIENCES 14 (2006).

4. *See generally* Allianz et al., *Small Sizes that Matter: Opportunities and Risks of Nanotechnologies* (2005).

5. Treye et al., *supra* note 3, at 19.

6. *Id.* at 18. CPSC Nanomaterial Statement, *available at* http://www.cpsc.gov/library/cpscnanostatement.pdf.

7. 15 U.S.C. §§ 1261–1278 (2006). The FHSA is one of five statutes administered by the CPSC. The FHSA was enacted in 1960 and has been amended several times, most

Until specific guidelines are issued to address nanomaterials, consumer products that contain them will be evaluated like any other consumer product that possibly contains hazardous substances (i.e., non-nanomaterial containing consumer products).[8]

Finally, this chapter discusses the CPSC's current efforts with respect to nanomaterials and the limitations of the CPSC's power. It should be noted that Congress enacted the Consumer Product Safety Improvement Act[9] that became effective on February 10, 2009 and which amended the CPSA. The law, which was enacted in the wake of a series of incidents concerning lead contamination in popular toys that were manufactured and imported from China, is specifically intended to address the presence of lead in toys and chemicals in plastics used by babies. But, importantly, as discussed below, the law generally strengthened CPSC's enforcement power, increased its budget, and increased penalties for violators.

II. THE CONSUMER PRODUCT SAFETY ACT

Congress established the CPSC as an independent regulatory commission in 1973 when it enacted the Consumer Product Safety Act[10] after an investigation and report to Congress indicated that a number of problems existed with consumer products.[11] The CPSC has jurisdiction over *consumer products*, defined as

> Any article, or component part thereof, produced or distributed (i) for sale to a consumer for use in or around a permanent or temporary household or residence, a school, in recreation, or otherwise, or (ii) for the personal use,

---

notably by the Labeling of Hazardous Art Materials Act (LHAMA), in 1988. The other four statutes administered by the CPSC are the Consumer Product Safety Act (CPSA), 15 U.S.C. §§ 2051–2084, the Poison Prevention Packaging Act (PPPA), 15 U.S.C. §§ 1471–1476, the Flammable Fabrics Act, 15 U.S.C. §§ 1191–1204, and the Refrigerator Safety Act, 15 U.S.C. §§ 1211–1214.

The PPPA allows the CPSC to establish special packaging standards for hazardous substances to prevent children from handling, using, or ingesting the substance. 15 U.S.C. § 1472 (2006). Some products regulated under the PPPA are aspirin, furniture polish, and prescription drugs. 16 C.F.R. § 1700.14 (2008) (listing all products regulated under the PPPA). Products that contain nanomaterials are also subject to regulation under the PPPA.

8. J. Bromme, *Nanotechnology and the CPSC*, BNA PRODUCT SAFETY & LIAB. REPORTER 5 (2005), *available at* http://www.arnoldporter.com/resources/documents/Article-Nanotechnology_and_the_CPSC(2005).pdf.

9. Consumer Product Safety Improvement Act of 2008, Pub. L. No. 110-314.

10. 15 U.S.C. §§ 2051–2084.

11. E. M. Felcher, *The Consumer Product Safety Commission and Nanotechnology*, PEN (Aug. 14, 2008), at 10–11.

consumption or enjoyment of a consumer in or around a permanent or temporary household or residence, a school, in recreation, or otherwise.[12]

This broad definition encompasses thousands of everyday items. However, the Commission's jurisdictional authority does not extend to household products that are regulated by other statutes and other agencies (e.g., tobacco, motor vehicles and equipment, pesticides, firearms and ammunition, aircraft and parts, boats and other vessels, drugs, devices, food, and cosmetics).[13] As indicated in Chapter 5, a petition has been submitted by the International Center for Technology Assessment to EPA to regulate nanosilver in consumer products as a pesticide. Given the prevalence of nanosilver in consumer products, how EPA addresses this petition will have a significant impact on the Commission's role.

The CPSA does not require premarket notification or permitting of the product. Rather, the CPSA, as amended after its original enactment, provides the CPSC with a limited set of regulatory tools. The 2008 amendment enhanced these tools, but it did not fundamentally alter them. Among the powers that are given to the CPSC are: (a) to develop a safety standard when it discovers unreasonable risk of injury associated with a product, (b) to ban a product if there is no feasible standard, (c) to recall a product that presents a *substantial product hazard*, and (d) to seek judicial intervention to seize a product. The term *substantial product hazard* means (a) the failure to comply with an applicable consumer product safety rule issued under the CPSA, which failure creates a substantial risk of injury to the public; or (b) a product defect which—due to a pattern of defect, the number of defective products, and other factors—creates a substantial risk of injury to the public.[14] The 2008 amendment expanded the scope of "substantial product hazard" to include the failure to comply with a similar rule, regulation, standard, or ban under *any* statute or regulation enforced by the Commission.

With respect to establishing safety standards (e.g., warnings, instructions, and performance standards), the CPSC can promulgate standards when it is "reasonably necessary to prevent or reduce an unreasonable risk of injury."[15] However, if a voluntary consumer product safety standard can eliminate or adequately reduce the risk of injury and it is likely that there will be substantial compliance with such voluntary standard, the CPSC will not impose a mandatory standard.[16] In fact, if in the course of creating a mandatory safety rule, the

---

12. 15 U.S.C. § 2052.
13. *Id.*
14. 15 U.S.C. § 2064(a).
15. 5 U.S.C. § 2056.
16. Procedurally, the steps for imposing a mandatory standard or accepting a voluntary one are outlined in 15 U.S.C. § 2058. With respect to voluntary standards, the CPSC is supposed to monitor compliance with the standard and/or provide technical support in formulating the standard. There are three coordinating organizations involved in formulating most voluntary standards: the American National Standards Institute (ANSI), the

Commission is informed of a voluntary standard already in existence, it may terminate its proceedings to issue a mandatory rule and instead adopt the voluntary rule (provided that the interested parties have an opportunity to comment).[17] The recent amendment to the act provides the Commission with the authority to adopt, by rule, voluntary standards for any consumer product or class of consumer products whose characteristics meet the definition of "substantial product hazard" if the Commission determines that (a) such characteristics are readily observable and have been addressed by voluntary standards; and (b) such standards have been effective in reducing the risk of injury from the consumer product and that there is substantial compliance with the voluntary standards.[18] An issue for nanomaterials would be whether the hazards are "readily observable."

There are two conditions under which a mandatory standard can be created: (a) if there is no voluntary standard, and (b) if industry fails to comply with voluntary standards. To create mandatory standards, the CPSC must demonstrate among other things: (a) the potential benefits and potential costs of the action, including an identification of those likely to receive the benefits and bear the costs; (b) the alternatives that were considered and the reasons those alternatives were not chosen, and (c) a summary of significant issues raised during the public comment period.[19] Once a safety standard is adopted, every manufacturer and private labeler must certify that the product conforms to all applicable standards.[20]

The CPSC has the power to recall or require notification for products that pose a "substantial product hazard." If the CPSC determines after giving interested parties an opportunity for an administrative hearing that a product presents a substantial hazard, the CPSC can order a manufacturer, distributor, or retailer to give public notice and specifically to notify those who have purchased the product.[21] Also, if the Commission, after notifying the manufacturer, determines that the product is an imminently hazardous consumer product and has filed action for seizure, then the Commission may order the manufacturer, retailer or distributor to take any of the following actions: cease distribution; give public notice; mail a notice to each person who is manufacturer, retailed or distributor; mail a notice to each person who purchased the product; and notify all transporters and state health officials as well.[22] In addition, if the CPSC determines (after affording interested parties a hearing) that a product in the marketplace poses a "substantial product hazard" and action is in the public

---

American Society for Testing and Materials (ASTM) International, and the International Underwriters Laboratories. The members of these entities (i.e., industry members and other stakeholders) develop the voluntary standards.

17. 15 U.S.C. § 2058(b).
18. 15 U.S.C. § 2064(j).
19. 15 U.S.C. § 2058(f)(2).
20. 15 U.S.C. § 2063(a)(1).
21. 15 U.S.C. § 2064(c).
22. *See id.*

interest, it can direct manufacturers, distributors and retailers to (a) bring the product into conformity with requirements of safety standard or to repair the defect, (b) replace such product with a like or equivalent product that complies with the applicable consumer product safety or that does not contain the defect, or (c) refund the purchase price with certain restrictions.[23]

If the CPSC provides the manufacturer with a complaint that the manufacturer's product poses a substantial product hazard, the CPSC may disclose any information it has obtained from the manufacturer to the public.[24] However, this ability to disclose is restricted to information that is nonconfidential business information. Moreover, if the information to be provided to the public allows the public to "readily ascertain" who the manufacturer is, the manufacturer will be afforded the opportunity to provide comments to the Commission.[25] However this opportunity is not available if the disclosure is being made because the product is imminently hazardous or the manufacture, distribution, or importation of the product has been prohibited.[26]

Finally, if a manufacturer, distributor, or retailer has information that *reasonably* supports a conclusion that a product (a) fails to comply with an applicable consumer product safety rule (a mandatory rule or a voluntary rule that was relied upon by the Commission) or any similar rules, standards and bans under any other statute governed by the Commission, (b) contains a safety defect that could create a "substantial product hazard," or (c) creates an unreasonable risk of serious injury or death, then the company is under an obligation to immediately report that to the CPSC.[27] The only exception to this requirement is if the company has actual knowledge that the CPSC has been adequately informed of the defect, failure to comply, or risk.[28] As the implementing regulations note, a company should not:

> delay reporting in order to determine to a certainty the existence of a reportable noncompliance, defect or unreasonable risk. The obligation to report arises upon receipt of information from which one could reasonably conclude the existence of a reportable noncompliance, defect which could create a

---

23. 15 U.S.C. § 2064(d).

24. 15 U.S.C. § 2055(b)(5). In addition, public disclosure can occur (a) in lieu of proceeding against a manufacturer regarding a "substantial hazard product" when the Commission has agreed to remedial settlement agreement dealing with the product; or (b) because the manufacturer agrees to its public disclosure. *See id.*

25. 15 U.S.C. § 2055(b)(1).

26. 15 U.S.C. § 2055(b)(4). The manufacture, sale, and distribution of products that are prohibited can result in civil penalties, criminal penalties and injunctive relief including seizure of the product. *See* 15 U.S.C. §§ 2069, 2070, 2071. Section 19 of the CPSA outlines what constitutes a prohibited product. 15 U.S.C. § 2068.

27. 15 U.S.C. § 2064.

28. *See id.* This notification provision is similar in scope as to the notification requirements under section 8(e) of the Toxic Substances Control Act.

substantial product hazard, or unreasonable risk of serious injury or death. Thus, an obligation to report may arise when a subject firm received the first information regarding a potential hazard, noncompliance or risk.[29]

Critics have challenged the adequacy of the CPSA as a regulatory tool to control products that contain nanomaterials. As noted above, the CPSC may impose consumer product safety standards when it is "reasonably necessary to prevent or reduce an unreasonable risk of injury." The vast majority of these standards, however, are voluntary. Industry—usually working through one of three coordinating organizations—has already agreed to these standards when they are adopted by the CPSC. Other stakeholders can have a role in developing these standards by participating in these coordinating organizations. Additionally, the CPSC can also be involved in developing these standards. But ultimately, given that these are voluntary standards, it is industry's role that is critical as to how they are formulated.

In its enforcement tool kit, CPSA provides the Commission with the power to impose civil and criminal penalties and to seek injunctive relief by, among other things, restraining a person from manufacturing or distributing a product in violation of the statute or by seizing the product. Section 19 of the statute provides an extensive list of prohibited acts for which enforcement may be undertaken. The CPSC also has the power to ban a product if it finds that (a) the consumer product is being, or will be, distributed in commerce; (b) the consumer product presents an unreasonable risk of injury; and (c) no feasible consumer product safety standard would adequately protect the public from the unreasonable risk of injury associated with the product.[30] Only a limited number of products have actually been banned. The procedural steps for banning a product are similar to the rulemaking process necessary for imposing a safety standard. Critics note that because of the relatively high standard for banning, it is not surprising that only a handful of products have been banned. One critic opined that the CPSA "has been crippled with amendments."[31] It should be noted that the CPSC could also ban a product under the FHSA provisions.

III. FEDERAL HAZARDOUS SUBSTANCES ACT

Although the enactment of FHSA precedes the creation of the CPSC by more than 10 years, administration of the FHSA was transferred from the Secretary of

---

29. 16 C.F.R. § 1115.12(a).

30. 15 U.S.C. § 2057.

31. C. J. Davies, *Nanotechnology Oversight: An Agenda for the New Administration*, PROJECT ON EMERGING NANOTECHNOLOGIES 16 (2008), *available at* http://www.nanotechproject.org/process/assets/files/6709/pen13.pdf.

Health, Education, and Welfare to the CPSC in 1973.[32] As with the regulation of cosmetics under the FFDCA, the FHSA (as amended) does not provide for pre-market review by the CPCS. Rather, a manufacturer has an obligation to either manufacture a nonhazardous product or to place a cautionary label on the product to provide adequate warning to purchasers and users indicating the hazards associated with the product.[33] Manufacturers face a substantial risk if a product that contains hazardous substances is put in the marketplace without an adequate label. If the label is inadequate or insufficient to protect to the human health or the environment, or if product is inherently dangerously to children, then CPSC has the authority to ban and seize the product as well as impose penalties and recalls.[34] Manufacturers who wish to avoid such possibilities should follow the CPSC's regulations set forth in 15 C.F.R. Part 1500 to determine for themselves if their product meets the standards for requiring a label.

A. Labeling Requirements

The burden is on the manufacturer to label the product consistent with the FHSA labeling requirements.[35] On the label, the manufacturer must provide its name and address, the name of the hazardous ingredients, and the signal word that provides notice of an appropriate level of caution. Thus, in order to draft such a label, a manufacturer must first determine whether a label is necessary (i.e., that the ingredient is, in fact, a hazardous substance). CPSC encourages manufacturers to use its guidelines to make this assessment.[36]

A substance is a hazardous substance if two conditions are met: (a) the substance or mixture of substances must be toxic, corrosive, an irritant, flammable or combustible, a strong sensitizer, or it must generate pressure through decomposition, heat, or other means;[37] and (b) it must have the potential to cause

---

32. 15 U.S.C. § 2079(a) (2006).

33. M. A. Babich, *Risk Assessment of Low-Level Chemical Exposures from Consumer Products under the U.S. Consumer Products Safety Commission Chronic Hazard Guidelines*, 106 ENV'T HEALTH PERSPECTIVES 387, 388 (February 1998).

34. Treye et al., *supra* note 3, at 19. Generally, any toy or other product intended for use by children, which is or contains a "hazardous substance," would be considered a "banned hazardous substance."

35. *See id.*

36. *Id.*

37. 15 U.S.C. § 1261(f)(1)(A) (2006). A corrosive substance will chemically destroy living tissue upon contact. 15 U.S.C. § 1261(i) (2006). An irritant will induce a local inflammatory reaction after repeated or prolonged contact with living tissue. 15 U.S.C. § 1261(j) (2006). A strong sensitizer means a substance that causes hypersensitivity on living tissue through an allergic or photodynamic process. 15 U.S.C. § 1261(k) (2006). A flammable material has a flashpoint between 20 degrees Fahrenheit and 100 degrees Fahrenheit, while combustible materials have flashpoints at or above 100 degrees Fahrenheit, with certain exceptions. 15 U.S.C. § 1261(l)(1) (2006); 16 C.F.R. § 1500.3(c)(6) (2008). A substance

"substantial personal injury or substantial illness during or as a proximate result of any customary or reasonably foreseeable handling or use, including reasonably foreseeable ingestion by children."[38] This two-part definition requires an analysis similar to the general risk assessment analysis set forth in Chapter 3— that is, it requires an examination of toxicity, bioavailability and exposure.[39]

1. Assessing Toxicity *Toxic substances* are those that (a) can cause death in laboratory animals within a set time period under certain concentrations[40] (i.e., acute hazards); or (b) can be demonstrated to be a carcinogen, a neurotoxin, or a developmental or reproductive toxin (i.e., chronic hazard).[41] If the substance does not fit into the first category, the product manufacturer must develop sufficient scientific evidence to determine whether it qualifies as a chronic hazard. Unlike acute hazards that should be apparent in the course of normal premarket testing, chronic hazards—and particularly latent chronic hazards— are difficult to detect in premarket testing, and may in fact not be detected for a number of years after the product is in the marketplace.

In testing for carcinogenicity, a manufacturer will need to examine the evidence developed through epidemiological and animal bioassay studies.[42]

---

may also be considered hazardous if generates pressure through decomposition, heat, or other means or explodes or erupts under certain circumstances. 15 C.F.R. § 1500.3(c)(7)(i) (2008).

38. 15 U.S.C. § 1261(f)(1)(A) (2006); 16 C.F.R. § 1500.3(b)(4)(i)(A) (2008).

39. CPSC, *Labeling Requirements for Art Materials Presenting Chronic Hazards; Guidelines for Determining Chronic Toxicity of Products Subject to the FHSA; Supplementary Definition of "Toxic" Under the Federal Hazardous substances Act; Final Rules,* 57 Fed. Reg. 46626, 46631 (1992) ("Chronic Hazard Guidelines"). Bioavailability is a term which indicates the extent to which a substance is absorbed by the body. Chronic Hazard Guidelines at 46648. A summary of the chronic hazard guidelines are at 16. C.F.R. § 1500.135. The full guidelines are at 57 Fed. Reg. 46626–46674 (1992).

40. A substance is *highly toxic* when a small amount (*e.g.,* 50 milligrams or less per kilogram of body weight if administered orally to white rats; concentration of 200 parts per million or less by volume of gas or vapor if inhaled continuously for 1 hour or less by white rats; 200 milligrams or less per kilogram when in direct contact with bare skin of rabbits for 24 hours or less) produces death within 14 days in at least half of tested laboratory animals. *See* 16 C.F.R. § 1500.3(c)(1)(ii), 1500.40, 15 U.S.C. § 1261(h). But substances can also qualify as *toxic* if death is produced within 14 days but the concentrations of achieving death are at higher (*e.g.,* single dose of from 50 milligrams to 5 grams per kilogram of body weight if administered orally to white rats; concentration above 200 parts per million and below 20,000 parts per million by volume of gas or vapor if inhaled continuously for 1 hour or less by white rats; more than 200 milligrams but not more than 2 grams per kilogram when in direct contact with bare skin of rabbits for 24 hours or less) than for highly toxic substances. 16 C.F.R. § 1500.3(c)(2)(i), 1500.40.

41. 16 C.F.R. § 1500.3(c)(2)(ii) (2008).

42. *See* 57 Fed. Reg. 46633.

These are supplemented with information on other factors such as absorption, distribution, metabolism, and elimination of substances.[43] The CPSC recognizes the limitations of epidemiological and animal bioassay studies (as discussed in Chapter 3). With respect to epidemiological studies (especially retroactive ones), it is difficult to establish a casual relationship between exposure and cancer because of confounding variables.[44] According to the CPSC, if one or more of the following criteria are met, there is a sufficient basis to conclude a causal relationship exists: (a) no identified bias that can account for the observed association has been identified, (b) all possible confounding factors that could account for the observed association can be ruled out with reasonable confidence, and (c) based on statistical analysis, the association has been shown unlikely to be due to chance.[45] If none of the criteria are met, there is insufficient evidence to establish that the substance will cause cancer in humans (not that the substance is, in fact, noncarcinogenic).[46]

In using animal studies (again, as noted in Chapter 3), the issue becomes whether the results in animals can be extrapolated to humans and what distinctions between animals and humans must be taken into account. Sufficient carcinogenicity evidence in animals requires that the substance was tested in well-designed and conducted studies and has been found to elicit a statistically significant exposure-related increase in the incidence of malignant tumors, combined malignant and benign tumors, or benign tumors that indicate an ability to turn malignant.[47] If such data is lacking, that product cannot be found toxic for the purposes of CPSC regulation.

As to testing for neurotoxicity or reproductive/developmental toxicity, the methods for testing (epidemiological and animal bioassay studies) are similar to the tests for carcinogenicity. The distinction lies in how much information is sufficient to find a causal relationship (for epidemiological studies) or to find a statistical significant relationship (for animal studies). For example, a neurotoxin link for a substance based on epidemiological studies may be established by concluding that two of the following conditions exist: (a) a consistent pattern of neurological dysfunction is observed in multiple studies; (b) adverse effects/lesions in the nervous system account for the neurobehavioral dysfunction with a reasonable degree of certainty; (c) all identifiable bias and confounding factors are discounted after consideration; and (d) based on statistical

---

43. *Id.* at 46634.
44. *Id.*
45. *Id.* at 46635.
46. *Id.*
47. *Id.* at 46636.

analysis, the association has been shown unlikely with reasonable certainty to be due to chance.[48]

2. Assessing Substantial Injury or Illness For something to be considered a *hazardous substance,* in addition to an assessment of its toxicity, there must be a substantial likelihood of personal illness or injury to occur as a result of any customary or reasonably foreseeable handling or use, including reasonably fore-seeable ingestion by children.[49] Thus, not only must a substance be toxic, it must also be demonstrated that: (a) persons will be exposed to the substance, (b) the substance can enter the body, and (c) there is significant risk of adverse health effects related to handling or use of the substance.[50]

As discussed in Chapter 3, there are primarily three routes of entry into the human body: inhalation, dermal absorption, and oral ingestion.[51] Methods to determine which method of entry is likely include direct monitoring of populations (e.g., general populations or subgroups), making predictions based on modeling, and using data based on a similar type of substance (e.g., similar structure, reactivity, and volatility).[52] Although each methodology has its issues, direct monitoring field studies are preferred over model predictions, and both are preferred over surrogate data.

A related issue to the mode of entry is whether differences exist between the absorption characteristics of a substance when it is exposed to humans as part of the consumer product as compared to the absorption of the substance when it is tested in human or animal toxicity studies.[53] These differences may arise due to the presence of constituents in the product of substances other than the substance of concern, and different dosages of the substance, as well as differences in the physical or chemical characteristics of the substance when it is part of the product matrix. There are two general approaches when considering absorption: (a) a default value may be assumed for the amount of substance absorbed, or (b) an assessment can be performed.[54] The default approach assumes a 100 percent absorption rate, allowing for a quick and easy determination of an upper

---

48. *Id.* at 46639. Sufficient evidence of a causal relationship between developmental or reproductive toxicity and exposure to a substance requires the following three criteria:
   (1) There should be no identifiable bias which can be introduced through a faulty design of the experiment . . . (2) Confounding factors such as socioeconomic status, age, smoking, alcohol consumption, drug use, environmental or occupational exposure, and other diseases should be adjusted for. (3) The association between an endpoint and a causal factor should not be due to chance; there must be a statistically significant association. *Id.* at 46642.
49. *Id.* at 46633.
50. *Id.* at 46644.
51. *Id.*
52. *Id.* at 46645.
53. *Id.*
54. *Id.*

bound on risk without the need for a time-consuming assessment.[55] A qualitative assessment, on the other hand, may be used to show that bioavailability of a substance is no greater than that shown in the toxicity studies or to demonstrate compelling evidence that surrogate bioavailability data may be used.[56]

3. Discretion of the CPSC The FHSA also allows the CPSC to declare by regulation something to be a hazardous substance whenever doing so will promote the objectives of the Act.[57] To date, the CPSC has used this discretion to promulgate regulations to cover only charcoal briquettes and metal-cored candlewicks.[58] Conversely, the CPSC also has discretion to exempt from the requirements of the FHSA any hazardous substance it finds is adequately addressed by another act of Congress (e.g., book matches, laboratory chemicals, ballpoint ink cartridges, paste shoe wax, cellulose sponges, etc.).[59]

4. Label Contents Once the manufacturer has determined whether the product contains hazardous substances and whether the product is intended for use in a household or by children, the manufacturer must properly label the product. Alternatively, if the CPSC determines that standard labeling is not adequate to protect human health and safety, it can require a special labeling[60] (e.g., turpentine, benzene, charcoal, and fireworks)[61] or possibly ban the product. Each label for a hazardous substance must contain:

1. the name and address of the manufacturer, packer, distributor, or seller;
2. the common, usual, or chemical name of the hazardous substance;
3. an appropriate signal word: (a) "POISON" and the skull and crossbones symbol for highly toxic substances; (b) "DANGER" for substances which are extremely flammable, corrosive, or highly toxic; and (c) "WARNING" or "CAUTION" for all other hazardous substances;
4. an affirmative statement of the principal hazard, such as "combustible" or "vapor harmful";
5. precautionary measures describing the action to be followed or avoided;
6. instruction, if appropriate, for first-aid treatment;
7. instructions for handling and storage if special care is required; and
8. the statement "Keep out of the reach of children" or an equivalent, unless the product is intended for children, in which case, directions for the protection of children.[62]

---

55. *Id.*
56. *Id.*
57. 15 U.S.C. § 1262(a).
58. 16 C.F.R. § 1500.12.
59. 15 U.S.C. § 1262(d); 16 C.F.R. § 1500.83.
60. 15 U.S.C. § 1262(b).
61. 16 C.F.R. § 1500.14.
62. 15 U.S.C. § 1261(p) (2006); 16 C.F.R. § 1500.121(b)(5) (2008).

This label must appear on the container in a location intended to be prominently displayed for retail sale.[63] To ensure that the placement of the label is consistent with the need to be conspicuous, a manufacturer should follow the requirements set forth at 16 C.F.R. 1500.121 (e.g., adequate type size, good contrast, not cluttered, and not interfered with by other graphics). Hazardous substances that fail to bear an appropriate, conspicuous label are deemed *misbranded hazardous substances.*[64] Moreover, an otherwise conspicuous label may be negated if there are deceptive disclaimers (e.g., the product is "Harmless" or "Safe around pets").[65]

B. Banning and Seizing Products

The CPSC has the power to ban a product from interstate commerce once it is deemed to be a *misbranded hazardous substance* or a *banned hazardous substance* if it follows the procedures outlined under the FFDCA.[66] If a product containing a hazardous substance is intended for use in the household or by children, and that product fails to bear a sufficient warning label, it will be deemed a *misbranded hazardous substance.*[67] A product will be declared a *banned hazardous substance* if it is: (a) a toy, or other product intended for use by children, that is, itself, a hazardous substance, or bears or contains a hazardous substance accessible to a child; or (b) a hazardous substance intended for household use about which the CPSC decides that notwithstanding a cautionary label, the public health and safety would be protected only by keeping the product out of the channels of

---

63. 16 C.F.R. § 1500.121(b)(2) (2008). The label must also appear on any outer container or wrapper. *See* 16 C.F.R. § 1500.121(b)(4). Additionally, the size of the cautionary label shall be reasonably related to the type size of other type appearing on the container; the height of capital or uppercase letters must be no less than three times the width; and all statements regarding the hazard must appear in the same size, style, color, and boldness. 16 C.F.R. § 1500.121(b) (2008).

64. 15 U.S.C. § 1261(p) (2006).

65. 16 C.F.R. § 1500.122 (2008).

66. 15 U.S.C. § 1262 (2006). Once the CPSC determines that a substance should be deemed a hazardous substance, banned hazardous substance, or misbranded hazardous substance, the Commission will institute rulemaking procedures. These steps include: (a) an advance notice of the rulemaking, (b) a delay in the effective date of the regulation, (c) the right to file objections within 30 days, (d) the right to automatic stay of the effective date of portions of any regulations to which objections are filed, (e) the right to a public hearing on such objections and a decision based on a fair evaluation of all the evidence of records at such hearing, and (f) a judicial review under 21 U.S.C. § 348(g)(2). *See* Spring Mills, Inc. v. CPSC, 434 F.Supp. 416, 428 (D.S.C. 1977) (listing the procedures of the FDCA relevant to the FHSA). The CPSC's failure to follow this rulemaking procedure will result in the violation of the due process rights of affected persons and nullification of the regulations. *See id.*

67. 15 U.S.C. § 1261(p).

interstate commerce.[68] A list of banned hazardous substances is available at 16 C.F.R. 1500.17 (e.g., carbon tetrachloride and cyanide salts).

CPSC may not classify a product as a banned hazardous substance unless it prepares a two-part regulatory analysis containing both a preliminary and final analysis. In both analyses, the CPSC must conduct a cost–benefit analysis that requires an evaluation of alternatives with the intent to select the least burdensome requirement that will adequately reduce the risk of injury. Under the preliminary analysis, the CPSC must: (a) identify the nature of the risk of injury, (b) summarize each regulatory alternative, (c) explain if there are any existing standards, and if so, why such standard does not eliminate or adequately reduce the risk of injury, (d) invite any interested person to submit comments on risks of injury identified or the alternatives offered by the CSPC, (e) invite any person to submit an existing standard, and (f) invite any person to develop a voluntary standard.[69] In the final analysis, the CPSC must provide: (a) a description of the potential costs and benefits (including nonmonetary costs and benefits) and identification of who will benefit and who will bear the costs, (b) a description of the considered alternatives to the final regulation and a cost–benefit analysis of each of these alternatives, and (c) a statement that the proposed regulation imposes the least burdensome requirement to prevent or adequately reduce the risk for which the regulation is being promulgated. [70]

A misbranded or banned hazardous substance is liable to be seized once it is placed in interstate commerce. The penalties for violating a product ban include fines and imprisonment.[71] For first-time offenses, individuals face fines of up to $5,000 ($10,000 for corporations or other organizations) and imprisonment for up to 90 days. Repeat offenses, or offenses with intent to defraud or mislead, lead to fines of up to $250,000 if the offense results in death ($500,000 for corporations or other organizations) and imprisonment for up to one year.[72]

In addition to seizure and penalties, if the article or substance sold in commerce is a banned hazardous substance, then CPSC after providing the manufacturer or distributor an opportunity for a hearing can order such a person to give notifications to customers, the general public, and/or other manufacturers, distributors, or dealers of the article or substances.[73] The CPSC may also order the manufacturer, distributor, or dealer of the article or substance to take one of the following actions of that person's choosing: (a) if feasible, repair or change the article or substance so it is no longer a banned hazardous substance; (b) replace the article or substance with a like or equivalent article or substance

---

68. 15 U.S.C. § 1261(q)(1).
69. 15 U.S.C. § 1262(f).
70. 15 U.S.C. § 1262(i).
71. 15 U.S.C. § 1264.
72. Id.
73. 15 U.S.C. § 1274(a).

that is not a banned hazardous substance; or (c) refund the purchase price of the article or substance (less a reasonable allowance for use under certain circumstances).[74] The FSHA specifically states that CPSC need not engage in a cost–benefit analysis before taking one of these above actions.[75]

C. Export and Import Requirements

The export and import rules of the CPSC are distinctive. There are export and import rules under both the CPSA and the FHSA, and there are distinctions between these laws. Given that consumer products containing nanomaterials may e imported into the country, this section will focus on import rules. We will briefly discuss the export rules under the FHSA only. Under the FHSA, it may be possible to export consumer products that are tagged as *misbranded* or *banned hazardous substance* under the CPSC rules if the exporter files notice with CPSC at least 30 days prior to the date of export.[76] CPSC will then notify the foreign government of the exportation and the basis upon which the substance is considered misbranded or banned.[77] If, at that time, the receiving country is willing to accept the product, it can be exported.

Importers, on the other hand, are subject to the full panoply of the CPSC's power. Under the CPSA, the Commission has the authority to refuse admission into the customs territory of the United States a product that fails to comply with an applicable consumer product safety rule; is not accompanied with a certificate as required; is or has been determined to be an imminently hazardous consumer products as determined in a proceeding; has a defect that constitutes a "substantial product hazard;" and was manufactured by a person who has violated the statute.[78] The importer would have an opportunity to modify the product under the supervision of the Commission and the Department of Treasury. But if the modification is not proceeding in a satisfactory manner, then the Commission can direct the Treasury Secretary to demand delivery of the product into customs custody or otherwise to seize the product. Any product that is refused admission may be destroyed, unless the importer or other party with an interest in the product requests that the Treasury Secretary export the product in lieu of destruction. Under the FHSA, the CPSC has the authority to sample hazardous substances that are being imported into the United States.[79] If it appears that these imported hazardous substances are either misbranded hazardous substances or banned

---

74. 15 U.S.C. § 1274(b). The CPSC may also take the same actions for articles or substances which are not banned hazardous substances, but are intended for use by children and have a defect which creates a substantial risk of injury to children. 15 U.S.C. § 1274(c).

75. 15 U.S.C. § 1274(g) (2006).

76. 15 U.S.C. § 1273(d) (2006).

77. *Id.*

78. 15 U.S.C. §2066(a).

79. 15 U.S.C. § 1273(a).

hazardous substances under the FHSA, the CPSC may destroy these hazardous substances unless they are exported within 90 days.[80] Due to the concern that the product may re-enter the United States, the Commission's practice is to destroy the product If the CPSC determines that relabeling of the hazardous substances would bring them into compliance with the FHSA, then the owner or consignee may complete such relabeling before the CPSC reaches a final determination.[81]

IV. CPSC ACTION AND NANOTECHNOLOGY

A policy statement entitled the "CPSC Nanomaterial Statement" was issued in 2005. In it, the CPSC stated that the safety and health risks associated with nanomaterials can be assessed under the existing CPSC statutes, regulations, and guidance. However, the CPSC acknowledged that due to variations in nanomaterials and the lack of scientific data on exposure and toxicity, the Commission could not make a "general statement" about the potential health and safety effects from exposure to nanomaterials.[82] Thus, the CPSC noted that research was needed to determine the "unique exposure and risk assessment strategies" necessary to identify the toxicity and exposure associated with a particular nanomaterial.

The policy statement discussed the CPSC's ability to regulate nanomaterials under both the CPSA and the FHSA. As noted above, under the CPSA the Commission has the authority to issue a safety standard. However, the CPSC noted it will not a priori issue a safety standard, but rather, as with other consumer products, it will "look to see" whether a defective product composed of or containing nanomaterials creates a substantial risk of injury and assess the severity of such risk. As to the FHSA, the CPSC took no definitive position on how it may use this statutory authority in noting that chronic hazards associated with nanomaterials have to be investigated. Some observers of the CPSC have criticized the policy statement on nanotechnology as being a reactive "wait and see" position that will result in action only after an incident has occurred.

There are some practical realities the CPSC must confront that make its "wait and see" position a necessity. The CPSC has only a limited budget and staff. Although the number of consumer products sold in the United States has proliferated, when accounting for inflation, CPSC's FY 2007 budget is nearly only half its FY 1980 budget.[83] The number of CPSC staff members has dwindled

---

80. *Id.*

81. 15 U.S.C. § 1273(b) (2006).

82. CPSC Nanomaterial Statement, *available at* http://www.cpsc.gov/library/cpscnanostatement.pdf.

83. Davies, *supra* note 29, at 10.

from 900 in FY 1981 to 393 in FY 2007.[84] The 2008 amendment does reverse the downward spiral of the Commission by increasing both funding and the number of individuals it may employ. However, even with this increased funding, its resources are still finite, and those resources will have to be deployed to address issues that are the forefront at the public's consciousness and have garnered the attention of the media and politicians—which has been toys and other products manufactured in foreign countries, not nanomaterials.

Even if the CPSC were interested in promulgating a mandatory rule, the Commission would have the burden of demonstrating that the benefits justify the costs and that the alternatives are not appropriate. Given the need for data on the health consequences of nanomaterials and how adverse impacts may be mitigated, the CPSC may not be able to meet its regulatory burden.

Given the constraints on the CPSC to act, but given the possibility that public concerns over nanomaterials will necessitate action, a question arises whether industry can forestall the need for a mandatory standard by creating a voluntary standard using the coordinating organizations (i.e., ASTM International, ANSI, and International Underwriters Laboratories). Incentives already exist for companies to create their own standards. Certain industry players may consider it in their best interests to set standards so that the quality of products is maintained and that other players do not try to free ride on their efforts to research and develop the best quality products. Moreover, by creating their own standards, companies may be able to assuage the concerns about nanomaterials that are being raised by members of nongovernmental organizations. Thus, at least for the near term, the more likely avenue will be for a voluntary standard to be developed.

---

84. *Id.* at 16.

10. CHEMICALS

I. THE USES OF NANOMATERIALS IN CHEMICALS

The chemical industry broadly defined is involved in the manufacture and development of virtually every commercial product. The preceding chapters focused on products that are impacted by chemicals or that are part of the larger chemical industry, including food packaging, processed foods, pesticides, cosmetics, medical devices, and consumer products. This chapter will discuss how manufacturers of basic and specialty chemicals (i.e., the manufacturers, importers, and processors of paints, dyes, lubricants, solvents, plastics, inks, catalysts, pigments, and a host of other products) who use or manufacture nanomaterials need to view their regulatory obligations.

Nanomaterials have a number of advantages that companies in the chemical industry would like to explore and commercialize upon. For example, incorporating nanoparticles into paint could reduce its solvent content and thereby lessen the adverse hazardous impacts that result. In addition, nanomaterials may reduce the weight of paint, thus making it more desirable for industries such as aircraft manufacturing.[1] Another example is the synthesization of nanoscale particles by sol-gel techniques (a process involving a transition from a liquid "sol" phase to a solid "gel" phase that is used with ceramic and glass) to make the next generation of light-emitting phosphors for television and computer displays.[2] EPA has identified over 200 existing chemicals that are produced at the nanoscale for commercial or R&D purposes.[3]

This chapter will view the obligation of the chemical manufacturer through the basic lifecycle of the product—from research in commercial settings, to start-up and scale-up operations, to manufacturing and production of products, and ultimately to disposal. Specifically, the research, start-up/scale-up, and manufacturing and production phases of a product are all potentially subject to the Toxic Substances Control Act, 15 U.S.C. §§ 2601 et seq., (TSCA). In 2008, EPA issued a number of documents attempting to clarify how nanomaterials may be governed under TSCA. Clearly this policy remains in flux. However, it is also clear that companies have already begun to—or are being required to—comply

---

1. The Royal Society and The Royal Academy of Engineering, *Nanoscience and Nanotechnologies* (June 2004) at 11.

2. *See id.*

3. EPA, *Nanoscale Materials Stewardship Program: Interim Report*, (January 2009) at 18.

with provisions of TSCA. Thus, the discussion in those sections can be based on actual application of this statute (albeit to a limited number companies). Similarly, even without issuing a specific guidance document on the applicability of the Occupational Health and Safety Act, 29 U.S.C. §§ 651 et seq., (OHSA), which governs, among other things, employee exposure to hazardous substances in a variety of contexts (including research, manufacture, and disposal), companies have begun to comply with its provisions.

Finally, this chapter will focus on other statute that could be relevant to the disposal of nanomaterials: the Solid Waste Disposal Recovery Act, 42 U.S.C. §§ 6901 et seq., more commonly referred to as the Resource Conservation and Recovery Act (RCRA), which governs the generation, transportation and disposal of hazardous waste. In discussing this statute, it remains an open question whether the standards that have historically been used to ensure compliance with it can actually be applied to nanomaterials.

II. TOXIC SUBSTANCES CONTROL ACT

The purpose of this section is to generally examine the provisions that may be applicable to the manufacture, processing, or distribution of nanomaterials and to provide analysis of recent interpretations or uses of those provisions. For example, some companies have already submitted applications and notifications in compliance with sections 5 and 8 of TSCA, and in one instance, negotiated a consent order to manufacture under TSCA section 5(e). Moreover, EPA has issued guidance on what constitutes a "new" chemical as well as notices as to what constitutes "significant new uses" for specific nanoscale materials.

A. Scope of TSCA

TSCA provides EPA with the authority to regulate the manufacture, importation, processing, and distribution of all nanoscale substances that meet the definition of "chemical substance."[4] The term *chemical substance* is broadly defined to mean any organic or inorganic substance of a particular molecular identity, including any (a) combination of such substances occurring in whole or in part as a result of a chemical reaction or naturally, and (b) any element or uncombined radical.[5] In enacting TSCA, Congress provided EPA with three major powers:

(a) The ability to require manufacturers, importers, and processors to test (as well as collect and submit data on) certain new and existing chemicals in order to determine their effects on human health and

---

4. EPA, TSCA Inventory Status of Nanoscale Substances—General Approach (Jan. 23, 2008) at 1.

5. 15 U.S.C. § 2602(2)(A).

the environment and report if there are substantial risks of injury to human health or the environment (sections 4 and 8 of TSCA).

(b) The ability to require any person who manufactures or imports for commercial purposes any "new" chemical substance or manufactures, imports, or processes an existing chemical substance for a "significant new use" to seek approval from EPA's Office of Pollution Prevention and Toxics (section 5 of TSCA).

(c) The authority to limit or ban the production, importation or use of a chemical substance if EPA determines those substances will cause or will present unreasonable risk of injury to human health or the environment (sections 5 and 6 of TSCA).

B. Statutory Exemptions

Any manufacturer, importer, or processor of any chemical substance (regardless of whether it is a nanoscale chemical substance) must first determine if TSCA is actually applicable to the chemical substance in question. The statute excludes items that are regulated under other federal statutes. Specifically, the statutory exclusions are: (a) any pesticide, as defined under FIFRA, when manufactured, processed, or distributed in commerce for use as a pesticide;[6] (b) tobacco and any tobacco product; (c) any source material, special nuclear material, or by-product material as defined by the Atomic Energy Act, 42. U.S.C. § 2011 et seq.,; (d) any article the sale of which is subject to the tax imposed by section 4181 of the Internal Revenue Code (i.e., firearms and ammunition); and (e) any food, food additive, drug, cosmetic, or device regulated under the FFDCA.[7]

Additionally, the statute exempts *mixtures*, which means "any combination of two or more chemical substances if the combination does not occur in nature and is not, in whole or in part, the result of a chemical reaction." Included in this definition is "any combination which occurs, in whole or in part, as a result of a chemical reaction *if* none of the chemical substances comprising the combination is a new chemical substance *and* if the combination could have been manufactured for commercial purposes without a chemical reaction at the time the

---

6. According to the agency, pesticide raw materials, intermediates, or inert ingredients that are not themselves pesticides are chemical substances subject to TSCA until they become actual components of registered pesticide products. 42 Fed. Reg. 64572, 74585 (Dec. 23, 1977). Additionally, those R&D chemicals being evaluated for pesticidal applications are substances subject to TSCA until their manufacturers or importers demonstrate intent to create a pesticide by submitting an application for an experimental use permit or an application of registration under FIFRA. 51 Fed. Reg. 15096, 15098 (Apr. 22, 1986).

7. 15 U.S.C. § 2602(2)(A). The agency interprets the FFDCA exemption as including substances intended for use as a component of a food, food additive, drug, cosmetic, or device within the meaning of those terms. The FDA considers intermediates and catalysts to be such components.

chemical substances comprising the combination were combined."[8] This exclusion refers only to the mixture itself and not to its chemical constituents. Thus, for example, if there is a new chemical substance in a mixture, although the mixture would not be subject to TSCA, the specific "new" chemical substance would be.[9]

In addition to the statutory exclusions, as discussed below EPA's regulations specifically exclude a number of different types of substances from specific regulatory requirements, such as the pre-manufacture notification.

C. Testing and Regulation of Risks

Under section 4 of TSCA, EPA may require a manufacturer, importer, processor, or producer to conduct testing (e.g., toxicological tests) on certain chemicals to evaluate their potential health and environmental effects. The purpose of this testing requirement is to provide EPA with a mechanism for obtaining data necessary to determine how the chemical should be regulated. The agency may impose such a demand once it has issued a rule or negotiated an enforceable consent agreement with the company.[10] Alternatively, a company may voluntarily request that EPA issue a rule to establish standards that should be used when the company conducts data tests. However, EPA may accept or reject such a petition request. EPA is currently evaluating whether it should issue a test rule for nanoscale materials in order to develop needed environmental, health and safety data.

In order to issue a rule, EPA must first make certain statutory findings. To reach its statutory threshold, EPA can use different mechanisms. One mechanism

---

8. 15 U.S.C. § 2602 (8) (emphasis added). Mixtures include alloys, inorganic glasses, ceramics, and cements.

9. 40 C.F.R. § 720.30.

10. EPA's decision to request that a company conduct testing is based on recommendations made by the Interagency Testing Committee (ITC). The ITC was created by Congress and is comprised of representatives from 16 agencies. The purpose of the ITC is to provide EPA with a list of chemicals (known as the Priority Testing List) for which EPA should request test information. In developing the Priority Test List, the ITC considers the quantities of the chemical substances that are or will be manufactured; the quantities of the chemical substances that enter or will enter the environment, the number of employees (as well as other people) that will be exposed and the duration of the exposure, the extent to which the substance or a closely related substance is known to present an unreasonable risk to health or the environment, the existence of data on health or environmental effects, a reasonable estimation of the degree to which testing is likely to produce data on health or the environment, and the reasonable foreseeable availability of facilities and personnel to conduct such testing. According to the statue, once a chemical is placed on the ITC Priority List, EPA must initiate a rulemaking proceeding (or indicate the reasons for not initiating such a proceeding). The agency, however, has consistently failed to conduct such rulemaking.

allows the agency to determine (a) that the chemical substance "may present an unreasonable risk of injury to health or the environment,"[11] (b) that there is insufficient information to permit a reasoned evaluation of the health and environmental effects, and (c) that insufficient data may be addressed through additional testing.[12] The other mechanism allows EPA to find (a) that the chemical is and will be produced in substantial quantities (100,000 kg/yr.); (b) that there are data gaps that may be addressed through additional testing, and that the manufacture, processing, use, disposal, or distribution in commerce without such data may present an unreasonable risk; and (c) either (i) there is or may be a significant exposure to humans, or (ii) the chemical is reasonably expected to enter the environment in substantial quantities.[13]

The rule shall include identification of the chemical substance for which testing is required as well as the standards that should be employed in developing the test data. In setting forth the standard by which the test data will be developed, EPA will need to examine the "relative costs of the various test protocols and methodologies which may be required under the rule and the reasonably foreseeable availability of the facilities and personnel needed to perform the testing required under the rule."[14] The characteristics of the chemicals that could be investigated include persistence, acute toxicity, subacute toxicity, chronic toxicity, and any other characteristics that may present risk. The methodologies for evaluating such characteristics include epidemiologic studies, serial or hierarchical tests, in vitro tests, and whole animal tests.[15]

Once EPA receives the data pursuant to a test rule, it will publish a Federal Register notice that (a) identifies the chemical substance for which data has been received, (b) lists the uses or intended uses for the chemical substance, and (c) states the nature of the test data that has been developed.[16] To date, EPA has not issued a section 4 rule regarding nanomaterials generally, or for any particular nanomaterial. However, if such a rule were to be imposed or proposed, among the issues that would be considered is whether the existing methodologies and testing equipment are sufficient to measure the impacts of nanomaterials. Moreover, even if the methodologies and technologies exist, the issue of cost arises.

Procedurally, the issuance of a rule must be in compliance with the Administrative Procedures Act. During the rulemaking process, any interested parties will have an opportunity for an oral presentation of data, views, or arguments

---

11. Chemical Mfrs. Ass'n v. EPA, 859 F.2d 977 (D.C. Cir. 1988).

12. 15 U.S.C. § 2603(1)(A).

13. 15 U.S.C. § 2603(1)(B).

14. 15 U.S.C. § 2603(b)(1).

15. 15 U.S.C. § 2603(b)(2)(A).

16. 15 U.S.C. § 2603(d).

as well as an opportunity for written comments.[17] Any person who is subject to the test rule may request an exemption, which EPA will grant if two conditions are met: (a) the chemical substance for which the data is requested is equivalent to a chemical substance for which data has previously been submitted; and (b) if data were to be submitted, it would be duplicative. In such limited cases, if an application for exemption is submitted during the *reimbursement period*, EPA will order that the person benefiting from the previously submitted data pay the original data submitter and any other contributors to the data a portion of their costs.[18] However, EPA's preferred alternative is to negotiate enforceable consent agreements whereby parties agree to provide information.

Companies also have the option of petitioning EPA to impose standards for conducting tests on a chemical. One of the reasons that a company may make such request is because it is obligated to provide data in order to satisfy the pre-manufacture notice requirements, and would like to know the standards for obtaining such data. However, it is discretionary on EPA's part to either accept or deny such a request.[19]

D. Notice Provisions

1. Exemptions Any person who intends to manufacture or import a *new chemical substance* for commercial purposes or who intends to manufacture, import, or process a chemical for a significant new use must notify EPA and seek approval at least 90 days before undertaking the activity, unless an exemption applies. As described below, certain nanomaterials are already subject to pre-manufacture notice. Thus, manufacturers must first determine if an exemption applies to their process before they consider the applicability of the pre-manufacture requirements. This section will examine these exemptions.

Some of the exemptions do not require EPA approval prior to the manufacture or import of chemicals, such as the exemption for substances that have no independent commercial purposes, the R&D exemption, or the polymer exemption. These exemptions do, however, require compliance with other requirements, such as record-keeping. Conversely other exemptions require EPA approval prior to manufacture or import under these provisions such as the low volume emission exemption and low exposure emission exemption. Of particular significance to manufacturers of nanomaterials may be the R&D exemption as well as the low volume emission exemption. In fact, EPA reports that it has already received a number of applications for this latter exemption.

(a) Commercial Purpose TSCA's section 5 reporting requirements are only applicable to manufacturing, importing, or processing for "commercial purposes."

---

17. 15 U.S.C. § 2603(b)(5).

18. 15 U.S.C. § 2603(c). As with FIFRA reimbursements, EPA has created a set of regulations governing the payment of the original data submitters.

19. 15 U.S.C. § 2603(g).

The following list contains substances that even though they are manufactured for commercial purpose are not manufactured for distribution in commerce as chemical substances per se and have no commercial purpose separate from the substance, mixture, or article of which they are a part. As such, these substances are exempt from pre-manufacture notification requirements:

1. Any substance that is an impurity—a substance that is unintentionally present in another chemical substance.
2. Any by-product which is not used for commercial purposes. A by-product is a chemical substance produced without separate commercial intent during the manufacturing or processing of another chemical substance or mixture.
3. Any chemical substance that results from a chemical reaction that occurs incidental to (a) exposure of another chemical substance, mixture, or article to environmental factors; or (b) storage or disposal of another chemical substance, mixture, or article.
4. Any chemical substance that results from a chemical reaction that occurs either (a) upon end use of another chemical substance, mixture, or article which is not itself manufactured or imported for distribution in commerce or for use as an intermediate (e.g., adhesive, paint, miscellaneous cleanser or other housekeeping product, fuel additive, water softening and treatment agent, photographic film, battery, match, or safety flare); or (b) upon use during the manufacture of an article destined for the marketplace without further chemical change of the chemical substance (e.g., curable plastic or rubber molding compounds, inks, drying oils, metal finishing compounds, adhesives, or paints).
5. Any chemical substance which results from a chemical reaction that occurs when (a) a stabilizer, colorant, odorant, antioxidant, filler, solvent, carrier, surfactant, plasticizer, corrosion inhibitor, antifoamer or defoamer, dispersant, precipitation inhibitor, binder, emulsifier, de-emulsifier, dewatering agent, agglomerating agent, adhesion promoter, flow modifier, pH neutralizer, sequesterant, coagulant, flocculant, fire retardant, lubricant, chelating agent, or quality control reagent functions as intended; or (b) a chemical substance, which is intended solely to impart a specific physiochemical characteristic, functions as intended.
6. Any non-isolated intermediate.[20]

---

20. 40 C.F.R. § 720.30(h). An "intermediate" is defined as "any chemical substance that is consumed, in whole or in part, in chemical reactions used for the intentional manufacture of another chemical substance(s) or mixture(s), or that is intentionally present for the purpose of altering the rates of such chemical reactions." 40 C.F.R. § 720.3(n).

(b) R&D Exemption The statute and regulations specifically exempt any chemical substance that is manufactured solely for noncommercial R&D purposes.[21] This exemption includes scientific experimentation, research, or analysis conducted by academic, government, or independent not-for-profit research organizations (e.g., universities, colleges, teaching hospitals, and research institutes) unless the activity has an eventual commercial purpose. In contrast, any research conducted at a company is considered to be for commercial purposes, but may be exempt if all of the following conditions are:

1. The chemical substance is manufactured or imported only in small quantities solely for research and development. This may include evaluation of physical, chemical, production or performance characteristics of a new or existing substance for the purposes of developing a new commercial product (e.g., synthesis of a new chemical; health or environmental effects testing; scale-up activities to determine whether the substance can be produced at a commercial scale; and testing of production capabilities such as process yield or uniformity in a now or modified production process). EPA has not yet established quantitative limits as to what constitutes small quantities. Rather, the agency has described such as being quantities reasonably necessary for R&D purposes.

2. The manufacturer or importer notifies all persons in its employ or to whom it directly distributes the chemical substance who are engaged in experimentation, research, or analysis regarding the chemical substance (including the manufacture, processing, use, transport, storage, and disposal of the substance associated with research and development activities) of any health risk that may be associated with the substance.[22] The notification must be made in accordance with EPA requirements[23] [hereinafter referred to as the "notification prong."]

3. The chemical substance is used by, or directly under the supervision of, a technically qualified individual.[24]

---

21. 15 U.S.C. § 2604(h)(3); 40 C.F.R. § 720.36.

22. 40 C.F.R. § 720.3(cc).

23. *See* 40 C.F.R. § 720.36(c) for an outline of the notification requirements.

24. 40 C.F.R. § 720.36(a). The term *technically qualified individual* means a person or persons (a) who because of education, training, or experience (or a combination of these factors) is capable of understanding the health and environmental risks associated with the chemical substance which is used under his or her supervision; (b) who is responsible for enforcing appropriate methods of conducting scientific experimentation, analysis, or chemical research to minimize such risks; and (c) who is responsible for the safety assessments and clearances related to the procurement, storage, use, and disposal of the chemical

With respect to the notification prong, to determine whether notification is required, the manufacturer or importer must review and evaluate the following information as to whether there is reason to believe there is any potential risk to health that may be associated with the chemical substance:

(i) Information in its possession or control concerning any significant adverse reaction by persons exposed to the chemical substance that may reasonably be associated with such exposure;

(ii) Information provided to the manufacturer or importer by a supplier or any other person concerning a health risk believed to be associated with the substance;

(iii) Health and environmental effects data in its possession or control concerning the substance;

(iv) Information on health effects that accompany any EPA rule or order issued that applies to the substance and of which the manufacturer or importer has knowledge.[25]

(c) Low Volume Exemption To obtain a *low volume exemption* (LVE), the manufacturer must intend to manufacture at an annual production rate of 10,000 kg or less.[26] A manufacturer must submit a notice to EPA at least 30 days before manufacture of the new chemical substance is to begin by filing an EPA Form No. 7710-25.[27] If no action is taken by EPA, the manufacturer may consider its exemption approved and may begin to manufacture the new chemical substance.

The notification must contain: (a) the manufacturer's identity; (b) chemical identity as outlined in 40 C.F.R. § 720.45(a); (c) impurities as outlined in 40 C.F.R. § 720.45(b); (d) known synonyms or trade names as outlined in 40 C.F.R. § 720.45(c); (e) by-products as outlined in § 720.45(d); (f) production volume as outlined in § 720.45(e);[28] (g) description of intended categories of use as outlined in § 720.45(f);

---

substance as may be appropriate or required within the scope of conducting a research and development activity. 40 C.F.R. § 720.3(ee).

25. 40 C.F.R. § 720.36(b).

26. 40 C.F.R. § 723.50(a)(1).

27. *See id.* at 723.50(a)(2).

28. The assumption is that the manufacturer will manufacture at an annual production volume of 10,000 kilograms. Manufacturers who intend to manufacture an exempted substance at annual volumes of less than 10,000 kilograms and wish EPA to conduct its risk assessment based upon such lesser annual production level rather than a 10,000–kilograms level may so specify by writing the lesser annual production volume in the appropriate box on the PMN form and marking the adjacent binding option box. Manufacturers who opt to specify annual production levels below 10,000 kilograms and who mark the production volume binding option box shall not manufacture more than the

(h) test data as outlined in § 720.50; and (i) a certification that indicates among other things that all exemption conditions have been met.[29]

EPA may determine during the review period that manufacture of the new chemical substance does not meet the criteria set forth above or that there are issues concerning toxicity or exposure requiring further review that cannot be accomplished within the 30–day review period. Accordingly, EPA will notify the manufacturer that the substance is not eligible.[30] If a substance is ineligible, the manufacturer may not begin to manufacture the new chemical substance without complying with pre-manufacture notice or submitting a notice seeking a new exemption. EPA will not allow the manufacture of a new chemical substance if the agency determines that the substance, any reasonably anticipated metabolites, environmental transformation products, by-products of the substance, or any reasonably anticipated impurities in the substance may under anticipated conditions of manufacture, processing, distribution in commerce, use, or disposal of the new chemical substance cause: (a) serious acute (lethal or sublethal) effects, (b) serious chronic (including carcinogenic and teratogenic) effects, or (c) significant environmental effects.[31]

Manufacturers must inform processors and industrial users that the substance can be used only for the uses specified in the notice provided to EPA and under the controls imposed on the substance by the agency.[32] This notification can be accomplished by means of a container labeling system, written notification, or any other method that adequately informs people of use restrictions or controls

The manufacturer has an obligation to take action if it learns that a direct or indirect customer is processing or using the new substance in violation of use restrictions or without imposing prescribed worker protection or environmental release controls.[33] If this is the first time, the manufacturer must cease distribution of the substance to the customer or the customer's supplier immediately, unless the manufacturer is able to document each of the following:

1. That the manufacturer has, within 5 working days, notified the customer in writing that the customer has failed to comply with the appropriate conditions/requirements; and
2. That, within 15 working days of notifying the customer of the noncompliance, the manufacturer received from the customer,

---

specific annual amount of the exempted substance unless a new exemption notice for a higher (up to 10,000 kgs) manufacturing volume is submitted and approved by EPA.

29. 40 C.F.R. § 723.50(e).
30. 40 C.F.R. § 723.50(h).
31. *See id.* 723.50(d).
32. *See id.* at 723.50(k)(1).
33. *See id.* at 723.50(k)(3).

in writing, a statement of assurance that the customer is aware of the appropriate conditions/requirements and will comply with those terms.[34]

If the manufacturer learns of another violation, the manufacturer must cease supplying the new chemical substance to that customer and report the compliance failure to EPA within 15 days.[35] Within 30 days of its receipt of the report, EPA will notify the manufacturer whether—and under what conditions—distribution of the chemical substance may resume.[36]

The manufacturer has an obligation to maintain records for five years on, among other things, annual production volume and import volume, compliance with the use restrictions, worker protection requirements, environmental controls, compliance with the notification requirements to processors and industrial users, and compliance with notification of EPA of customer violations.[37] The agency may request such records at any time, and manufacturers have an obligation to provide such records within 15 working days of receipt of such a request.[38]

(d) *Low Release and Exposure Exemption* EPA also exempts from the premanufacture requirements any chemicals that have low environmental releases and human exposures. To be eligible for this exemption, a manufacturer must satisfy all of the following conditions:

1. For consumers and the general population, no dermal exposure or inhalation exposure (except as described with respect to incineration) and exposure in drinking water no greater than a 1 milligram per year (estimated average dosage resulting from drinking water exposure in streams from the maximum allowable concentration level from ambient surface water releases).
2. For workers, no dermal exposure and no inhalation exposure. This criterion is met if adequate exposure controls are used in accordance with applicable EPA guidance.
3. For ambient surface water releases, no releases resulting in surface water concentrations above 1 part per billion, unless EPA has approved a higher surface water concentration supported by relevant and scientifically valid data submitted to EPA in the manufacturer's notice that demonstrates that the new substance will not present an unreasonable risk of injury to aquatic species or human health at the higher concentration.

34. *See id.*
35. *See id.* at 723.50(k)(4).
36. *See id.*
37. *See id.* at 723.50(n).
38. *See id.*

4. For ambient air releases from incineration, no releases of the new chemical substance above 1 microgram per cubic meter maximum annual average concentration.

5. For releases to land or groundwater, no releases to groundwater, to land, or to a landfill unless the manufacturer has demonstrated to EPA's satisfaction in the manufacturer's notice that the new substance has negligible groundwater migration potential.[39]

The notice that a manufacturer must submit is similar to the one done in the context of a low volume emission exemption. The only difference is that for a low volume emission exemption, the manufacturer must be provided the estimated maximum amount to be manufactured during the first year of production and the estimated maximum amount to be manufactured during any 12–month period during the first three years of production.[40]

The approval process described above for LVE is equally applicable to this type of exemption. Furthermore, the notification and record-keeping requirements of the low volume emission exemption are also equally applicable. Manufacturers who use this exemption may distribute the chemical substance only to other persons who agree in writing to refrain from further distribution of the substance until it has been reacted, incorporated into an article, or otherwise rendered into a physical form or state in which environmental releases and human exposures thresholds noted above are not likely to occur.[41]

(e) *Test Market Exemption* Any person may apply for an exemption to manufacture or import a new chemical substance solely for test-marketing purposes.[42] Test-marketing refers to the distribution in commerce of no more than a predetermined amount of chemical substance, mixture, or article containing that chemical substance or mixture, by a manufacturer or processor, to no more than a defined number of potential customers, for the purpose of exploring market capability in a competitive situation during a predetermined testing period prior to the broader distribution of that chemical substance, mixture, or article in commerce.[43] To obtain the exemption, the person must demonstrate that the substance will not present an unreasonable risk of injury to health or the environment as a result of the test-marketing. To make such a demonstration, the applicant must submit all existing data on health and environmental impacts associated with the chemical, a description of the proposed test-marketing activity, and the number of people that will be exposed to the chemical.[44]

---

39. *See id.* at 723.50(c)(2).
40. *See id.* at 723.50(e)(2)(vi)(B).
41. 40 C.F.R. § 720.50(k)(2).
42. 15 U.S.C. § 2604(h)(1).
43. 40 C.F.R. § 720.3(gg).
44. 40 C.F.R. § 720.38(b).

2. Pre-Manufacture Notification Requirements Any person who intends to manufacture or import[45] for commercial purposes any "new" chemical substance must notify EPA 90 days before doing so.[46] To determine what constitutes a new chemical substance, EPA requires the person to review the TSCA Chemical Substances Inventory. Any chemical substance not on the Inventory is "new" for the purpose of pre-manufacture notification. There are two portions to the TSCA Inventory: (a) a nonconfidential, publicly available and searchable database, and (b) a confidential portion of the database that is only searchable by EPA upon a request by a party that demonstrates it has a bona fide interest in manufacturing or researching the chemical substance.

When searching for a chemical substance on the Inventory, two classification systems can be examined. Class 1 chemical substances are represented by a distinct chemical structure and specific molecular formula.[47] Class 2 chemical substances are substances that have unknown or variable composition, complex reaction products, and biological materials.[48] As these chemicals cannot be represented by chemical structure, they are described using either partly indefinite names indicating variable structures (e.g., heptene), or names that are descriptive of complex compositions (e.g., tall-oil fatty acids), or names based on compositional characteristics (e.g., C15-18.alpha.–alkenes).[49]

A significant issue with nanoscale chemical substances is whether they are *new* chemical substances.[50] In 2008, EPA issued a guidance document that stated that the physical characteristics (e.g., particle size and shape) of a nanoscale chemical is not relevant to determining whether it is considered a new chemical substance for TSCA notification purposes.[51] Rather, as EPA noted, it is the *molecular identity* of the nanoscale chemical that will determine if it is considered

---

45. The term *manufacture or import* means (a) to sell or to offer for sale the substance, mixture, or article in commerce; to introduce or deliver for introduction into commerce, or the introduction or delivery for introduction into commerce of the substance, mixture, or article; or to hold (or the holding of) the substance, mixture, or article after its introduction into commerce, including for test-marketing purposes; or (b) for use by the manufacturer, including for use as an intermediate. 40 C.F.R. § 720.3(r)

46. 15 U.S.C. § 2604(a)(1)(A).

47. EPA, TSCA Inventory Status of Nanoscale Substances—General Approach (Jan. 23, 2008) at 4.

48. *See id.*

49. *See id.*

50. EPA claims that it has received and reviewed more than 50 *new* nanoscale chemicals. Moreover, EPA has identified 18 nanoscale materials as potentially being *new* chemicals, including some that are at the R&D stage. *See* EPA, *Nanoscale Materials Stewardship Program: Interim Report,* (January 2009).

51. EPA, TSCA Inventory Status of Nanoscale Substances—General Approach (Jan. 23, 2008) at 3–4.

a new or an existing chemical.[52] To determine a chemical's molecular identity, it is necessary to examine the types and number of atoms in a molecule, the types and number of chemical bonds, the connectivity of the atoms in the molecule, and the spatial arrangement of the atoms within the molecule.[53] If the nanoscale chemical differs from a conventional chemical that is already registered on the TSCA Inventory in any of these categories, it would be considered a new chemical requiring pre-manufacture notification.

The first application of this policy document was EPA's announcement in October 2008 that carbon nanotubes would not be considered identical to graphite or other allotropes of carbon.[54] EPA explained that there may have been a misunderstanding based on the agency's communication with a single company a few years prior that a substance now considered a carbon nanotube was already on the Inventory. EPA clarified its position to that company and to the larger industry by stating that carbon nanotubes would be considered to be new chemical substances and thus subject to pre-manufacture notice unless they had the same molecular identity as non-nanoscale allotropes of carbon.[55]

In some cases, companies have submitted a *bona fide intent to manufacture letter*. Such a letter is typically transmitted when a potential applicant has reason to believe that a similar chemical may be listed in the confidential, nonpublicly available portion of the TSCA Inventory or alternatively, if the identity of a particular substance is claimed as a trade secret. By submitting a bona fide intent to manufacture letter, the applicant is requesting that EPA search the confidential database and inform the manufacturer if, in fact, the chemical substance is listed in the TSCA Inventory.[56] The letter must demonstrate a genuine interest in developing the chemical, and thus the requester must provide detailed information on the chemical's identity, probable manufacturing site, process used, and date when the PMN would be submitted.[57] EPA is supposed to provide a conclusive determination within 30 days after receipt of a complete submission.[58]

(a) Notice Contents As noted above, a pre-manufacture notice (PMN) must be submitted to EPA 90 days prior to the intended date of the activity. The PMN Form 7710-25 requires, to the degree such information is known or reasonably ascertainable, such information as: (a) the common or trade name, the chemical identity, or molecular structure of the chemical; (b) the categories or proposed categories of use; (c) the total or reasonable estimate of the amount that will be

---

52. *Id.* at 3.

53. *Id.*

54. 73 Fed. Reg. 64946, 65947 (Oct. 31, 2008).

55. Pre-manufacture notice requirements would naturally not be applicable if one of the preceding exemptions applied.

56. 40 C.F.R. § 720.25.

57. 40 C.F.R. § 720.25(b)(2).

58. *See id.* at 720.25(b)(8).

manufactured, used, or processed; (d) a description of the by-products; (e) the number of or reasonable estimate of the number of people who be exposed and the duration of such exposure; and (f) the manner of its disposal.[59]

With respect to toxicity data, a manufacturer is not required to conduct tests prior to submission of a PMN.[60] However, a manufacturer is required to provide any test data that it knows of or is reasonable ascertainable and that is in the applicant's possession or control.[61] The test data would include any health effects data, ecological effects data, physical and chemical properties data, monitoring data on human exposure or environmental impacts, and environmental fate and transport data.[62] Within the context of nanomaterials, data on items such as particle size, surface charge and area, diffusion, mobility, dispersion, and crystal structure may be more relevant than data that is typically sought with conventional chemicals.

EPA claims that it has received and reviewed a number of PMN applications, including for carbon nanotubes and fullerenes.[63] Moreover, EPA recently entered into a consent decree under section 5(e) allowing the manufacture of multiwalled carbon nanotubes.[64] A redacted copy of the consent decree was made available to the public, and it provides insight as to the direction of EPA's review of nanomaterials under section 5 (see below for further discussion).

(b) Responses to a Notice Once EPA receives a complete PMN, it has a number of different options:

First, EPA may not act on the PMN. If the 90-day review period expires without EPA indicating it has an objection, the applicant may manufacture or import the chemical.[65] The applicant must then file a Notice of Commencement (NOC) after it has begun manufacturing. Specifically, on or within 30 days after the first day of manufacturing or importation, an EPA Form 7710-56 must be filed. Once EPA receives the NOC, it will place the chemical listed therein on the TSCA Inventory.

Second, EPA may extend the review period for good cause (e.g., the PMN being incomplete),[66] but such an extension constitutes final agency action and will be subject to judicial review. In reviewing an application, EPA must determine

---

59. 15 U.S.C. § 2604(d), 15 U.S.C. § 2607(a)(2).

60. 40 C.F.R. § 720.50

61. 40 C.F.R. §§ 720.50, 720.40(d).

62. 40 C.F.R. § 720.50(a)(2).

63. 73 Fed. Reg. 64947.

64. *See* EPA, Consent Decree and Determinations Supporting Consent Decree, P-08-0177 (undated and redacted). The redacted Consent Decree does not identify the company. However, Thomas Swan & Co. Ltd. in a press release noted that it had entered into the first consent decree for the manufacture of its high purity multiwalled carbon nanotubes and also entered into a consent decree for single-walled carbon nanotubes.

65. 40 C.F.R. § 720.70.

66. *Id.*

whether the information is sufficient to conclude that the manufacture, processing, distribution, use, and disposal of the chemical or any combination of such activities will not present an unreasonable risk of injury to human health or the environment.[67] EPA will have to examine the toxicity of the substance both to humans and the environment as well as the magnitude of both human and environmental exposure.[68]

Third, if EPA determines the information is not sufficient to permit a reasoned evaluation of the health and environmental effects and either (a) in the absence of such information, the manufacture, processing, use, disposal, or distribution in commerce presents an unreasonable risk; or (b) the substance may be produced in substantial quantities (100,000 kg/yr.) and (i) have a *significant exposure* to humans, or (ii) may reasonably be expected to enter the environment in substantial quantities, then EPA may issue an section 5(e) order.[69] The order can prohibit or limit certain activities associated with a new chemical substance. Typically, a section 5(e) order is in the form of a consent decree that has been negotiated between EPA and the company.[70] However, because a consent decree is only binding upon the company that enters into it, and because once the chemical is placed on the Inventory as a result of the consent decree another company can manufacture the chemical without notifying EPA, typically the agency will issue a Significant New Use Rule for this chemical.

The consent order entered into by EPA regarding multiwalled carbon nanotubes illustrates the issues that may be addressed through a consent decree. In it, EPA acknowledges that it lacks the information necessary to make a "reasoned evaluation" of human health effects, and because there is potential risk of human health, that it cannot permit uncontrolled manufacturing or processing of the substance. To address this issue, EPA required the manufacturer to develop and submit the results of a 90-day inhalation study in rats with an observation period of three months, at least two weeks before manufacturing or importing a specified kilograms of the substance or after a specified period of time after commencing nonexempt commercial manufacture of the substance, whichever

---

67. 15 U.S.C. § 2604(b)(2).

68. 15 U.S.C. § 2604(b)(4).

69. 15 U.S.C. § 2604(e)(1)(A). EPA has established thresholds for what constitutes a significant exposure. *See* www.epa.gov/oppt/newchems/expbased.htm However, those thresholds are based on conventionally sized chemicals, and thus it remains to be seen if they will applied to nanoscale chemicals.

70. In those rare instances in which the notified party and EPA do not reach a consent agreement, EPA may seek a judicial injunction to prohibit or limit the manufacturing, processing, or use. 15 U.S.C. § 2604(e)(2)(B), (C), (D). For instance, if there is a possibility that the manufacturer will be able to move forward with manufacturing or importation because EPA's notification period was going to expire, then EPA may seek (and a court may impose) a temporary restraining order or preliminary injunction.

comes first.[71] EPA also requested a one-gram sample of the sample. Additionally, the agency requested certain material characterization data, including: type of multiwalled carbon nanotube (concentric cylinders or scrolled tubes, number of walls/tubes), configuration of nanotube ends, description of any branching, width/diameter of innermost wall/tube, carbon unit cell ring size and connectivity, alignment of nanotube along long axis, hexagonal array orientation when rolled up, particle size of catalyst used in the manufacture of the nanotube, molecular weight (average and range), and particle properties (shape, size, weight, count, surface area, surface to volume ratio, and aggregation/agglomeration).[72]

In addition, because there is an unreasonable risk of exposure, under the consent decree, EPA requires compliance with certain OSHA requirements. Any person reasonably likely to be dermally exposed to the substance (by directly handling it, by contact through equipment on which the substance may exist, or by the substance becoming airborne) is required to wear personal protective equipment. Specifically, the person must wear gloves and full body clothing that are impervious to the substance. Any person who is reasonably likely to inhale the substance must be provided with a NIOSH-certified respirator as described in the consent decree. The consent order requires that any person to whom the substance is distributed agrees to these terms and will limit the uses of the substance to undisclosed uses (but presumably the uses intended by the manufacturer who would otherwise not agree to the consent order).

Fourth, if the agency has a reasonable basis to conclude that the manufacturing, processing, distribution in commerce, or disposal of the chemical substance presents or will present an unreasonable risk before EPA can issue a rule under section 6 of TSCA (see below), the agency can regulate the chemical under section 5(f) of TSCA. Under this section, EPA may limit the amount of production or impose other restrictions as authorized under section 6 on the substance via an immediately effective proposed rule, issue a proposed rule that would prohibit the activity, or seek an injunction issued through the U.S. District Court.[73] If the agency issues a proposed order, the applicant may object, and, the agency may seek an injunction from a federal court. The procedures and standards that a court will use in evaluating a proposed order that has been objected to or an initial application for injunctive relief from the agency are similar to those outlined for section 5(e).[74]

---

71. *See* Consent Decree, *supra* note 64. The Consent Decree provides for a waiver of this provision if the company submits toxicity data under the Nanoscale Stewardship Program (discussed in Chapter 11) and if other certain conditions are met.

72. *See* Consent Decree, *supra* note 64, at 6–7.

73. 15 U.S.C. § 2604(f)(1),(2).

74. 15 U.S.C. § 2604(f)(3)(C).

Finally, under section 6, if EPA has a reasonable basis to conclude that the manufacture, processing, distribution in commerce, use, disposal, or combination thereof would or will present an unreasonable risk of injury to health or the environment,[75] the agency has the authority, among other things, to prohibit or limit the amount of the chemical substance for a particular use, to require an approved label or warning be placed on the chemical substance, to give notices to potential distributors or the general public of an unreasonable risk of injury, to make and maintain records of the processes used to manufacture or process the substance, to monitor or conduct tests that are reasonable and necessary to assure compliance with the rule issued under section 6, to prohibit or regulate any method of commercial use, to prohibit or regulate the manner of disposal (but in conformity with state or local laws), and to direct manufacturers and processors to give the public notice of risks.[76] EPA may, however, only use the least burdensome mechanism that adequately addresses the risk; thus, there must be a balancing of different factors.[77]

In promulgating a rule to impose of any of these restrictions, EPA must consider four factors: (a) the effects of a chemical substance on human health and the magnitude of the exposure to human beings; (b) the effects of a chemical substance on the environment and the magnitude of the exposure on the environment; (c) the benefits of such substance for various uses and the availability of substitutes for such uses; and (d) the reasonably ascertainable economic consequences of the rule, after consideration of the effect on national economy, small business, technological innovation, the environment, and public health.[78]

Procedurally, in issuing a section 6 rule, EPA will need to publish a notice of proposed rulemaking that is subject to public notice and comment, provide for an opportunity for an informal hearing that is available to any interested party that allows for oral and documentary evidence submissions as well as cross-examinations, and then publish a final rule based on the record established.[79]

---

75. 15 U.S.C. § 2605(a).

76. *See id.*

77. *See id.*

78. 15 U.S.C. § 2605(c)(1). In the event EPA determines that the health or environmental risk can be eliminated or sufficiently reduced through compliance with the requirements set forth in another federal statute(s), then the agency may not promulgate a section 6 rule unless it finds, in its own discretion, that it is in the public interest to issue such a rule. However, EPA is required to make certain comparisons as to cost and efficiency in issuing a section 6 rule as opposed to compliance with other statute.

79. 15 U.S.C. § 2605(c)(2),(3). The statute provides that if a person represents an interest that would "substantially contribute to a fair determination of the issues to be resolved in the proceeding," and if that person's economic interest is small in comparison to costs for effective participation, and if the person has a demonstrated lack of resources, then

Because the statute requires a balancing of different interests, it is usually a complicated, long, and drawn-out process that is subject to legal challenges from different sides. Moreover, the burden is higher when EPA seeks to partially or totally ban a product rather than to merely regulate it. Courts have examined section 6 rules under the *substantial evidence* test, which is less deferential to the agency than the arbitrary and capricious standard. Thus, while the provision provides EPA with a significant tool, it is not a tool that EPA can readily employ.

3. Significant New Use If EPA considers nanomaterials as being equivalent to conventionally sized counterparts that are already on the Inventory, it is still possible for the agency to impose virtually all the obligations a manufacturer has under a PMN by issuing a "Significant New Use Rule" (SNUR). A SNUR can be issued for a single chemical, or alternatively, it can be issued to cover a category or class of chemicals. Thus, with respect to nanomaterials, it is possible that EPA could issue a SNUR to cover an entire group of nanomaterials. However, in order to do so, EPA must determine that the activity, in fact, constitutes a *significant new use*. The agency examines factors such as, projected production and processing volume; the anticipated extent to which the new use increases the type, form, magnitude and duration of human exposure or environmental impacts; and the reasonably anticipated manner or methods of manufacturing, processing, distribution in commerce, and disposal of a chemical substance.[80]

The agency has already issued SNURs for certain nanomaterials. For example, in November 2008, EPA issued SNURs for two nanomaterials: siloxane modified silica nanoparticles (generic) and siloxane modified alumina nanoparticles (generic).[81] EPA noted that based on the physical properties of both of these substances and on test data from "analogous respirable, poorly soluble particulates," it had concerns about the potential systemic effects from dermal exposure and about lung effects.[82] Even though EPA noted that it had not determined that the manufacture, processing, or use of either substance might present "an unreasonable risk," it determined that "use without impervious gloves or a NIOSH-approved respirator with an APF of at least 10; the manufacture, process, or use of the substance as a powder, or uses of the substances other than as described in the PMN may cause serious health effects."[83] Additionally, the agency recommended that manufacturers undertake a 90-day inhalation toxicity test that would assist in characterizing the human health effects of the substance. As noted above, with the entering of a section 5(e) consent order for multiwalled

EPA will allow reasonable attorneys' fees, expert fees, and other costs of participation. 15 U.S.C. § 2605(c)(4).

80. 15 U.S.C. § 2604(a)(2).

81. 3 Fed. Reg. 65763 (Nov. 5, 2008).

82. *See id.* at 65751–65752.

83. *See id.*

carbon nanotubes, EPA has signaled that it will be examining—and potentially issuing—a SNUR to cover those substances as well. Moreover, in that consent order, EPA also asked for a 90-day inhalation study to be completed.

A person who manufactures, imports, or processes[84] a chemical substance that is under the SNUR must file a Significant New Use Notice (SNUN) at least 90 days before undertaking the new uses.[85] Thus, manufacturers, importers, or processors who want to use the generic versions of the nanomaterials noted above as an additive will need to submit a SNUN. The exemptions from the PMN requirements are equally applicable to the SNUN requirements,[86] and once the SNUN is submitted, EPA has the same options for regulatory approval as it does with a PMN such as section 5(e) orders, section 5(f) orders, or section 6 rulemaking. If EPA does not take action, the agency is required to explain in the Federal Register its reasons for not doing so.[87]

Critics of EPA's efforts to regulate nanomaterials through SNURs have noted the disadvantages of the process. Specifically, unlike PMNs, companies have to comply with a SNUR only after EPA has had to propose a rule, provide the public with notice and comment, and then issue a final rule. Also, there is a substantial evidentiary burden in issuing a SNUR for an existing chemical because EPA has to know which uses of the material are new and which are not (e.g., EPA would have to know the extent to which the use changes the manufacturing or processing and their impacts on the magnitude, duration, or type of exposure).[88]

E. Reporting Requirements

Section 8 of TSCA identifies the various mechanisms that can be used by EPA to obtain additional information from companies on toxicity and human exposure. Specifically, section 8(a) provides EPA with the authority to issue rules that can broadly ask for information, section 8(b) requires EPA to compile and maintain the TSCA Inventory database and provides the agency with the authority to

---

84. Unlike the PMN requirements, the SNUR requirements apply as well to those who process. *Process* means to (a) use as a part of a chemical reaction to produce another substance, (b) add stabilizers or additives to a substance, (c) repackage a substance, or (d) use a substance to produce an article that contains either the substance or another substance from it during the production of the article. For example, EPA considers a company to be a processor if it uses a TSCA-regulated substance to manufacture an article which, that distributed into commerce, contains either (a) the substance, (b) a mixture containing the substance, or (c) a reaction product of the substance.

85. 15 U.S.C. § 2604(a)(1)(B).

86. The exemptions authorized by TSCA section 5 (h)(1), (h)(2), (h)(3), and (h)(5).

87. 15 U.S.C. § 2605(g).

88. R. Denison, Statement of Richard A. Denison, Ph. D., Senior Scientist, at USEPA's Public Meeting in the Development of a Voluntary Nanoscale Materials Stewardship Program (Aug. 2, 2007), *available at* http://www2.envirionmentaldefense.org/article.cfm?contentID=6748.

request information from manufacturers to update the Inventory, section 8(c) provides EPA with authority to require the reporting of any allegations of significant adverse reactions, and section 8(d) provides EPA with the authority to require the submission of any ongoing and completed unpublished health and safety studies that are known or available. Additionally, section 8 also creates certain obligations on manufacturers, importers, distributors, or processors regardless of whether EPA issues any additional rules or requests. Specifically, section 8(c) creates an obligation to make a notation of any allegation of significant adverse reaction that must be maintained internally (regardless of whether EPA makes a request that such information be reported to it), and section 8(e) creates an obligation to immediately notify EPA of any new, unpublished information (e.g., preliminary results of animal bioassay studies or epidemiological studies) that reasonably supports a conclusion that there is a substantial risk associated with the chemical. Given the issues addressed in the preceding chapters about the lack of scientific information on health and environmental impacts, EPA may use its authority under sections 8(a) or 8(d) to request information from manufacturers. However, as discussed in Chapter 11, EPA thus far has not decided to use these regulatory tools. Rather, EPA launched a two-year voluntary reporting program referred to as the Nanoscale Materials Stewardship Program. The first part of the program has been completed, and EPA issued a report in January 2009. In that report, EPA noted that it is still considering how to apply its authority under section 8(a) to address the data gaps it identified through the use of the voluntary program.

Section 8(a) provides EPA with broadly worded authority to require by issuing a rule that manufacturers, importers, and processors retain such records and report such information as EPA may require to carry out its mandate. The statute provides that EPA may request information on: (a) the common or trade name, the chemical identity, and molecular structure of each chemical or mixture; (b) the categories or proposed categories of use; (c) the total quantities manufactured or processed or reasonable estimates of future manufacturing or processing, (d) the by-products that have been generated from the production process or usage; (e) all existing data on impacts to human health and the environment; (f) the estimates on the number of employees exposed and reasonable estimates on the number that will be exposed; and (g) the method for disposal.[89]

Under section 8(b) of TSCA, EPA is required to compile a list of each chemical substance that is manufactured or processed in the United States, except those manufactured or processed in small quantities. Section 8(b) states that the agency

---

89. 15 U.S.C. § 2607(a)(2). EPA has used its section 8(a) rulemaking authority to issue a general reporting rule entitled the Chemical Assessment Information Rule (CAIR), which is applicable to persons who manufacture or import one or more listed substances at any plant site; it requires them during the reporting period to provide information including general production, use, and exposure information on Form 7710-35.

must "compile, keep current, and publish a list."[90] This list is referred to as the TSCA Inventory—the use of which has already been discussed in the preceding sections. Prior to the creation of the TSCA Inventory, the federal regulatory agencies had no idea how many existing chemicals were in commerce.[91] The Inventory contains more than 80,000 organic, inorganic, and other chemical substances that have been manufactured, imported, or processed for commercial purposes in the United States since 1976.[92]

Section 8(c) requires manufacturers, importers, processors, and distributors[93] to record, retain, and, when requested by EPA, report "allegations of significant adverse reactions" to human health or to the environment for any substance or mixture.[94] The term *allegation* is broadly defined as "a statement, made without formal proof or regard for evidence, that a chemical substance or mixture has caused a significant adverse reaction to health or the environment."[95] Records must be maintained on all consumer complaints of personal injury or harm to health (e.g., substantial impairment of normal activities, or long-lasting or irreversible damage), reports of occupational disease or injury, and reports of complaints of injury to the environment (e.g., abnormal death of certain organisms).[96]

---

90. 15 U.S.C. § 2607(b).

91. Testimony of J. Clarence (Terry) Davies at EPA Public Meeting on Nanoscale Materials Stewardship Program (Aug. 2, 2007).

92. In 1986, EPA promulgated the Inventory Update Rule (IUR), which required manufacturers of nonpolymeric organic chemicals to report on the production volume and plant site information if the production or importation of a chemical on the TSCA Inventory is at levels in excess of 10,000 pounds per year at the site. Such reports were required every four years until 2003 when EPA amended the IUR to require reporting for the manufacture or importation of inorganic chemicals as well. A list of chemicals is noted at 40 C.F.R. § 710.45, with exemptions listed at 40 C.F.R. § 710.46. EPA also changed the reporting threshold to 25,000 pounds per year at a single location or importation and manufacture beginning in 2005 and every five-year interval thereafter, and added that if the substance was processed or used in quantities of above 300,000 pounds per year at a single site, that would also need to be reported. The regulation exempts those who manufacture or import the chemical substance solely for the purposes of scientific experimentation, those who manufacture or import fewer than 1100 lbs at a single plant site, and those who qualify as a "small" manufacture (i.e., total sales below $30 million and total production of the listed substance is below 100,000 lbs for the reporting period at the plant site).

93. 15 U.S.C. § 2607(c). TSCA § 8(c) applies to "any person," which includes distributors of chemicals in commerce, except for retailers and other companies who solely distribute chemical substances. *See* 40 C.F.R. § 717.7(c).

94. 15 U.S.C. § 2607(c). The term *significant adverse reactions* is defined as "reactions that may indicate a substantial impairment of normal activities, or long-lasting or irreversible damage to health or the environment." 40 C.F.R. § 717.3(i).

95. 40 C.F.R. § 717.3(c).

96. 15 U.S.C. § 2067(c).

Once an allegation is made, a notation of the allegation must be maintained for 5 years (except for employee allegations which must be maintained for 30 years). If the recipient disagrees with the allegations, then along with the notation this disagreement and the reasons for it should not be noted.

EPA can also issue rules that require chemical manufacturers, importers, processors, or distributors in commerce to submit lists and/or copies of ongoing and completed unpublished health and safety studies that are known or available.[97] The agency has used this provision to request information to develop industry standards. Among the information that can be sought are toxicological and epidemiological studies, clinical and ecological effect studies, studies of occupational exposure, studies based on environmental monitoring data, data on physical and chemical properties, bioconcentration, and other data that bear on the effects of a chemical on health and the environment.[98] EPA can ask a company to produce reports in its possession even if the company is not the entity that is actually engaged in manufacturing, processing, distributing, or importing the chemical. Additionally, a company can be asked to produce studies on chemicals that are manufactured in small quantities solely for the purposes of research.

TSCA Section 8(e) requires that a person who commercially engages in the manufacturing, process, or distribution of a chemical substance must notify EPA immediately of any new, unpublished information (e.g., preliminary results of animal bioassay or epidemiological studies) on a chemical that reasonably supports a conclusion that there is a substantial risk.[99] A person who is potentially subject to this provision should consult with EPA's revised guidance statement concerning reporting under section 8(e).[100] To determine if the information justifies making a submission under section 8(e), a company must evaluate the toxicity and exposure data (both these factors are weighted) and make a determination of whether there is substantial risk of injury to humans or the environment. If any of the data indicates any instance of, or evidence suggesting the possibility of cancer, birth defects, mutagencity, death, or serious or prolonged incapacitation, EPA must be notified.[101] In providing the notice to EPA, the person should include toxicity data, exposure, environmental persistence, and actions being taken to reduce human health and environmental risks.

The notification is not necessary if the company has actual knowledge that EPA is adequately informed of the risk.[102] For example, information obtained from the following sources is not reportable: (a) an EPA study or report, (b) an

---

97. 15 U.S.C. § 2607(d).

98. 40 C.F.R. § 716.3. Reporting requirements under Section 8(d) will terminate 60 days after the substance at 40 C.F.R. § 716.120 unless otherwise extended.

99. 15 U.S.C. § 2607(e).

100. 68 Fed. Reg. 33129 (June 3, 2003).

101. See id. at 33138.

102. See id.

official publication of another federal agency, (c) radio or television broadcasts, (d) recorded public scientific conferences held in the United States, (e) scientific databases, and (f) scientific conferences sponsored and cosponsored by EPA.[103] Thus far, at least three section 8(e) notices have been filed with regard to nano-materials: specifically, for single-walled carbon nanotubes, multiwalled carbon nanotubes, and carbon nanotubes. For example, the BASF submitted the results of a subchronic inhalation study conducted on rats that were exposed to aerosol dust of carbon nanotubes for a specified duration of time. The results indicated that while there was no observed effect at 0.1 mg/m^3, there were effects on the lung at .5 mg/m^3 and 2.5 mg/m^3.[104]

F. Penalty Policy and Citizen Suit Provision

TSCA identifies a broad range of prohibited acts, including failure or refusal to comply with (a) any rule or order issued under section 4, or (b) any rule, order, or requirement issued under sections 5 or 6. It is also unlawful for any person to use for commercial purposes a chemical substance or mixture which the person knew or had reason to know was manufactured, processed, or distributed in commerce in violation of sections 5 and 6 or a rule or order already issued under sections 5, 6, or 7. It also unlawful for a person to fail or refuse to establish or main-tain records, submit reports or notices, permit access to or copying of records, or to refuse entry or inspection to EPA. Any person who violates any of these pro-hibited acts described is subject to a civil penalty. Additionally, if the person knowingly or willing violated any of these provisions, he or she is subject to crim-inal penalties. There is significant case law examining imposition of civil and criminal penalties as well as penalty policies that outline the factors that will be used in determining an appropriate penalty. However, to date these provisions have not been applied against any companies or individuals with respect to manufacturing, processing, or importing of nanomaterials.

Section 21 of the TSCA provides that any person may petition the agency to initiate a proceeding for the issuance, amendment, or repeal of a rule under sections 4, 6, or 8, or an order under sections 5(e) or 6(b)(2). The petitioner must set forth the facts it claims establishes the need for such an action by the agency. Within 90 days of the filing of the petition, EPA must either grant or deny it. If the agency agrees to the petition, it must promptly commence an appropriate proceeding in accordance with sections 4, 5, 6, or 8. If EPA denies the petition, it must publish its reasons in the Federal Register. Within 60 days after the

---

103. See id. at 33139.

104. Letter from BASF to United States Environmental Protection Agency (dated July 8, 2008) (on file at http://www.epa.gov/oppt/tsca8e/pubs/8emonthlyreports/2008/8eaug 2008.htm).

denial, or the expiration of the 90-day review period, the petitioner may have its petition reviewed de novo by a court.

III. OCCUPATIONAL SAFETY AND HEALTH ACT

In 1970, Congress enacted the Occupational Safety and Health Act (OSH Act).[105] Under OSH Act, employers across industries are required to maintain a safe and healthful workplace "free from recognized hazards likely to cause death or serious physical harm."[106] The OSH Act created two different agencies: the Occupational Safety and Health Administration (OSHA) and the National Institute of Occupational Safety and Health (NIOSH). OSHA is within the jurisdiction of the Department of Labor; it is charged with setting and enforcing standards for occupational safety; providing training, outreach, and education on workplace hazards; and continuing to foster improvement in workplace safety by supervising ongoing research and promulgating compliance and enforcement mechanisms.[107] NIOSH operates as part of the Centers for Disease Control and Prevention (CDC) in the Department of Health and Human Services. NIOSH conducts research and provides recommendations to OSHA regarding health and safety standards, but it cannot issue regulations or impose requirements.

With respect to nanomaterials, NIOSH is conducting extensive testing on health effects and the appropriate control mechanisms. These efforts are outlined in further detail in Chapter 11. OSHA, on the other hand, as of December 2008, had not published any specific new regulations or even guidance documents on the applicability of existing regulations to nanomaterials. However, companies are being required to comply (or have already been in compliance with) the various OSHA requirements. For example, companies have issued Material Safety Data Sheets; used dermal protection, respirators or other devices to limit exposure; and adopted on their own permissible levels for airborne nanoparticles that are lower than permissible levels for their conventionally sized counterparts. The question that follows is whether what has been done adequately addresses the occupational health and safety risks associated with the particular nanomaterial being manufactured.

A. Hazard Communication

OSHA's Hazard Communication Standard (HCS) requires: (a) chemical manufacturers or importers to assess the hazards of chemicals that they produce

---

105. 29 U.S.C. §§ 651 *et seq.*

106. OSHA Mission Statement, *available at* http://www.osha.gov/oshinfo/mission.html.

107. *See id.*

or import, and (b) all employers to provide information to their employees about the hazardous chemicals to which they are exposed and the hazards associated with those chemicals.[108] The HCS is not applicable to laboratories as they are subject to their own set of communication requirements as discussed below.

The HCS applies to any chemical known to be present in such a way that employees may be exposed to it under normal conditions of use or in a foreseeable emergency.[109] The HCS suggests various methods for communicating with a company's employees about the hazards associated with the chemical, including container labeling and warnings, material safety data sheets (MSDS), and employee training.[110] Of these various methods, the most significant one for employers and employees is the MSDS. A chemical manufacturer and importer must create an MSDS for each hazardous chemical it produces or imports. Employers that use these hazardous chemicals must obtain and maintain a copy of the MSDS at their premises.[111]

The OSHA regulations define the term *hazardous chemical* to mean "any chemical which is a physical hazard or a health hazard."[112] The term *physical hazard* is defined as:

a chemical for which there is scientifically valid evidence that it is a combustible liquid, a compressed gas, explosive, flammable, an organic peroxide, an oxidizer, pyrophoric, unstable (reactive) or water-reactive.[113]

The term *health hazard* is broadly defined as any "chemical for which there is statistically significant evidence based *on at least one study* conducted in accordance with established scientific principles that acute or chronic health effects may occur in exposed employees"[114] (emphasis added). Thus, employers must

---

108. 21 C.F.R. § 1910.1200(b)(1). OSHA has published guidance documents to assist employers to comply with the requirements, including the "Hazard Communication Guidelines for Compliance," Publication 3111 (2000) and "Chemical Hazard Communication," Publication 3084 (1998).

109. The HCS requirement contains some exceptions, such as notably exempting laboratories from its requirements. Separate requirements for laboratories are discussed in Part (d).

110. 29 C.F.R. § 1910.1200(a)(1).

111. 29 C.F.R. § 1910.1200(g)(1). The term *use* is broadly defined as "to package, handle, react, or transfer." 29 C.F.R. 1910.1200(c).

112. 29 C.F.R. § 1910.1200(c).

113. *See id.*

114. *See id.* The term *health hazard* includes chemicals that are carcinogens; toxic or highly toxic agents; reproductive toxins; irritants; corrosives; sensitizers; hepatotoxins; nephrotoxins; neurotoxins; agents that act on the hematopoietic system; and agents that damage the lungs, skin, eyes, or mucous membranes. Appendix A provides further definitions and explanations of the scope of health hazards covered by this section, and

recognize that notwithstanding the limited information regarding the health impacts of various nanomaterials, if a single scientific study establishes that employees may be subject to acute or chronic effects as to their health, a MSDS must be used to communicate those effects to the employees.

The MSDS shall contain, among other things, the chemical and common names (for nonmixtures), chemical and common names of ingredients of a certain percentage (for mixtures), physical hazards (such as "potential for fire, explosion, and reactivity"),[115] health hazards ("including signs and symptoms of exposure, and any other medical conditions that are generally recognized as being aggravated by exposure to the chemical."),[116] primary routes of entry, OSHA permissible exposure limits (or any other exposure limits used or recommended by the manufacturer or employer preparing the sheet), and safe handling and first aid procedures.[117] With nanomaterials, given that research is still ongoing, it is important for manufacturers, importers, and employers to realize that if they become aware of "any significant information regarding the hazards of the chemical, or ways to protect against the hazards," this new information must be added to the MSDS within three months.[118]

There are already MSDSs for nanomaterials such as, single-wall carbon nanotubes, quantum dots, multiwall carbon nanotubes, copper powders, nano-cobalt phosphorus, zinc oxide, nano silver and 40 nm titanium dioxide. However, because of the paucity of data on specific nanomaterials, the data or information contained in these MSDS are mainly based on their conventionally sized counterparts. For example, OSHA has not promulgated any permissible emission limits for any nanomaterial, yet certain MSDSs refer to permissible emission limits. In such instances, the limits are usually referring to the permissible emission limits for the respective conventionally sized counterpart.

Additionally, once an employer has been given an MSDS or otherwise informed about hazardous chemicals being used at its facility, the employer is charged with developing an employee information program to educate employees on the known hazards regarding these chemicals.[119]

B. Chemical-Specific Standards and Permissible Exposure Limits

OSHA establishes the permissible exposure limit (PEL)(i.e., the maximum amount of safe exposure to an air contaminant based on an eight-hour time weighted average). Subpart Z of the CFR contains three tables identifying various

---

Appendix B describes the criteria to be used to determine whether a chemical is to be considered hazardous for purposes of this standard.

115. 29 C.F.R. § 1910.1200(g)(2)(iii).
116. 29 C.F.R. § 1910.1200(g)(2)(iv).
117. 29 C.F.R. § 1910.1200(g).
118. 29 C.F.R. § 1910.1200(g)(5).
119. 29 C.F.R. § 1910.1200(h).

exposure limits for about 400 substances.[120] As noted above, to date there is no permissible exposure limit for any type of nanomaterial. Thus, as explained above, when companies typically note a PEL, it is the PEL for the bulk form (e.g., using the PEL for graphite in an MSDS for carbon nanotubes). Some research has indicated that the use of graphite PEL may not be appropriate. In other circumstances, companies refer to the OSHA permissible limits for Total Dust (15 mg/m3) or Respirable Dust (5 mg/m3). However, again these limits may not be at the appropriate levels.

Once a PEL is established, employers have an obligation to comply with it. To achieve compliance, the employer must first undertake administrative (e.g., substituting a less hazardous chemical for a more hazardous one, enclosing the manufacturing in a closed system, etc.) or engineering controls (e.g., use of a portable HEPA-filtered vacuum cleaner, ventilated enclosure for weighing/ mixing, fume hood, etc.) whenever feasible. If these control mechanisms are not sufficient to achieve compliance with the PEL, then protective equipment or other protective measures shall be used to keep the exposure of employees to air contaminants within the limits.

C. Personal Protective Equipment

As noted in the preceding section on TSCA, EPA is requiring companies to use personal protective equipment for their employees who may have dermal contact with or who may inhale the nanoparticles. As a general matter, under the OSHA regulations, personal protective equipment (PPE) (e.g., dust masks, gas masks, protective clothing, gloves, and other devices) are provided and used "wherever it is necessary by reason of hazards of processes or environment, chemical hazards, radiological hazards, or mechanical irritants encountered in a manner capable of causing injury or impairment in the function of any part of the body through absorption, inhalation or physical contact."[121]

Even if EPA were not requiring the use of such equipment, employers have an obligation to protect the health of their employees. As such, an employer shall prepare a written assessment of the workplace "to determine if hazards are present, or are *likely to be present*, which necessitate the use of PPE"[122] (emphasis added). At this stage, nanomaterials are likely to fall under the designation of "likely to present" a risk. OSHA's recommendations as to how to conduct a hazard

---

120. 29 C.F.R. § Subpart Z. The three tables can be accessed via this Web site: http:// www.osha.gov/pls/oshaweb/owastand.display_standard_group?p_toc_level=1&p_part_ number=1910. Permissible exposure limits were first promulgated in 1989 and were developed from a list of Threshold Limit Values, which had been published by the American Conference of Governmental Industrialists, and the NIOSH Respiratory Exposure Limit list. The information in these lists was supplemented by and reviewed against existing scientific literature and opinion, leading to the formation of the PELs.

121. 29 C.F.R. § 1910.132(a).

122. 29 C.F.R. § 1910.132(d)(1).

assessment are set forth in 29 C.F.R. Part 1910, Subpart I, Appendix B, which outlines, among other things, how a survey should be conducted, how the findings should be analyzed, and what various PPE options should be considered. Following this assessment, the employer must make a professional judgment as what (if any) would be the appropriate PPE for the existing hazards.[123] For example, employers could require the use of gloves, NIOSH-approved respirators, safety glasses with side shields, and impervious clothing. However, research is ongoing as to whether these items will actually be protective for nanomaterials, and thus, what is appropriate gear is subject to change as new data is analyzed and new equipment is developed.

D. Respiratory Protection Standard

Under OSHA Regulations, the use of respirators is necessary when engineering control measures (e.g., enclosure or confinement of the operation, general and local ventilation, and substitution of less toxic materials) do not adequately keep employee exposure below the regulatory limits or internal control targets.[124] If the workplace needs respirators, an employer must establish and implement a written respiratory protection program with worksite-specific procedures.[125] The program, which shall be updated as necessary to reflect those changes in workplace conditions, includes the following elements: (a) an evaluation of the worker's ability to perform the work while wearing the NIOSH-approved respirator; (b) regular training of personnel; (c) periodic environmental monitoring; (d) respirator fit testing; and (e) respirator maintenance, inspection, cleaning, and storage.[126] The standard further requires that the respirator program be designed by personnel knowledgeable about the properties of the respirators and the nature of the hazards.[127]

As noted above, there is no PEL for any particular nanomaterial. As a result, if a company is interested in adopting a respiratory protection program, the employer may choose to fashion a voluntary respirator program using the framework provided by existing regulations. In the event the employer does so, the employees have to be advised of the following four points:

1. To read and heed all instructions provided by the manufacturer on use, maintenance, cleaning, and care as well as warnings regarding the respirator's limitations.

---

123. NIOSH Nanotechnology Research Center, *Progress towards Safe Nanotechnology*, (June 2007) at 28.

124. 29 C.F.R. § 1910.134(a)(1); *See* NIOSH, *Approaches to Safe Nanotechnology: An Information Exchange with NIOSH* 2 (2006) *available at* http://www.cdc.gov/niosh/topics/nanotech/safenano/ [hereinafter, "Information Exchange"] at 23.

125. 29 C.F.R. § 1910.134(c)(1).

126. 29 C.F.R. § 1910.134(c)(1).

127. 29 C.F.R. § 1910.134.

2. To ensure that the chosen respirator is certified for use by NIOSH to protect against the contaminant of concern, and to review the label or statement of certification on the respirator or respirator packaging so as to know what the respirator is designed for and to what degree it will protect.

3. To not wear the respirator into atmospheres containing contaminants the respirator is not designed to protect against.

4. To keep track of the respirator so that there is no mistaken use of another person's respirator.[128]

With respect to respirators for nanomaterials, NIOSH and its affiliated entities have been conducting studies on the effectiveness of respirators to protect against the inhalation of nanoparticles. For example, a NIOSH-funded study determined that smaller-sized nanoparticles were not able to evade respiratory filters with greater success than larger-sized particles. Specifically, the study examined whether nanoparticles that size of 3 nm to 20 nm were able to penetrate through various filter media. The research concluded that there is no evidence that nanoparticles the size of 3 nm are able to pass through the filter media at higher rates than larger particles.[129]

E. Laboratory Standards

As a supplement to the HCS standards, which exempt laboratories from its requirements, OSHA has promulgated a specific set of regulations relating to the use of hazardous chemicals in laboratories. To qualify for the laboratory standard, two conditions must be met: the facility must be a laboratory, and the use of the chemical must be for a laboratory use. The term *laboratory* is defined as including, "a workplace where relatively small quantities of hazardous chemicals are used on a non-production basis."[130] *Laboratory uses* is defined as "handling or use of such chemicals in which all of the following conditions are met: (i) chemical manipulations are carried out on a 'laboratory scale;' (ii) multiple chemical procedures or chemicals are used; (iii) the procedures involved are not part of a production process, nor in any way simulate a production process; and (iv) 'protective laboratory practices and equipment' are available and in common use to minimize the potential for employee exposure to hazardous chemicals."[131] The term

---

128. 29 C.F.R. § 1910.134, Appendix D.

129. *See NIOSH, supra* note 125, at 29.

130. 29 C.F.R. § 1910.1450(b).

131. *Id.* The term *health hazard* includes chemicals which are carcinogens; toxic or highly toxic agents; reproductive toxins; irritants; corrosives; sensitizers; hepatotoxins; nephrotoxins; neurotoxins; agents that act on the hematopoietic systems; and agents that damage the lungs, skin, eyes, or mucous membranes. Appendices A and B of the Hazard Communication Standard (29 C.F.R. § 1910.1200) provide further guidance in defining

laboratory scale means work with substances in which the containers used for reactions, transfers, and other handling of substances are designed to be easily and safety manipulated by one person;[132] it specifically excludes those workplaces whose function is to produce commercial quantities of materials.[133] Moreover, as with HCS, a substance is considered a *hazardous chemical* if there is statistically significant evidence based on at least one study conducted in accordance with established scientific principles that acute or chronic health effects may occur in exposed employees.[134] Thus, as with HCS, if studies indicate that a particular nanomaterial poses an acute or chronic health impact on employees, the laboratory standards must be adhered to.

OSHA regulations state that laboratories that work with hazardous substances must create a "Chemical Hygiene Plan" to be formally overseen by a "Chemical Hygiene Officer." The Chemical Hygiene Plan must effectively: (a) protect employees from the known hazards of chemicals in the workplace, and (b) implement controls to regulate the amount of exposure employees have to hazardous substances. Measures necessary to protect employees from known dangers include: training workers on appropriate methods of handling hazardous chemicals; making PPE available, disseminating the Chemical Hygiene Plan, developing standard operating procedures relevant to the safe handling of chemicals, and allowing exposed employees the opportunity to seek medical attention if so required.[135] Finally, similar to the HCS provisions, the laboratory standards require that employers maintain the MSDS and other communications that may arrive from any chemical manufacturer or importer and to provide such information to the workers.

F. General Duty Clause

The "General Duty" clause as outlined in section 5 of the Act is a catch-all provision providing that "each employer shall furnish to each of his employees employment and a place of employment which are free from recognized hazards that are causing or are likely to cause death or serious physical harm to his employees." The clause can only be used if no standard has been issued with respect to that particular hazard (e.g., permissible exposure levels for certain types of nanomaterials). The clause is violated if the following four conditions are present: (a) the employer failed to prevent or remove a hazard to which

---

the scope of health hazards and determining whether a chemical is to be considered hazardous for purposes of this standard.

132. *See* 29 C.F.R. § 1900.1450(b).
133. *See id.*
134. *See id.*
135. 29 C.F.R. § 1910.1450(e).

employees of that employer were exposed; (b) the hazard is recognized (i.e., the employer has to know or should have known about the hazard); (c) the hazard is causing or was likely to cause death serious physical harm; and (d) there is a feasible and known way to eliminate or at least materially reduce the hazard through either physical means, administrative controls, or safety training. Presently, given the lack of toxicological and exposure, it would be difficult for OSHA to assert that a company has violated the General Duty clause. However, as the data develops, companies have to remain cognizant of their obligation under this provision and constantly evaluate and reevaluate whether they need to undertake any action to satisfy it.

IV. RESOURCE CONSERVATION AND RECOVERY ACT

In the process of manufacturing nanomaterials, a waste stream may be generated containing nanomaterials. One statute that covers the disposal of wastes is the Resource Conservation and Recovery Act (RCRA). Under the statute, EPA is authorized to manage the generation, transport, and disposal of solid waste, and specifically hazardous waste. EPA's position thus far toward nanomaterials is that "[n]anomaterials that meet one or more of the definitions of a hazardous waste potentially would be subject to" hazardous waste requirements.[136] Thus, the fundamental question is whether the nanoparticle wastes will meet the definition of *hazardous waste* and thus, be subject to the requirements under RCRA.[137] As of December 2008, EPA has not indicated that any nanoparticle actually meets the definition of "hazardous waste."

---

136. U.S. Environmental Protection Agency, Nanotechnology White Paper (February 2007) at 68.

137. In 1976, Congress significantly amended the Solid Waste Disposal Act to address the growing volume and improper management of solid and hazardous waste. These amendments were referred to as the Resource Conservation and Recovery Act of 1976 (RCRA). Specifically, the Subtitle C of RCRA addresses hazardous waste management and seeks to regulate hazardous waste from the time it is generated until its ultimate disposal—what is commonly referred to as "from cradle to grave." In order to create this cradle-to-grave system, Congress requires generators, transporters, and disposal sites to track the wastes and to maintain records. Additionally, there are obligations to ensure that hazardous wastes are treated, stored, or disposed of in such a manner as to minimize the present and future threat to human health and the environment. This section will mainly address the requirements applicable to generators of wastes.

A. Defining *Hazardous Waste* and the Scope of Exceptions

In determining if a nanomaterial would be regulated as a RCRA hazardous waste, three issues must be addressed. First, because hazardous waste is a subset of solid waste, does the waste qualify for the definition of solid waste? Second, even if it qualifies for a solid waste designation, is it excluded from the regulatory program? Third, if it is not excluded, does it qualify as a listed or characteristic hazardous waste?

Hazardous waste is considered a subset of solid waste. The term is defined as "a solid waste, or combination of solid wastes which because of its quantity, concentration, or physical, chemical, or infectious characteristic may (a) cause or significantly contribute to an increase in mortality or an increase in serious irreversible, or incapacitating reversible, illness; or (b) pose a substantial present or potential hazard to human health or the environment when improperly treated, stored, transported, or disposed of, or otherwise managed."[138]

The statute defines *solid waste* in very broad terms to include any "garbage, refuse . . . or any other discarded material."[139] The term *discarded material* is further defined in the implementing regulations as any material that is "abandoned," "recycled," or considered "inherently waste-like."[140] A material is *abandoned* if it is "disposed of," "burned or incinerated," or "accumulated, stored, or treated (but not recycled) before or in lieu of being abandoned by being disposed of, burned or incinerated."[141] In comparison to the definition of *abandoned*, the term *recycled* is more complex. Specifically, materials are solid waste if they are recycled—or accumulated, stored, or treated before recycling—through any of the following mechanisms: (a) materials used in a manner constituting disposal or used to produce products that are applied to the land; (b) materials burned for energy recovery, used to produce a fuel, or contained in fuels; (c) materials accumulated speculatively (except for commercial chemical products under certain circumstances); or (d) materials that are reclaimed.[142] However, in order to encourage recycling, RCRA exempts three types of wastes from the definition of solid waste (and thereby from the definition of hazardous waste): wastes that are used or reused as ingredients in an industrial process to make a product; wastes that are used or reused as effective substitutes for commercial products; or wastes that are returned to the original process from which they are generated, without first

---

138. 42 U.S.C. § 6903(5).

139. 42 U.S.C. § 6903(27).

140. 40 C.F.R. § 261.2.

141. 40 C.F.R. § 261.2 (b).

142. *See id.* at 261.2(c). These are exempt from the solid waste definition when they are reclaimed: sludges, by-products that exhibit a characteristic of hazardous waste, and commercial chemical waste.

being reclaimed or land disposed.[143] With respect to the third category, in determining whether a material is "inherently waste-like," EPA will consider whether the material is ordinarily disposed of, burned, or incinerated, or whether the material contains toxic constituents that are not ordinarily found in raw materials and are not used or reused during recycling. EPA will also consider whether the material may pose a substantial hazard to human health and the environment even when recycled.[144]

The next step is determining whether the solid waste is excluded from the regulatory program. There are 22 categories of materials that are not considered solid wastes for regulatory purposes under the statute.[145] Of particular relevance to those companies that use nanomaterials is an exclusion from the definition of solid waste for materials that are reclaimed and returned to the original process. This exclusion only applies if the production process is in a closed-loop system, reclamation does not involve controlled flame combustion, the secondary material is not allowed to accumulate in tanks for more than 12 months without being reclaimed, and the reclaimed material is not used to produce a fuel or used in a manner constituting disposal.[146] Another exemption from the solid waste definition applies to domestic sewage. That is, wastes that are passed through the sewer system and are treated by a publicly owned treatment works (POTW) are excluded. As discussed in Chapter 5, it was California associations representing POTWs who raised the issue of silver ions from Samsung washing machines entering their systems and ultimately reaching San Francisco Bay. Thus, when applied to certain nanomaterials, this exemption may come under greater scrutiny.

---

143. *See id.* at 262.2(e)(1).

144. 40 C.F.R. § 261.2(d).

145. With regard to materials that are not solid wastes, RCRA regulations provide 22 exclusions including: domestic sewage; industrial wastewater discharges that are point source discharges subject to regulation under the Clean Water Act; irrigation return flows; radioactive waste regulated under the Atomic Energy Act; materials subject to in-situ mining techniques that are not removed from the ground as part of the extraction process; pulping liquors; spent sulfuric acid; secondary materials that are reclaimed and returned to the original process, provided that the entire process is enclosed; spent wood-preserving solutions; coke by-product waste, nonwastewater splash condenser dross residue; oil-bearing hazardous secondary materials that are generated at a petroleum refinery and are inserted into the petroleum-refining process; processed scrap metal; shredded circuit boards; condensates derived from the overhead gases from kraft mill steam strippers; comparable fuels or comparable syngas fuels; spent materials generated within the primary mineral processing industry; petroleum-recovered oil from an associated organic chemical manufacturing facility; spent caustic solutions from petroleum-refining liquid treating processes; hazardous secondary materials used to make zinc fertilizers; zinc fertilizers made from hazardous waste; and used cathode ray tubes (CRTs). 40 C.F.R. § 261.4(a).

146. *See id.* at 262.4(8).

If a nanomaterial falls within the definition of solid waste, the next issue is whether it falls within one of the 18 exclusions for the definition of hazardous waste.[147] Of particular relevance for nanomaterials is the exclusion for *household wastes*. Thus, consumer products containing nanomaterials that are discarded would not qualify as *hazardous wastes*.

If a given nanomaterial falls within the definition of solid waste and is not excluded via one of the 18 standard exclusions, the next issue is whether it qualifies as a hazardous waste. EPA has taken the broad statutory definition of *hazardous waste* and created two categories: listed wastes and characterized hazardous waste. Listed wastes are found at 40 C.F.R. § 261.30–35 where they are organized into four lists: the F list (hazardous waste from nonspecific sources),[148] the K list (hazardous waste from specific sources),[149] and the P and U lists (discarded commercial chemical products).[150] Each waste is assigned an identification number, which is significant for the management standards and requirements that apply to the waste.[151] To be a listed waste, the particular waste must meet any of the following criteria: (a) the waste has characteristics of ignitability, corrosivity, reactivity, or toxicity; (b) the waste is an acute hazardous waste, or one that has been found to be fatal to humans in low doses, or is otherwise capable of causing or significantly contributing to an increase in serious illness; or (c) the waste is a toxic waste, or capable of posing a substantial present or potential hazard to human health or the environment when improperly treated, stored, transported, disposed of, or otherwise managed.[152] To date, no nanomaterials have been specifically listed as a hazardous waste.

---

147. With regard to the second category of exclusion (solid wastes that are not hazardous wastes), there are 18 exclusions, including: household waste; agricultural waste; mining overburden returned to the mine site; flash ash waste, bottom ash waste, slag waste, and flue gas emission control waste; drilling fluids and other wastes associated with the development or production of crude oil, natural gas, or geothermal energy; trivalent chromium wastes; solid waste from the extraction and processing of ores and minerals; cement kiln dust; solid waste that consists of discarded arsenical-treated wood; petroleum-contaminated media and debris from Underground Storage Tanks (USTs); hazardous injected groundwater that is reinjected through an underground injection well pursuant to free-range hydrocarbon recovery operations undertaken at petroleum refineries and plants; used chlorofluorocarbon refrigerants; non-tern plated used oil filters; used oil re-refining distillation bottoms; and leachate or gas condensate collected from landfills in which certain solid wastes have been disposed. 40 C.F.R. § 261.4(b).

148. 40 C.F.R. § 261.31.
149. 40 C.F.R. § 261.32.
150. 40 C.F.R. § 261.33.
151. *See* 40 C.F.R. § 261.30(c).
152. 40 C.F.R. § 261.11. Even if a facility's waste is listed, the facility may seek delisting by submitting a petition to EPA demonstrating that the waste does not pose a sufficient hazard to merit RCRA regulation. To submit a petition, the applicant must comply with the requirements of 40 C.F.R. § 260.22. To be successful, an applicant must satisfy two

If a waste is not listed, a facility must determine whether the waste exhibits characteristics of a hazardous waste such as ignitability, corrosivity, reactivity, or toxicity. These characteristics are described in further detail in Subpart C of the 40 C.F.R. Part 261.

In brief, and with further elaboration in the regulations, solid waste exhibits the characteristic of: (a) ignitability, if a representative sample of the waste has a flash point of 140° F;[153] (b) corrosivity, if a representative sample of the waste has a pH less than or equal to 2 or greater than or equal to 12.5;[154] (c) reactivity, if a representative sample of the waste is normally unstable or readily undergoes violent changes or generates toxic gases when heated, compressed, or comes in contact with water;[155] and (d) toxicity, if, when using the Toxicity Characteristic Leaching Procedure, the extract from the representative sample of waste contains a containment listed in 21 C.F.R. § 261.24. Table 1 wastes (e.g., arsenic, barium, cadmium, etc.) are considered toxic at concentrations at or above the levels set forth in that Table.[156]

For waste streams that contain nanomaterials, the question EPA and stakeholders must consider is whether these tests are appropriate mechanisms when dealing with nanomaterials for classifying a waste as hazardous. For instance, the TCLP test is based on concentrations. However, as discussed in Chapter 3, mass and concentrations may not be an appropriate basis for evaluating the toxic hazards posed by nanomaterials. Moreover, even if it is concluded that these tests are adequate and appropriate mechanisms, data must be generated about specific nanomaterials rather than assuming that data for bulk-sized counterparts would be applicable.

B. Regulation of Generators

If the waste stream with nanomaterials is considered a hazardous substance, the next issue is: what are the obligations of the generator? A *generator* is defined as "any person, by site, whose act or process produces hazardous waste identified or listed in Part 261 of this chapter or whose act first causes a hazardous waste

---

criteria: (a) that the waste produced by the particular facility does not meet any of the criteria under which the waste was listed as a hazardous or an acutely hazardous waste; and (b) EPA Administrator has a reasonable basis to believe that factors other than those for which the waste was listed do not warrant the waste being considered as a hazardous waste. If EPA grants the delisting, the waste will not be regulated as a hazardous waste at that *particular facility*. If the facility does not seek to have the waste delisted, or if EPA denies the request to have the waste delisted, then the waste and the entities that generate, transport, treat, store, or manage the waste will be subject to RCRA regulation.

153. *See id.* at 261.21.
154. *See id.* at 261.22.
155. *See id.* at 261.23.
156. *See id.* at 262.24.

to become subject to regulation."[157] There are three categories of hazardous waste generators: large quantity generators (LQGs), small quantity generators (SQGs), and conditionally exempt small quantity generators (CESQGs). The following general requirements are applicable to LQGs and SQGs; however, these requirements are not applicable or are applicable in a less stringent form to CESQGs:[158]

- Generators must obtain an EPA identification number, which helps EPA monitor and track generators. To obtain an EPA identification number, a generator must submit EPA Form 8700-12 to EPA. The generator must also not offer hazardous waste to transporters or to treatment, storage, or disposal facilities that have not received an EPA identification number.
- A generator that transports or offers for transport a hazardous waste for offsite treatment, storage, or disposal must prepare a manifest on EPA Form 8700-22.[159] The manifest permits parties involved in hazardous waste management to track the movement of hazardous waste from generation to treatment, storage, or disposal (i.e., RCRA's cradle-to-grave system).[160] The generator must sign the manifest certification, obtain the signature of the initial transporter, retain a copy, and give the transporter the remaining copies of the manifest. Once the waste is delivered to the designated facility, the owner and operator of the facility must sign and return a copy of the manifest to the generator. An exception report must be sent to EPA if a manifest is not returned to the generator within 45 days of the date the waste was initially accepted for transport.
- Generators must comply with pre-transport requirements. Before transporting hazardous waste or offering hazardous waste for transportation offsite, a generator must package the waste, label each package, mark each package appropriately, and placard it or offer the initial transporter the appropriate placards in accordance with applicable Department of Transportation regulations.[161]

In addition to the general requirements set forth above, LQGs and SQGs have other specific obligations described below. Due to the regulatory structure's focus on mass, a facility's nanomaterial waste stream (by itself) may not qualify

---

157. 40 C.F.R. § 260.10.

158. All generators have an obligation to identify their waste stream and to determine if they are generating hazardous wastes.

159. 40 C.F.R. § 262.20.

160. Id.

161. 40 C.F.R. §§ 262.30–33.

the facility for either LQG or SQG status. However, generators should recognize that the facility's other operations may generate sufficiently larger waste streams with the result that the facility may be subject to such requirements, including regarding its nanoscale wastes, even if those wastes are by mass only a small amount of the total waste stream.

1. **Large Quantity Generator** LQGs generate 1,000 kilograms per month or more of hazardous waste, or greater than 1 kilogram per month of acutely hazardous waste. A LQG may accumulate hazardous waste onsite for 90 days or less without a permit or without having interim status provided that: the waste is placed in conforming containers, tanks, or drip pads; the date upon which each period of accumulation begins is clearly marked and visible for inspection on each container; and while being accumulated onsite, each container and tank is labeled or marked with the words "Hazardous Waste."

LQGs have the following obligations:

- Submit a biennial report to EPA by March 1 of each even numbered year.[162] The report must be submitted on EPA Form 8700-13A; cover generator activities during the previous year; and include information such as EPA Identification number and name and address for each transporter and offsite treatment, storage, or disposal facility to which waste was shipped during the year, a description of hazardous waste shipped offsite, and a description of efforts undertaken to reduce the volume of waste generated.[163]
- Have a personnel training program in place for the proper handling of hazardous waste[164] and develop and follow a written waste analysis plan that describes the procedures the facility will carry out to comply with the treatment standards.[165]
- Prepare a contingency plan designed to minimize hazards to human health or the environment from fires, explosions, or any unplanned sudden or non-sudden release of hazardous waste to air, soil, or surface water, as well as maintain emergency procedures with at least one employee designated as an emergency coordinator.[166]
- Maintain and test equipment; make arrangements with local emergency Authorities; and be equipped with an alarm system, radio, or telephone to summon emergency assistance, fire control equipment, and an adequate water system to supply water hose streams or a sprinkler system.[167]

---

162. 40 C.F.R. § 262.41.
163. Id.
164. 40 C.F.R. § 262.34(a)(4); 40 C.F.R. § 265.16.
165. 40 C.F.R. § 262.34(a)(4); 40 C.F.R. § 268.7(a)(5).
166. 40 C.F.R. § 262.34(a)(4); 40 C.F.R. § 265.50–56.
167. 40 C.F.R. § 262.34.

2. **Small Quantity Generator** SQGs generate between 100 kilograms and 1,000 kilograms of hazardous waste per calendar month and accumulate less than 6,000 kilograms of hazardous waste at any time.[168] SQGs may accumulate hazardous waste onsite for 180 days or less without a permit or without having interim status, or 270 days if the generator must ship the waste greater than 200 miles. [169] Similar to LQGs, SQGs must properly identify waste containers with the words "Hazardous Waste" and the date accumulation began. Although SQGs are not required to have a written emergency plan, they are required to ensure that an emergency coordinator is available at all times.[170] Although SQGs are also not required to have a formal personnel training program, they must ensure that employees handling such hazardous waste are familiar with proper handling and emergency procedures.[171]

3. **Conditionally Exempt Small Quantity Generator** CESQGs are those generators that produce 100 kilograms or less of hazardous waste per calendar month, 1 kilogram or less of acutely hazardous waste per calendar month, and 100 kilograms or less of any acute spill residue.[172] CESQGs may accumulate 1,000 kilograms or less of onsite hazardous waste, 1 kilogram or less of acutely hazardous waste, and 100 kilograms or less of acute spill residue or soil. Unlike LQGs and SQGs, CESQGs are not included in the Subpart C requirements and thus do not have to engage in such things as obtaining an EPA identification numbers; using a manifest, package, or label; or engaging in record-keeping as these other generators.[173] CESQGs may treat or dispose of hazardous waste to a state-approved, RCRA-permitted, or RCRA-interim status facility.

C. Transporter and Owners and Operators of Treatment, Storage, and Disposal Facilities

RCRA also imposes requirements on those who transport hazardous wastes as well as owners and operators of facilities that are used for treatment, storage, and/or disposal facility (TSDF) of hazardous waste. Given that the obligations of the parties handling nanomaterials will not be triggered until nanomaterials are considered as hazardous waste, this section will merely outline what the general obligations of transporters and TSDFs would be should that designation be made.

Transporters must comply with specific manifest requirements to ensure that the hazardous waste can be tracked while also complying with Department

---

168. 40 C.F.R. § 262.34(d).
169. *Id.*
170. 40 C.F.R. § 262.34.
171. *Id.*
172. 40 C.F.R. § 261.5(e).
173. 40 C.F.R. § 261.5.

of Transportation requirements.[174] In the event of a hazardous waste discharge during transportation, the transporter must take appropriate and immediate action to protect human health and the environment by notifying local authorities and diking the discharge area.[175] A transporter must clean up the hazardous waste discharge or take such action as may be required or approved by federal, state, or local officials so that the hazardous waste discharge no longer presents a hazard to human health or the environment.[176] If immediate removal of the waste is necessary, the regulations permit federal, state, or local government officials to authorize the removal of the waste by transporters who do not have EPA identification numbers and a manifest.[177] Thus, for generators who ship wastes containing nanoscale materials that are hazardous and for haulers who accept such wastes for transport, it is necessary that both parties communicate prior to the shipment about any special needs that such a transport may have with respect to cleanup or handling and that such special instructions are noted on the manifest.

For owners and operators of TSDFs, there is an extensive set of requirements including (but not limited to) general facility standards, manifest requirements, contingency planning, financial assurance, emergency procedures, and closure requirements. In addition, owners and operators of TSDFs are required to obtain a detailed chemical and physical analysis of a representative sample of its wastes before treating, storing, or disposing of any hazardous waste.[178] At a minimum, the analysis must include all of the information that must be known in order to properly treat, store, or dispose of the waste.[179] It must be repeated as necessary to ensure that it is accurate and up-to-date.[180] The facility must develop a written waste analysis plan that describes the procedures to carry out the analysis.[181] Thus, when nanomaterials are regulated as hazardous substances, owners and operators will need to acquire sufficient data to perform such analyses.

D. Enforcement Provisions

This section will briefly examine the enforcement authority given to EPA under RCRA. There is extensive case law on this subject matter, and thus this section provides only a brief sketch.

RCRA provides EPA with the authority to inspect facilities and to demand samples in order to develop or assist in the development of any regulation or to

---

174. 40 C.F.R. § 260.10.

175. 40 C.F.R. § 263.30(a).

176. 40 C.F.R. § 263.31.

177. 40 C.F.R. § 263.30(b).

178. 40 C.F.R. § 264.13(a)(1).

179. *Id.*

180. 40 C.F.R. § 264.13(a)(3).

181. 40 C.F.R. § 264.13(b).

enforce any provisions of the statute. Thus, EPA could use this statutory provision to gather preliminary data to make a determination regarding how nanomaterials should be regulated.

For violations of the statute or implementing regulations, EPA has the authority (except when the RCRA program is delegated to the state)[182] to seek civil penalties for past and current violations, seek injunctive relief, and/or require compliance immediately or within a specified period of time.[183] The statute also provides EPA with the authority to pursue criminal prosecution for *knowing* violations of the law,[184] as well as enhanced punishment for *knowing endangerment*.[185]

---

182. If the program is delegated to the state, then EPA will notify the state of the alleged violation.

183. 42 U.S.C. § 6928 (a).

184. 42 U.S.C. § 6928(d). For example, it is violation to (a) knowingly transport or cause to be transported hazardous waste without a permit; (b) knowingly treat, store, or dispose of any hazardous waste without a permit, or in violation of a material condition or requirement of a permit; (c) knowingly omit material information or make any false statement in any application, label, manifest, permit, or other document; (d) knowingly generate, store, treat, transport, dispose of, or otherwise handle hazardous wastes and knowingly destroy, alter, conceal, or fail to file any record, application, manifest, or other document; (e) knowingly transport without a manifest or cause to be transported without a manifest; (f) knowingly export a hazardous waste without receiving appropriate approval for export or providing the appropriate notification; and (g) knowingly store, treat, transport, dispose of, or otherwise handle a hazardous waste in material violation of an applicable regulation or standard.

185. If a person violates any of the conditions that would constitute a criminal violation and that person *knows* at the time that he is placing another person in imminent danger of death or serious bodily injury, this would constitute *knowing endangerment*. There is a separate provision that addresses *imminent hazard*. If EPA receives evidence that the past or present handling, storage, treatment, transportation, or disposal of any solid or hazardous waste may present an imminent or substantial endangerment to the health or the environment, the agency may bring suit in federal district court against any person (including any past or present generator, past or present transporter, or past or present owner or operator of TSDF) who has contributed or who is contributing to such handling, storage, treatment, transportation, or disposal so as to order the person to take such action as EPA deems necessary.

11. EVOLUTION OF BIOTECHNOLOGY REGULATION AND THE FUTURE OF NANOTECHNOLOGY REGULATION

I. LESSONS LEARNED FROM THE REGULATION OF BIOTECHNOLOGY

As laid out in the previous chapters, the U.S. government's policy toward the regulation of biotechnology is based on the perspective that as the process of biotechnology is not conceptually different from traditional methods of selective breeding and cross-hybridization, the new process does not per se pose new hazards or risks. What this means in practice is that regulation of biotech products—and now of products using nanotechnology as well—is shaped by the risks associated with a particular product, not the process that created it. Moreover, even though the current regulatory framework preceded the advent of biotechnology (and, of course, nanotechnology as well) because it is based on regulating products as opposed to the process by which they were created, it has been assumed that this framework with some modification would provide an adequate means of ensuring public heath and protection of the environment.

As a result, as concerns have escalated regarding the new risks posed by biotechnology and nanotechnology, a highly contentious debate has arisen regarding the adequacy of the existing regulatory structure. This has been particularly the case with regard to GM crops and foods. Given that this experience with agricultural biotechnology evolved over a period of two decades preceding the commercialization of nanomaterials, a number of lessons about the appropriate regulation of nanotechnology can be—and have been—drawn from it. Among these are: (a) the importance of developing toxicity, exposure, bioaccumulation, ecological impact, and other data in order distinguish risks between products; (b) the need to address labeling issues in a manner that satisfies the public's need for information to make informed decisions; and (c) the need for agencies to formulate policies in a timely manner to help companies understand what is expected of them. In the course of examining these lessons, this chapter will also discuss the policy issues that now confront the development of both biotechnology and nanotechnology.

A. Lesson 1: Developing Environmental, Health, and Safety Data

The regulation of any product is based on an assessment of the risks it poses and how those risks should be managed. The procedural and substantive requirements for assessing risk differ among regulatory programs based on the particular statute that is applicable (e.g., FIFRA requires an analysis of risk while also considering economic benefits as part of the overall assessment). However, all risk assessment fundamentally requires scientific data on impacts on human health and the environment by examining factors such as toxicity to human health, exposure pathways, persistence in the environment, and direct and indirect ecological impacts. Thus far, with regard to research on nanomaterials, the focus of both federal and corporate monies has been primarily on development of the technologies, with only a small fraction put toward understanding environmental, health, and safety risks. This allocation of resources has been bemoaned by all stakeholders, but, nonetheless it remains. However, those who are involved in the allocation of funds would do well to remember the Monarch Butterfly case as a example of how the lack of comprehensive scientific data can lead to adverse consequences for proponents of the technology—even if ultimately their position is scientifically vindicated.

As discussed in Chapters 3 and 5, the experience with the monarch butterfly underlines the need for a comprehensive set of data. As a result of a single laboratory experiment on the mortality of monarch butterfly larvae, coupled with media coverage that simplified the assertions made, GM foods and crops were— and to this date still are—perceived by a substantial segment of the population as a threat to the environment and human health. Because EPA had not required manufacturers to provide ecological data prior to registering their products, sufficient data was not immediately available to counter the monarch butterfly study when it was issued. Rather, EPA had to request that such studies be conducted prior to the reregistration of the GM corn implicated in the experiment. These later studies, along with other research, indicated that GM corn did not significantly adversely impact monarch butterflies. Nonetheless, the cost to the manufacturers went beyond research dollars and the withdrawal of registrations as serious damage was also done to public perception.

No study pertaining to nanomaterials has had a similarly significant impact as the Monarch Butterfly experiment. However, recent studies have suggested that multiwalled carbon nanotubes can act in the body similarly to asbestos, and older studies have questioned whether nanoscale titanium dioxide can breach the skin's protective defenses and have suggested that uncoated fullerenes can cross the blood–brain barrier. Thus, as more products with nanoscale products are being introduced into the marketplace, researchers, agencies, and stakeholders—all of whom acknowledge the need for data—must take on this task of addressing basic scientific questions and establishing research priorities with a greater sense of urgency.

Among these questions are: Can the effects and properties demonstrated by one form of nanomaterial be extrapolated to the entire class of nanomaterials or to a subgroup of nanomaterials? Are existing testing methodologies and technologies adequate for testing nanomaterials? What are the costs associated with such testing, and what are the costs associated with creating or adopting new testing technologies and methodologies? Are there thresholds at which a particular nanomaterial is toxic? Is the mass–dose metric effective in determining the toxicological impact of nanomaterials? What are the properties of a nanomaterial that impact the rate of absorption, distribution, metabolism, or excretion from the body?[1]

For proponents and users of nanomaterials, there must be an added urgency as one of their foremost concerns must be how this information is being relayed to the public. If test results are analogized to well-known hazardous compounds such as asbestos, then distinctions between the impacts (or the lack thereof) among various types, uses, and structures of nanomaterials may become academic as the stigma of a highly recognizable pollutant (e.g., asbestos) is imputed to the entire genre of nanomaterials.[2]

B. Lesson 2: Addressing the Need for Labels and Traceability

The public's confidence that a product will not pose an unreasonable risk is central to the ultimate way agencies will regulate a particular product. As technologies have advanced, so has the public's desire to know more about the products available in the marketplace so that they may make informed choices. Thus, another key issue that has presented a challenge to the federal government with regard to agricultural biotechnology is that of labeling or some other form of distinguishing identification. This is also likely to become an issue with consumer products that contain nanoparticles.

The history of FDA's labeling requirements for biotech foods illustrates how an agency's attempt to address the public's concerns may create uncertainty. FDA's initial position with respect to GM foods was that no specific labels were necessary to identify the production process. The rationale for this position was based on the view that genetic modification was simply the next technological advancement in the process of selective breeding. Thus, labels were necessary only if there were safety issues with the particular food product or if the food differed from its nongenetically modified counterpart in such a manner that its common name would confuse consumers. This policy came under withering criticism from environmental and consumer groups and was challenged in the courts. But, more importantly, it was clear from FDA focus groups comprised of ordinary citizens that, apart from interest group pressure, there were widely

---

1. *See* United States Environmental Protection Agency, Nanotechnology White Paper (Science Policy Council, February 2007) 72–81.

2. In this context, the fact that certain commercial entities and associations are currently developing (and to an extent have already developed) databases that will maintain toxicological and other health and safety data will assist in quickly analyzing results and making comparisons between studies whose results may be contradictory, and thus, possibly prevent overgeneralization of the impact of nanomaterials on human health or the natural environment.

held suspicions about the safety of biotech foods and concerns about whether regulatory agencies were really interested in protecting the public.

In response, FDA proposed guidelines for voluntary labeling that made distinctions between GM and nongenetically modified foods—a concession to the concept that food containing genetically engineered ingredients was inherently different from food that did not. However, FDA's draft guidance document was as much focused on giving assistance to companies who wanted to advertise the use of biotechnology as it was on admonishing those companies who sought to distinguish their products by asserting that they were "Genetically Modified Organism-free" or "GMO-free." Thus, critics could fault the agency not only for proposing a voluntary system, but also for developing guidelines that—in their view—were still aimed at blurring any distinction between GMO-containing and non-GMO containing foods. In the end, because it remains a draft guidance, as discussed below, it remains subject to reappraisal and adjustments. Moreover, even if it were finalized, it still would not be binding on the agency.

For nanomaterials, the question of whether labels are necessary for products that contain nanoscale substances has moved beyond demands in public policy papers. Environmental organizations and consumer groups have now called for such labels through the filing of petitions with EPA and FDA challenging those agencies' regulatory policies toward nanoscale silver, titanium dioxide, and zinc oxide. As with GM foods, the fundamental question that must be resolved is whether the nanotech process itself justifies the need for a label, or whether the nanoscale material must possess significantly different characteristics from its conventionally sized counterparts for a label to be needed. For those who oppose placement of a label identifying a product as containing nanomaterials because of the manufacturing process, the lesson from the biotech experience is that, while this may be a scientifically sound position, the public may not understand or appreciate it. Moreover, without a specific regulatory policy toward labeling of products that contain nanomaterials, competitors may be able to distinguish products using the absence of containing nanoscale materials as a selling point, thus helping to stigmatize products that contain nanomaterials.

Another issue that has become more prominent in the public consciousness is traceability. As discussed in the preceding chapters, the European Union has issued a directive on traceability for GM foods, but there is no such requirement in the United States. However, the issue of traceability has gained traction not due to GM foods, but because of significant media coverage of illnesses caused by the consumption of non-GM but pathogen-carrying vegetables (e.g., spinach and jalapeño peppers). If traceability is imposed on all foods, this requirement would naturally also cover all GM foods.

Similarly, the discovery of lead in popular children's toys manufactured in China resulted in the recently enacted Consumer Product Safety Improvement Act explicitly requiring that toy labels contain traceability information in order that recalled products can be tracked to their source. Thus, as with GM, traceability may be imposed to products that contain nanomaterials for reasons unrelated to the fact that they contain nanomaterials.

C. Lesson 3: Need for Issuance of Timely Enforceable Rules and Regulations

As technology advances, stakeholders need clear direction from regulatory officials as to how they intend to regulate these new developments. The lack of regulatory guidance can generate uncertainty that in turn can drive investment decisions and skew who participates in the market. For example, researchers who had wanted to develop transgenic animals in the past for human consumption complained that the lack of clear guidance hindered their ability to obtain corporate investment. Moreover, companies with research laboratories may be able to experiment with technologies that may or may not face regulatory hurdles while small companies that may be interested in the same technology may not be able to find funding to conduct research because investors are often unwilling to provide funds if there may ultimately be stringent regulatory requirements that would significantly impact profits.

The agencies can use a variety of implementing their policy positions. The biotechnology experience indicates that the agencies will often rely on interpretative statements or guidance documents as a means of communicating "firm" agency policy rather issuing a rule or regulation. While these regulatory tools have the benefit of assisting the regulated industry and the general public in clarifying an agency's position on how the statute and regulations are going to be interpreted and implemented, they also pose drawbacks. The most significant of these is that such documents are not legally binding on the agency. Thus, the regulated industry cannot claim that an agency has failed to comply with the guidance document because the agency always retains the option of not following such documents in favor of different procedures, practices, or protocols as long as they are consistent with existing statue or regulations. The second issue with guidance documents is that, unlike rule making, there are no formal procedures for public participation in developing such documents. For some agencies and for some guidance documents there has been public participation in drafting the documents. These agencies held workshops and public meetings, and asked individual and experts for comments. But for other agencies or other documents, there has been little or no public participation. Moreover, and more importantly, under formal rule-making, the agency not only has to solicit comments, but also has to address those comments when it issues its final rule. With guidance documents, the agency may take into account the public comments but it does not have to specifically address them. Finally, the failure of any agency to follow an interpretative statement or guidance document is not subject to judicial review, but the failure to follow a final rule is subject to such review. Thus, while stakeholders need information from agencies as to how the statutes will be interpreted when dealing with new technologies, this information should be in the form of enforceable rules and regulations.

II. PROPOSALS AND INITIATIVES FOR REGULATING NANOTECHNOLOGY

Considering some of the lessons learned from the agricultural biotechnology experience, it is not surprising that environmental and consumer groups, manufacturers, and agencies have attempted to craft different mechanisms for addressing certain concerns. The main lesson that has been absorbed is the need for acquiring data, and a number of initiatives have been undertaken to achieve this goal. These initiatives as described below are ongoing and therefore, subject to change. Moreover, the question remains as to whether these programs will generate sufficient data in a timely manner, and if not, what other programs or tools will be needed to acquire the necessary data.

A. EPA's Nanoscale Materials Stewardship Program

In 2008, EPA initiated a two-year pilot program for voluntary reporting called the Nanoscale Materials Stewardship Program. When the program was launched, critics compared it with a similar two-year pilot voluntary reporting program that began in 2006 in the United Kingdom. Therefore, it may be useful to first note how the British program has worked and the limitations of that program.

In September 2006, the British Department for Environment, Food and Rural Affairs (DEFRA) launched a two-year pilot voluntary reporting program for engineered nanomaterials.[3] The program is designed to enable DEFRA to obtain information on the characteristics, hazards, use, exposure, potential risks, and risk management techniques. The purpose of collecting this data was to determine appropriate regulatory controls.[4] The program is open to any company or organization that manufactures, uses, imports, researches, and/or manages deliberately engineered nanomaterials (e.g., excluding nanomaterials that are unintentional by-products of other processes) that have two or more dimensions up to 200 nm and that are "free" within any environmental media at any stage in the product's life cycle. There was no mandatory requirement to submit information, and as of August 2008, there were only 11 submissions—9 from companies and 2 from academia.[5]

It was anticipated that their involvement in determining the appropriate regulatory structure would entice companies would join the program. However, due to concerns about the confidentiality of the information they would be providing, and about how the information might be used by the government or third parties

---

3. DEFRA "UK Voluntary Reporting Scheme for Engineered Nanomaterials Materials" (September 2006), *available at* www.defra.gov.uk.

4. Specifically, DEFRA sought, among other things, information on use patterns, exposure pathways, water solubility, inhalation toxicity, dermal toxicity, bioaccumulation, degradation, and effects on organisms.

5. DEFRA, "UK Voluntary Reporting Scheme for Engineered Nanomaterials Materials: Seventh Quarterly Report" (August 2008).

(i.e., concerns about prosecution or third-party actions), the participation rate as noted above has been minimal. DEFRA recognized that the fear of public disclosure was a stumbling block and tried to placate potential submitters by promising that the data owner would be involved from the outset in responding to any request to review a submitter's information.[6] At the same time, DEFRA had to ensure the sharing of scientific information and public access as otherwise the purposes of the program would be undermined and public confidence in the ultimate regulatory method would be compromised.[7]

In January 2008, EPA issued its own version of a two-year pilot voluntary reporting scheme. The concept for this voluntary program was first conceived in June 2005 when EPA held its first public meeting regarding engineered nanomaterials. Shortly thereafter, EPA established a working group to advise it on an overall approach for addressing the potential risks associated with nanotechnology. However, it took EPA two more years before it issued a concept paper in 2007 and then launched the voluntary program in January 2008. The program is referred to as the "Nanoscale Materials Stewardship Program."[8] As with the rationale behind the British program, EPA noted that companies would participate because it would allow them to share data on nanoscale materials and receive assistance and guidance with developing appropriate risk management practices and plans.

Similar to the British program, the intent of the Nanoscale Materials Stewardship Program is to: (a) obtain, on a completely voluntarily basis, existing data and information to build EPA's knowledge based on hazard, exposure, and fate and transport; (b) identify risk management methods being utilized; and (c) encourage the development of additional test data in order to develop appropriate regulations.[9] The program is open to companies that manufacture or import engineered nanomaterials, physically or chemically modify or process an engineered nanomaterial, physically or chemically modify or process a non-nanomaterial to create an engineered nanomaterial, and/or use nanomaterials in the manufacture of a product.[10]

The program had two components: the "Basic Program" and the "In-Depth Program." The "Basic Program" participants were asked to voluntarily report existing data on such areas as material characterization, toxicity and fate studies,

6. DEFRA, "UK Voluntary Reporting Scheme for Engineered Nanomaterials Materials: Fifth Quarterly Report" (December, 2007).

7. R. Denison, Statement of Richard A. Denison, Ph. D., Senior Scientist, at USEPA's Public Meeting in the Development of a Voluntary Nanoscale Materials Stewardship Program (Aug. 2, 2007), *available at* http://www2.envirionmentaldefense.org/article.cfm?contentID=6748.

8. 73 Fed. Reg. Lat 4861 (Jan. 28, 2008).

9. 73 Fed. Reg. 4861.

10. *See id.*

uses, potential exposures, and risk management practices.[11] Enrollment in the Basic Program was supposed to close in July 2008. However, based on the analyses in the interim report, EPA accepted information after that date and continues to ask for additional information. The purpose of the In-Depth Program is to better characterize hazards, risks, and exposure issues. If a company agrees to participate in the In-Depth Program, EPA and the company will develop a plan of action that could include testing for health and environmental hazards, determining fate and transport, monitoring exposures, and evaluating effectiveness of engineering controls and worker protection equipment.[12] Thus far, only four manufacturers have indicated a willingness to participate in the In-Depth Program.

The interim report is based on an evaluation of the data provided by companies under the Basic Program. Apparently, 29 companies submitted information on approximately 123 nanoscale materials based on 57 different chemicals as of December 2008. However, the report examines only the data provided as of September 2008, and thus, involves data from 16 companies and trade associations regarding 91 different nanomaterials based on 47 different chemicals. This level of participation was significantly greater than the participation rate in the DEFRA program.

EPA has noted that it now has better insight into the physical and chemical properties of these nanomaterials, their uses, the methods of manufacturing them, and the risk management strategies employed for them. However, EPA recognizes the limitations of the information they have received. EPA noted that apparently two-thirds of the chemical substances from which commercially available nanoscale materials are based were not reported, approximately 90 percent of likely commercially available nanoscale materials were not reported, and little toxicological and exposure data was actually provided. Moreover, the scant participation in the In-Depth Program has led EPA to conclude that most companies are not inclined to voluntarily test their nanoscale materials. As a result, EPA is considering whether to issue a mandatory reporting obligation under section 8(a) and whether to issue a test rule under section 4. Critics of how the program was established have noted that the failure to require reporting under section 8 was a crucial design failure because if companies knew that they had a mandatory reporting obligation under the TSCA rules, they would have been more likely to participate in the voluntary program as they would have had to provide the data in any event.[13]

---

11. 73 Fed. Reg. 4863; Concept Paper for the Nanoscale Materials Stewardship Program under TSCA.

12. *See id.* at 4863–4864.

13. *See* J. Davies, Testimony of J. Clarence (Terry) Davies at EPA Public Meeting on Nanoscale Materials Stewardship Program at 4 (Aug. 2, 2007 comments); 73 Fed. Reg. 4862.

B. NIOSH

As discussed in Chapter 10, NIOSH as the sister agency to OSHA is the research arm of occupational health and safety. NIOSH can provide OSHA with data on health and safety and make recommendations based on said data as to appropriate mechanisms of control for ensuring appropriate worker safety (e.g., the creation of health-based exposure limits). However, it is not automatically a given that NIOSH recommendations will actually be implemented by OSHA as actual PELs or control mechanisms.

As researchers and workers who are involved in the production of nanomaterials are potentially the first to have exposure to nanomaterials, the standards and issues that NIOSH raises may not only apply to that particular nanomaterial or production process, but also may have a broad impact on public perceptions of nanomaterials and how stringently other agencies decide to regulate nanomaterials. As of December 2008, NIOSH had issued two major documents related specifically to nanomaterials: the Strategic Plan (2008 updated draft) and the Information Exchange (draft 2006). The Strategic Plan laid out four key goals and plans for addressing them:

1. Determine if nanomaterials pose risks for work-related injuries and illnesses by examining their toxicity and exposure through innovative research strategies.
2. Conduct research on and develop a roadmap for preventing work-related injuries and illnesses.
3. Promote healthy workplaces through developing guidance based on a review of the science and an evaluation of current best practices, available knowledge, and professional judgment.
4. Enhance global workplace safety and health through national and international collaborations on nanotechnology research and guidance.[14]

In issuing the draft Information Exchange document in 2006, NIOSH's aim was to provide an overview of what was known at the time about nanomaterial hazards as well as suggest the interim measures that can be taken to minimize workplace exposure as the scientific information on nanomaterials continues to develop.[15] In practical terms, due to the absence of a more complete knowledge of the hazards associated with nanomaterials, NIOSH adopted a position that

---

14. In 2004, NIOSH established the NIOSH Nanotechnology Research Center (NTRC) comprised of NIOSH scientists to implement the strategies necessary to achieve these goals. The NTRC has identified 10 critical research areas that will be examined to address the strategic goals and has set forth time frames for achieving these goals.

15. NIOSH, Approaches to Safe Nanotechnology: An Information Exchange with NIOSH 2 (2006), *available at* http://www.cdc.gov/niosh/topics/nanotech/safenano/, at Executive Summary. [hereinafter "Information Exchange"].

precautionary measures should be developed to reduce occupational exposure to them.[16] From NIOSH's perspective, it is likely that production processes that generate nanomaterials in the gas phase or that use or produce nanomaterials as powders or slurries/solutions pose the most significant occupational risks.[17] Specifically, NIOSH has recommended that employers take the following actions:

- Determine worker exposure. NIOSH notes that exposure assessments can be performed using traditional industrial hygiene sampling and by examining various potential metrics of nanomaterials that could influence exposure (e.g. particle size, mass, surface area, concentration, composition, and surface chemistry).[18]
- Minimize worker exposure. Employers should take steps to do so even if these steps may need to be altered as additional information is developed. With respect to nanomaterials, NIOSH adopts the traditional hierarchy of preferences: first engineering controls, then administrative controls (i.e., good work practices), then personal protective equipment and respirators. Engineering controls include implementing source enclosures that would isolate the source of hazardous materials from the worker and using local exhaust ventilation systems. Good work practices include cleaning work areas on a frequent basis, effectively disposing of waste in a sanitary manner, and providing adequate hand washing and shower facilities for employees working with hazardous substances.[19] As to personal protective equipment and respirators, items such as gloves, impervious clothing, and adequate respirators would be the appropriate mechanisms for reducing or mitigating against exposure.
- Base cleanup strategies on current good practices until specific guidance on cleaning up nanomaterial spills or contaminated surfaces emerge. Such good practices include using filtered vacuum cleaners, wetting powders down, using dampened cloths to wipe up powders, and applying absorbent materials and liquid traps.[20] NIOSH further cautions that employers should remain cognizant of any potential exposure that could occur during cleanup such as inhalation or dermal exposure.[21] Once the nanomaterials have been effectively cleaned from the workplace, NIOSH recommends that the waste be handled and disposed of according to relevant federal, state, and local regulations.[22]

---

16. *See id.* at 14.

17. *See id.*

18. *See id.* at 16. For a comprehensive discussion of various types of sampling methods, *see id.* at 16–21.

19. *See id.* at 22–23.

20. *See id.* 25.

21. *See id.* at 30.

22. *See id.*

- Develop an occupational health surveillance program while NIOSH formulates the relevant guidance for such a surveillance program.[23]

NIOSH"s recommendations will need to evolve as more information on the characteristics, uses, toxicological and exposure data become available.

C. FDA

FDA has also been investigating the way it should regulate nanomaterials. In August 2006, FDA formed an internal Nanotechnology Task Force (NTF).[24] The Task Force issued a report the following year in which it concluded that, unlike EPA's determination of what constitutes "new" under TSCA, the physical characteristics of nanoscale materials (e.g., size, surface area, shape, etc.) are in fact relevant in determining how FDA should regulate the material.[25] As discussed above, FDA's jurisdiction over areas where nanomaterials may be used is broad, ranging from drugs to food additives and from food packaging to medical devices and cosmetics. Thus, the Report summarized the state of the science in understanding how nanoscale materials will react with the human body and examined the adequacy of the regulatory structure given this stage of knowledge.[26] As a follow-up to the Report, FDA is considering issuing guidance documents that cover the agency's view of how nanoscale materials in a particular context (e.g., drugs, medical devices, food packaging, and cosmetics) should be regulated.[27] In order to develop these guidance documents, FDA has begun to hold public meetings, with the first held in September 2008.

One of the areas that will come under increasing scrutiny is FDA's regulation of nanomaterials in cosmetics. As discussed in Chapter 8, cosmetic regulations are based on the premise that lotions, sprays, and other applications do not affect the structure or function of the body. However, the precise issue with nanomaterials in cosmetics is whether they have physiological effects as some research has indicated. This places increased pressure on FDA to consider whether such products should be regulated as drugs rather than as cosmetics. With regard to cosmetics, FDA is currently focusing on the following questions:

(a) What characteristics or types of nanoscale materials would be important to focus on when considering the potential risks of cosmetic products?
(b) If your company markets a cosmetic product with nanoscale particles, what function do these particles perform, at what concentration are they used, and how stable are they in the formulation?

---

23. *See id.*
24. FDA News. *FDA Forms Internal Nanotechnology Task Force* (Aug. 9, 2006), *available at* http://www.fda.gov/bbs/topics/NEWS/2006/NEW01426.html.
25. Compare EPA, *TSCA Inventory Status of Nanoscale Substances—General Approach* (Jan. 23, 2008) with FDA, *Nanotechnology: A Report of the U.S. Food and Drug Administration, Nanotechnology Task Force* (July 25, 2007).
26. *See* 73 Fed. Reg. 46022–46024, 46022 (Aug. 7, 2008).
27. *See id.*

(c) What, if any, additional studies must be done for a product containing nanoscale particles to prove that this type of formulation is safe? What differences in safety or absorption have you observed between products formulated with nanoscale particles versus those that are formulated with non-nanoscale materials?

(d) Are safety assessments being done at the bulk ingredient level or final formulation or both? How do these assessments differ?

(e) What is the effect on bioavailability of making conventional particles nanoscale? Would you expect to see increased absorption/toxicity?

As this is the beginning of the process, FDA will need to address such issues as how to acquire this information, how to analyze it, and how to formulate a policy response based on it. In the interim, the agency also needs to address a petition filed by a coalition of nongovernmental organizations that asks FDA to look at all nanomaterials, but specifically focuses on FDA regulation (or in their opinion, the lack thereof) of titanium dioxide and zinc oxide in sunscreen products.

D. Nongovernmental Organizations

In addition to governmental agencies attempting to garner further information about nanomaterials, a number of environmental and consumer groups have taken positions on how nanotechnology generally—and certain nanomaterials particularly—should be regulated while the data is being gathered and even thereafter. At least one, the Canadian-based ETC Group, a few years ago advocated a strong use of the precautionary principle and argued that there should be a moratorium on all nanotechnology research and manufacturing until regulations have been instituted to specifically regulate nanomaterials. However, given that U.S. regulations are based on quantitative risk assessment, a proposal such as this one—imposing regulations before analyzing risks—is unlikely to be implemented. The only circumstance in which such a moratorium or ban would be realistically feasible would be if there were a major health incident or environmental disaster that caused public perception to turn so adversely against nanomaterials that a moratorium or ban seems the most appropriate mechanism to protect the public. Moreover, were such a scenario to occur, it is likely that those who manufactured or distributed the products involved would simply halt production and distribution because they would recognize that the potential for lawsuits alleging injury, along with their being subject to intense regulatory scrutiny, would militate against developing a product infused with nanoscale materials.

Other groups have taken the approach of petitioning different agencies to regulate certain commonly used nanomaterials. Specifically, a coalition of environmental and consumer groups have filed petitions with FDA and EPA to regulate the use of nanoscale titanium dioxide and zinc oxide in sunscreen as well as nanoscale silver in consumer products. In both instances, the aim is to require the respective agencies to consider how they regulate products with nanoscale materials generally. For example, in FDA petition, it is noted that petitioners would like FDA to develop a set of regulations under which nanoparticles are treated as new substances, subjected to nano-specific paradigms of health and safety testing, and labeled to delineate all nanoparticle ingredients.[28] EPA has asked for public comment on the nanosilver petition. Although FDA did not request specific comments on the petition, it did ask generally for comments on its sunscreen requirements. By requiring the agencies to directly confront the issue of regulating these ingredients, petitioners can accomplish multiple objectives, including raising public awareness while also potentially having an impact on the shape of regulatory policy.

E. Business/NGO Alliances

In September 2005, the nonprofit organization Environmental Defense and the chemical manufacturer DuPont entered into a partnership to develop "a comprehensive, practical and flexible system" for evaluating environmental, health and safety risks of nanotechnology. In 2007, they launched a six-step program (referred to as the "Nano Risk Framework") for identifying, evaluating, managing, and reducing the risks involved over the life cycle of the nanomaterial—that is, from research to manufacturing to distribution and sale to disposal.[29] As indicated at the launch, the Nano Risk Framework was not intended to be a substitute for federal regulation. Rather, it was to serve as a platform that agencies could use for acquiring data to formulate their regulations. The framers of the program assumed that other companies in addition to DuPont would join the Framework; however, to date, no other companies have done so.

Notwithstanding the lack of broader industry participation, the program may still influence how the federal agencies ultimately regulate the manufacturing, use, and disposal of nanomaterials. The reason is two-fold. First, the Framework represents a collaboration between two parties that are typically confronting one another, and thus provides agencies a new perspective as to what can be achieved. Second, the Framework follows the traditional risk-assessment paradigm used by EPA in evaluating chemicals.[30] While it accepts the premise that nanomaterials are different from their bulk counterparts, the program is designed to evaluate each particular nanomaterial and determine how that particular nanomaterial should be controlled so as to avoid adverse impacts.

---

28. The International Center for Technology Assessment, et al., Petitioners, Filed With: Andrew C. Von Eschenbach in his official capacity as, Acting Commissioner, Food and Drug Administration (May 16, 2006). *Petition Requesting FDA Amend Its Regulations for Products Composed of Engineered Nanoparticles Generally and Sunscreen Drug Products Composed of Engineered Nanoparticles Specifically* ("Petition") (May 16, 2006) at 3–4.

29. *See* Environmental Defense—DuPont Nano Risk Framework (Feb. 26, 2007) at 4.

30. *See id.* at 6.

The Framework is an interactive step-by-step process:

- Development of a general description of the nanomaterial and its intended uses based on tests and other information either generated by the company or by third parties.
- Development of a profile of the nanomaterial's physical and chemical properties, toxicity, exposure pathways, and fate and transport. During this step, the company may discover that it is necessary to rely on data on ultrafine particles of the same material because of the paucity of data on the particular nanomaterial. Additionally, if the company receives the nanomaterial from a third party, it may be dependent on the third party's willingness to share information regarding its physical and chemical properties.
- Evaluation of risk by using the information garnered and identifying and characterizing the "nature, magnitude and probability of risks presented by this particular nanomaterial and its anticipated application."[31]
- Development of a risk-management strategy. In formulation of a risk management strategy while addressing the uncertainty problem generated by the lack of data, the program calls for the use of a "reasonable worst-case assumption." Because the program assumes that more information will be developed and that the company will make adjustments based on that additional knowledge, the "reasonable-worst" scenario is seen as a temporary status. This standard was designed as a response to those critics who assert that because of the lack of data and the resulting uncertainty there should be moratorium on manufacturing. For example, one case study that DuPont outlined used an occupational exposure limit of titanium dioxide air particles substantially below the OSHA permissible exposure limit for conventionally sized titanium oxide. However, others, such as the U.S. Department of Defense, have noted that using a "reasonable worst scenario" is unnecessarily stringent and could substantially increase costs of compliance, thereby making it cost-prohibitive for certain companies to enter the marketplace. The issue of the costs of compliance is central to whether such a program could be instituted on a wider scale. The incremental costs for a company that routinely handles hazardous substances and has the equipment and personnel to implement such a program would be significantly less than those for small or medium-sized companies who would have ramp-up costs. As a result, there may be a substantial barrier to market for these companies.

---

31. *See id.* at 14.

- Determination based on the risk management alternatives as to which of these alternatives should be adopted. This decision must then be reviewed and adapted to changing circumstances as the company continues to develop or manufacture the nanomaterial.

III. CONCLUDING THOUGHT ON NANOTECHNOLOGY

Having addressed some of the lessons from biotechnology and the way different entities have tried to incorporate these lessons in current actions, there is just one final thought. As discussed in this book, nanotechnology has an impact a number of different fields, and therefore nanomaterials are potentially subject to a plethora of laws administered by different agencies. The important point that must be stressed is the *urgent* need to develop scientific data based on an examination of particular nanomaterials in specific applications. It is only by obtaining such data that an adequate regulatory structure that builds up public confidence can be developed. There are clearly costs associated with developing such data, which given the difficult economic conditions would be particularly tasking on small and medium-sized companies that have limited resources and are uncertain about the prospects of their technology or product. Thus, it is incumbent on the Government to allocate monies already committed to nanotechnology research to pay for the costs of specifically environmental, health, and safety research.

INDEX

Page numbers including *n* indicate notes.